“This book illuminates the critical yet underexplored interplay between addictive and neurodegenerative disorders. Comprehensive, easy-to-read, and insightful, it is an invaluable resource for health professionals and academics alike.”

Professor Mark S. Gold, Washington University, St. Louis, Missouri

“Dr. El Hayek has masterfully woven together cutting-edge research and practical insights on addiction and neurodegenerative disorders. This book is an invaluable guide for navigating the challenges of comorbid conditions in clinical practice and research.”

Dr. Firas Kobeissy, University of Florida, Gainesville, Florida

“This groundbreaking book on neurodegenerative disorders and addiction provides essential insights for psychiatrists and healthcare providers seeking a deeper understanding of the complex relationship between these conditions. Prepare to deepen your knowledge of diagnostic and treatment strategies, explore the effects of substances on neurodegeneration, and enhance your clinical practice.”

Dr. Vanessa Padilla, University of Miami, Miami, Florida

Addiction and Neurodegenerative Disorders

The brain regions and neuronal processes that underlie addiction extensively overlap with those needed for cognitive functions, including learning, memory, and reasoning. Addiction negatively impacts cognitive functions, making it more challenging to maintain abstinence and potentially leading to progression into neurodegenerative disorders. Furthermore, individuals living with neurodegenerative disorders can engage in behaviors, such as self-medication, due to impaired judgment and impulsivity associated with their condition. Addressing either of these conditions becomes markedly more intricate when they co-occur. This book explores the intricate and often overlooked connection between neurodegenerative disorders and addiction. It delves into the complex interplay between the two areas, discussing preclinical to clinical findings that highlight commonalities, shared risk factors, causality, comorbidity, management, and evidence-based interventions. By examining all these factors, an international team of contributors offer a comprehensive understanding of this underexplored intersection and its profound implications on affected individuals and their families, the healthcare system, and treatment strategies.

Neuroscience of Neurodegenerative Disorders

Series Editor
Wael Mohamed Yousef Mohamed
BBCH (Egypt), MMSc (Egypt), MD/PhD (USA)

Dr. Mohamed earned his PhD from Penn State University (USA) and currently is an Associate Professor at IIUM Medical School in Malaysia. He serves as an Editor for *Frontiers in Neurology* and *PLOS One* journals. His research is well supported by extramural funding. He is the Founder and Chair of the AfrAbia-PD-Genomic Consortium.

About This Series

This book series aims to publish volumes focusing on both basic and clinical research in the field of Neurodegenerative Disorders with a focus on delineating the recent advances in various disorders including Parkinson's disease, Alzheimer's disease, ALS, Huntington's disease, among other common neurodegenerative disorders. The books will examine different aspects of disorders including pathophysiology, diagnostic, treatment, and precision medicine. Volumes in the series will also highlight the recent advances in the management of neurodegenerative disorders especially in the OMICS and Genomics era and precision medicine. Volumes will be useful for neuroscientists, neurologists, psychiatrists, psychologists, neurosurgeons, and those interested in preventive medicine.

Integrating Neuroimaging, Computational Neuroscience, and Artificial Intelligence
Edited by Indranath Chatterjee and Nasrallah Moradikor

Addiction and Neurodegenerative Disorders: Unraveling the Nexus
Edited by Samer El Hayek

More information on this series can be accessed at:
https://www.routledge.com/Neuroscience-of-Neurodegenerative-Disorders/book-series/NND

Addiction and Neurodegenerative Disorders

Unraveling the Nexus

Edited by
Samer El Hayek

CRC Press
Taylor & Francis Group
Boca Raton London New York

CRC Press is an imprint of the
Taylor & Francis Group, an **informa** business

Designed cover image: Shutterstock.

First edition published 2026
by CRC Press
2385 NW Executive Center Drive, Suite 320, Boca Raton FL 33431

and by CRC Press
4 Park Square, Milton Park, Abingdon, Oxon, OX14 4RN

CRC Press is an imprint of Taylor & Francis Group, LLC

ISBN: 978-1-032-76242-5 (hbk)
ISBN: 978-1-032-76455-9 (pbk)
ISBN: 978-1-003-47848-5 (ebk)

DOI: 10.1201/9781003478485

Typeset in Times New Roman
by Deanta Global Publishing Services, Chennai, India

To my incredible parents, Linda and Atef.

Your unwavering belief in me throughout every step of my journey has been my greatest source of strength. From my earliest dreams to my professional milestones, you have stood by me, offering boundless encouragement and unconditional love.

This book reflects your continuous support and is a testament to the inspiration you have given me. With all my love and gratitude, I am honored to dedicate this work to you.

Contents

Foreword

In this groundbreaking work, Dr. El Hayek and a distinguished group of international contributors provide a comprehensive exploration of the interface between addictions and neurodegenerative disorders.

The convergence of addiction psychiatry and neurodegenerative disorders represents a complex and fascinating frontier in medical science. The intricate interplay between these domains has been rarely explored. This book masterfully presents a much-needed synthesis of evidence-based research and clinical insight, useful for clinicians, educators, students, and researchers alike.

This book elucidates the complex and multifaceted relationships between addiction and neurodegenerative disorders, exploring shared biological pathways, potential risk and protective factors, and their critical implications for patient care. By integrating cutting-edge science with an evidence-based and holistic approach to treatment, Dr. El Hayek delivers a resource that is both intellectually rigorous and clinically relevant.

What sets this work apart is its international perspective and capacity to bridge disciplines, connecting addiction psychiatry with neurology, public health, and policymaking. The book deepens our understanding of these interconnected conditions while providing actionable strategies to improve outcomes for affected individuals and communities.

This book is a collaborative project featuring 17 contributions from clinicians and academics worldwide, offering a comprehensive examination of the relationship between neurodegenerative disorders and addiction. Several illuminating chapters address major neurocognitive disorders, such as Alzheimer's disease, Huntington's disease, and Parkinson's disease, and their relationship to addictive disorders, including behavioral addictions. The text provides a comprehensive and practical approach to the screening and management of this complex comorbidity, including novel psychedelic-inspired approaches for treating neurodegenerative disorders. It also addresses the crucial topics of ethics and the role of environmental factors. A special chapter presents a cogent perspective on the challenges of neurocognitive disorders in low- and middle-income countries.

It is a privilege to introduce this outstanding and much-needed contribution to the literature. I am confident it will inspire clinicians, researchers, and policymakers to push the boundaries of our knowledge—and to transform these insights into meaningful, person-centered care.

Ihsan M. Salloum, MD, MPH, DLFAPA
Sujata G. Krishnan Endowed Chair
Joe R. and Teresa Lozano Long Distinguished Professor
University of Texas Rio Grande Valley School of Medicine
and Director of the Institute of Neuroscience and
Founding Chair of the Department of Neuroscience
Harlingen, Texas
Professor Emeritus
University of Miami Miller School of Medicine
Miami, Florida

Preface

The greatest wealth is health.

—Virgil

Welcome to this comprehensive work on addiction medicine and neurodegenerative disorders. In recent years, the intersection of these fields has become a focal point of scientific research and clinical interest. As our understanding of the biological, psychological, and social factors driving these conditions grows, we are increasingly aware of their complex interplay and the challenges they pose when co-occurring. This book seeks to bridge the gap between theoretical knowledge and real-world clinical practice, offering insightful perspectives and practical guidance on managing these disorders concurrently.

The inspiration for this work stems from my experience over the past three years treating patients who struggle primarily with substance-related challenges, often accompanied by comorbid neurological conditions. These issues present profound difficulties for individuals, families, and societies worldwide. With an evolving appreciation of their interconnected nature, the urgency for holistic approaches, cutting-edge research, and innovative therapeutic strategies that address both conditions simultaneously becomes ever more pressing.

This resource is designed for a diverse audience of healthcare professionals—psychiatrists, psychologists, neurologists, and other specialists—as well as researchers, residents, and students involved in treating populations affected by either condition. Each chapter brings together experts from various disciplines, providing a comprehensive and nuanced understanding of how neurodegenerative disorders and substance-related challenges can influence one another.

I hope this book will foster continued collaboration and progress in the fields of addiction medicine and neuroscience, bringing us closer to effective treatments and ultimately improving the lives of those affected.

Editor

Samer El Hayek, MD, is a Specialist Psychiatrist at the American Center for Psychiatry and Neurology, Dubai, United Arab Emirates, and an Adjunct Assistant Professor of Clinical Psychiatry at the American University of Beirut Medical Center (AUBMC), Beirut, Lebanon. An accomplished adult and addiction psychiatrist, Dr. El Hayek brings a robust clinical, academic, and research background to his work.

In 2022, Dr. El Hayek completed a fellowship in addiction psychiatry at the University of Miami-Jackson Health System in Florida, following his clinical psychiatry residency at AUBMC. With over three years of focused experience in addiction psychiatry, Dr. El Hayek has developed expertise in managing dual diagnoses and implementing patient-centered, multidisciplinary approaches. His clinical practice is informed by his extensive work with individuals from various cultural and socioeconomic backgrounds, underscoring his global perspective and dedication to reducing mental health disparities.

A passionate researcher and thought leader, Dr. El Hayek has authored over 85 peer-reviewed publications, with research interests spanning global mental health, addiction, and stigma. His work has been cited more than 2,000 times, earning him an *h*-index of 22 by 2024. In recognition of his research achievements, he was inducted into Sigma Xi, the Scientific Research Honor Society, in 2023. He is also the recipient of several awards, including the 2023 Yearly Research Contribution Award from the Erada Center for Treatment and Rehabilitation in Dubai, the 2019 Early-Career Psychiatrists Award from the 15th International Congress on Psychiatry at Ain Shams University in Cairo, and the 2019 Asian Federation of Psychiatric Associations Fellowship Award for Young Psychiatrists at the 7th World Congress of Asian Psychiatry in Melbourne, Australia.

Dr. El Hayek is a board member and the Eastern Mediterranean Region representative for the Network of Early-Career Professionals Working in Addiction Medicine (NECPAM). He also holds editorial roles in several scientific journals, including *Frontiers in Public Health* and *Current Psychopharmacology.*

Contributors

Oyeyemi Afolabi
Department of Psychiatry
University College Hospital
Ibadan, Nigeria

Kirtika Aggarwal
Department of Psychiatry
All India Institute of Medical Sciences
Dehli, India

Morufat Abosede Alabi
Department of Nursing
College of Medicine
University of Ibadan
Ibadan, Nigeria

Chadi Antoun
Department of Psychiatry
American University of Beirut Medical Center
Beirut, Lebanon

Elie Atallah
Department of Psychology and Social Sciences
Faculty of Arts and Sciences
Holy Spirit University of Kaslik
Kaslik, Lebanon
and
Department of Psychiatry
Bellevue Medical Center
Mansourieh, Lebanon

Dennis Buwembo
Brain Health Program
Makerere University College of Health Sciences
School of Medicine
Kampala, Uganda

Ankita Chattopadhyay
Department of Psychiatry
Jawaharlal Institute of Postgraduate Medical Education and Research
Puducherry, India

Shinjini Choudhury
Department of Psychiatry
All India Institute of Medical Sciences
New Dehli, India

Lea Daou
Department of Psychiatry
American University of Beirut Medical Center
Beirut, Lebanon

Ismail Deniz
Department of Addiction Psychiatry
University of California at Los Angeles
Los Angeles, California

Tala Diab
Faculty of Medical Sciences
Lebanese University
Beirut, Lebanon

Mario Eid
Department of Psychiatry
Lebanese American University
Beirut, Lebanon

Sarah El Halabi
Department of Psychiatry
Westchester Medical Center
Westchester, New York

Myriam El Khoury-Malhame
Department of Psychology and Education
School of Arts and Sciences
Lebanese American University
Beirut, Lebanon

Youssef Ernez
Department of Psychiatry D
Razi Hospital, Faculty of Medicine
University of Tunis El Manar
Tunis, Tunisia

Mounir Fakhouri
Department of Psychiatry
University of Rochester
Rochester, New York

Yara Ghanem
Gilbert and Rose-Marie Chagoury School of Medicine
Lebanese American University
Beirut, Lebanon

Abhishek Ghosh
Drug Deaddiction and Treatment Center
Postgraduate Institute of Medical Education and Research
Chandigarh, India

Joy Louise Gumikiriza-Onoria
Makerere University College of Health Sciences
School of Medicine
Kampala, Uganda

Sasidhar Gunturu
Bronx Care Health System
Department of Psychiatry
Icahn School of Medicine at Mount Sinai
New York, New York

Ahmad Hassan
Department of Anesthesiology
Washington University
St. Louis School of Medicine
St. Louis, Missouri

Noreen Hassan
Department of Psychology
Weinberg College of Arts and Sciences
Northwestern University
Evanston, Illinois

Wassim Hassan
Department of Anesthesiology
University of Illinois College of Medicine
Chicago, Illinois

Hala Itani
Department of Psychiatry
American University of Beirut Medical Center
Beirut, Lebanon

Chonnakarn Jatchavala
Department of Psychiatry
Faculty of Medicine
Prince of Songkla University
Songkla, Thailand

Karine Kahil
Department of Psychiatry
American University of Beirut Medical Center
Beirut, Lebanon

Lwere Kamada
Makerere University College of Health Sciences
School of Biomedical Sciences
Kampala, Uganda

Shyam Sundar Kanagarajan
Department of Psychiatry
Saveetha Institute of Medical and Technical Sciences
Chennai, India

Ruta Karaliuniene
Rhein Mosel Clinic
Academic Hospital at Mainz University
Mainz, Germany

Karl Khalil
Gilbert and Rose-Marie Chagoury School of Medicine
Lebanese American University
Blat, Lebanon

Rawan Koleilat
Department of Psychiatry
American University of Beirut Medical Center
Beirut, Lebanon

Rania Lansari
Department of Psychiatry D
Razi Hospital, Faculty of Medicine
University of Tunis El Manar
Tunis, Tunisia

Amine Larnaout
Department of Psychiatry D
Razi Hospital, Faculty of Medicine
University of Tunis El Manar
Tunis, Tunisia

Tiwatayo Olufolahan Lasebikan
New World Psychiatric Hospital
Ibadan, Nigeria

Victor Olufolahan Lasebikan
Department of Psychiatry
College of Medicine
University of Ibadan
Ibadan, Nigeria

Victor Makanjuola
Department of Psychiatry
College of Medicine
University of Ibadan
Ibadan, Nigeria

Wahid Melki
Department of Psychiatry D
Razi Hospital, Faculty of Medicine
University of Tunis El Manar
Tunis, Tunisia

Awanish Mishra
Department of Pharmacology and Toxicology
National Institute of Pharmaceutical Education and Research
Lucknow, India

Sachin Nagendrappa
Fiona Stanley Hospital
Perth, WA, Australia

Rita Nassanga
Makerere University College of Health Sciences
School of Medicine
Makerere, Uganda

Maryam Noroozian
Tehran University of Medical Sciences
Memory and Behavioral Neurology Division
Roozbeh Hospital
Roozbeh, Iran

Sahar Obeid
Department of Psychology and Education
School of Arts and Sciences
Lebanese American University
Sabtiyeh, Lebanon

Margaret Isioma Ojeahere
Jos University Teaching Hospital
Plateau, Nigeria

Oluremi Oladipo Oladele
Department of Psychiatry
College of Medicine
University of Ibadan
Ibadan, Nigeria

Kavitha Parthiban
Department of Psychiatry
Saveetha Institute of Medical and Technical Sciences
Tamil Nadu, India

Victor Pereira-Sanchez
Department of Psychiatry
Columbia University
New York, New York

Luke Petry
Department of Psychiatry
University of Illinois College of Medicine
Chicago, Illinois

Ramdas Ransing
Department of Psychiatry, Clinical Neurosciences, and Addiction Medicine
All India Institute of Medical Sciences
New Dehli, India

Ahmad Sabalbal
Department of Psychiatry
American University of Beirut Medical Center
Beirut, Lebanon

Mahdieh Saeidi
Bronx Care Health System
Department of Psychiatry
New York, New York

Mona Salehi
Department of Psychiatry
University of Minnesota School of Medicine
Minneapolis, Minnesota

Michel Sfeir
Department of Clinical Psychology and the Distress Lab
University of Mons
Mons, Belgium

Mohammadreza Shalbafan
Department of Psychiatry
School of Medicine
Mental Health Research Center
Psychosocial Health Research Institute
Iran University of Medical Sciences
and
Brain and Cognition Clinic
Institute for Cognitive Sciences Studies
Tehran, Iran

Farid Talih
Department of Psychiatry
American University of Beirut Medical Center
Beirut, Lebanon

Serkan Turan
Department of Child and Adolescent Psychiatry
Faculty of Medicine
Bursa Uludağ University
Nïlüfer/Bursa, Turkey

Ramyadarshni Vadivel
Mental Health and Addictions Health
Hamilton, New Zealand

Bita Vahdani
Cognitive Neurology, Dementia and Neuropsychiatry Research Center
Roozbeh Hospital
Tehran University of Medical Sciences
Tehran, Iran

Sanya Virani
Indiana University School of Medicine
Indianapolis, Indiana

Celal Yesilkaya
Department of Child and Adolescent Psychiatry
Konya Ereğli Public Hospital
Ereğli/Konya, Turkey

Odel Zadeh
University of Massachusetts Chan Medical School
Worcester, Massachusetts

Common Abbreviations

β-amyloid	Aβ
AD	Alzheimer's disease
ALS	Amyotrophic lateral sclerosis
AUD	Alcohol use disorder
BBB	Blood–brain barrier
DSM	*Diagnostic and Statistical Manual of Mental Disorders*
FTD	Frontotemporal dementia
HD	Huntington's disease
ICD	International Classification of Diseases
LMICs	Low- and middle-income countries
NDD	Neurodegenerative disorder
OUD	Opioid use disorder
PD	Parkinson's disease
SSA	Sub-Saharan Africa
SUD	Substance use disorder
WHO	World Health Organization

1 Animal Models of Substance Use and Neurodegeneration

Amine Larnaout, Youssef Ernez, Rania Lansari, and Wahid Melki

INTRODUCTION

Aging is increasingly becoming a worldwide challenge. The World Health Organization (WHO) estimates that, between 2015 and 2050, the percentage of the world's population over 60 years will rise from 12% to 22% (World Health Organization, n.d.). The emergence of several complex health issues, including dementia, characterizes aging. According to a report published in March 2023, the WHO estimates that more than 55 million individuals live with dementia globally, and there are nearly 10 million new cases annually (World Health Organization, n.d.). Substance use disorders (SUDs) are biopsychosocial diseases. They are chronic, relapsing, and debilitating disorders marked by compulsive drug-seeking and persistent use despite harmful consequences. Long-lasting behavioral changes and indelible brain changes are observed in SUDs (National Institute on Drug Abuse, n.d.).

While dementia and addiction seem to be distinct conditions, new research indicates that they may have overlapping neurobiological mechanisms, especially when it comes to synaptic plasticity, neurotransmitter systems, and neuroinflammation. Animal models offer a platform for examining these connections and identifying similar pathways that might guide the development of innovative therapeutic approaches for both conditions.

Animal models of diseases (Table 1.1) have long played a role in solving many mysteries in medicine. They have significantly contributed to the understanding of the pathophysiological and biological mechanisms of diseases and the development of different therapeutic techniques.

In this chapter, we describe the animal models used in neurodegeneration and addiction research, and we identify the common points between these two models to unveil their intersecting pathophysiology.

ANIMAL MODELS OF NEURODEGENERATIVE DISORDERS

Neurodegenerative disorders (NDDs) are a category of illnesses characterized by the progressive loss of neuronal function and structure. The complex and unique structure of the human brain makes it difficult to fully understand the underlying causes of these complex disorders (Tello et al., 2022). In this regard, animal models (Table 1.2) are essential as they offer invaluable insights into the course of disease and potential therapeutic interventions.

Cell Cultures

Several cell culture models have been used to study NDDs. The cell types include immortalized cell lines, induced pluripotent stem cell-derived cell lines (iPSCs), patient-derived iPSCs, and organ-like models (Slanzi et al., 2020).

Neuronal cell culture remains the gold standard for studying NDDs at a cellular level. Neuronal cells can be acquired directly from patients. Their primary advantage consists of carrying the cluster

DOI: 10.1201/9781003478485-1

TABLE 1.1
Main Types of Animal Models in Research

The most common animal models involve administering medications or other biologicals and exposing animals to injuries, stress, or other environmental factors to induce a disease that resembles those affecting humans or other animals.
In models based on spontaneous changes, animals are predisposed to certain diseases due to the natural progression of their lives.
Some models are genetically modified with knock-in or knockout genes or proteins.
Healthy animals are also animal models.
Some animals are not susceptible to developing certain diseases and serve as negative models to evaluate susceptibility to specific pathologies.

Source: Domínguez-Oliva et al., 2023.

of genes that would have caused the disease. The other significant advantage of human cell cultures is that they abolish species-specific barriers (Slanzi et al., 2020). Among the disadvantages of cell line cultures is that they are traditionally cultured in a 2D monolayer system and do not replicate the complex architecture of tissues and the different relationships of cells within the tissue (Langhans, 2018). There have been many attempts to create 3D cultures, and some promising models will soon be available (Korhonen et al., 2018; Medda et al., 2016).

For NDDs and SUDs, cell models have played a significant role in understanding the disease at a molecular level but have failed to replicate the disorder in totality or account for other factors in its complex pathophysiology.

Most human studies on Alzheimer's disease (AD) use patient-derived iPSCs or human neuron cultures modified by CRISPR-Cas9 technology to generate knock-in cellular models (Penney et al., 2020). Mini-brains based on 3D culture techniques, such as organoids, have been among the most promising techniques. Such co-culture allows the creation of a vascular system that mimics the blood–brain barrier (BBB). This has been an enormous step toward creating a reliable model for research on the treatment and pathophysiology of NDDs, including AD (Slanzi et al., 2020).

Yeast Models

Yeast models have widely contributed to the understanding of many NDDs. In this model, disease-associated genes with unknown functions are efficiently transfected into the yeast genome to unravel their function. This method has been instrumental in understanding protein misfolding and subsequent toxicity in Huntington's disease (HD) (huntingtin), Parkinson's disease (PD) (α-synuclein), and AD (tau, β-amyloid [Aβ]) (Delenclos et al., 2019; Liu et al.,2017; Winderickx et al., 2008)

Saccharomyces cerevisiae is the most widely utilized yeast model. It has the advantage of genome similarity with humans since it shares about 23% of its genes with humans, and a high degree of conservation of gene function is found among the shared genes (Botstein et al., 1997).

Yeast models have several limitations, however. Yeasts are single-celled organisms that lack tissue organization and signaling pathways; thus, they cannot provide clear visibility into the different pathophysiological and inflammatory mechanisms involved in the pathophysiology of NDDs (Bilinski et al., 2017).

Nematodes

In 1992, Link et al. developed *Caenorhabditis elegans* as a model to study NDDs. Since then, *C. elegans* has been the most widely used nematode in NDD research (Link, 2006). Its genome has at least 83% homology with the human genome (Lai et al., 2000; Kim et al., 2018), while its neuronal

TABLE 1.2
Advantages and Limitations of Animal Models of Neurodegenerative Disorders

Animal Model	Cost-Effectiveness	Requirements	Main Advantages	Main Limitations	Example of Diseases
Cells cultures	Good	Few ethical considerations	No species-specific barrier bias.	Does not replicate the complex condition of cells in tissues and their signaling and connections.	AD, PD, HD, ALS (amyotrophic lateral sclerosis)
Yeast models	Good	Short lifespan and generation cycles	Strong conservation of gene function between yeast and humans.	Lack of tissue organization. Lack of intercellular signaling system.	AD, PD, HD, ALS, Prion disease
Nematodes	Good	Short lifespan Many worms are needed to obtain sufficient material for biochemical analyses	High homology with human genes. Easy genomic manipulation. Ability to perform unbiased genetic screens.	Limited ability to replicate human pathophysiology. Inability to replicate inflammatory processes seen in some NDDs.	AD, PD, HD, ALS
Zebrafish	Good	Short generation time	High homology with human disease-associated genes. Mimics AD features in humans.	Zebrafish can regenerate neurons in some parts of the brain. Compounds are added to water and absorbed randomly through the skin, making quantitation challenging.	AD
Fruit flies	Good	Fast reproducibility	High homology between human-disease-related genes and *D. melanogaster* genes. Short lifetime and a large number of progenies. Many structural and functional similarities between the human and *Drosophila* nervous systems.	Differences in BBB permeability between flies. and humans. Some differences in neurotransmitter systems and neurotransmitters' functions between flies and humans (e.g., flies do not have an adrenergic system).	AD, PD, ALS, HD
Mice and rats	Good	Long generation time	It could show age manifestations similar to those seen in humans. A well-conserved gene expression, brain connectivity, and network modules. Easiness of genetic manipulation.	Some aspects of human disease are not well replicated (RNA-binding sites implicated in some NDDs are not well-conserved). Weak translatability of findings from rodent models to humans.	AD, PD
Non-human primates	Very costly	Long longevity	The closest animal models to study human disease.	Artificially induced models employ surgical procedures that could harm the animal. Very few sources of non-human primates. Difficulty of creating a transgenic model.	AD, PD, HD

signaling system, made up of 118 neuron types with different neurotransmitters and peptides, makes it a reliable model for studying molecular pathways underlying NDDs (Li et al., 2013; Hobert, 2010). The quick aging of this model is another significant advantage that has allowed the rapid construction of numerous transgenic models and efficient assessment of experimental strategies (Link, 2006). *C. elegans* invertebrates have enabled researchers to conduct genetic screening, allowing the identification of genes involved in AD. Notably, using *C. elegans* led to the discovery of two novel factors involved in tau pathology in AD: SUT-1 and SUT-2 (Alexander et al., 2014).

Nevertheless, this model has some limitations. First, *C. elegans* does not have a defined central nervous system with individualized tissues and structures, making it impossible to study the impact of NDDs on different central nervous system structures. Second, these nematodes do not have an adaptive immune system, which could play a role in some NDDs (Li & Le, 2013).

Zebrafish

Zebrafish (*Danio rerio*), a vertebrate model, is suitable for studying NDDs. The Zebrafish genome has about 70% homology with the human genome (Kalueff et al.,2014; Saleem & Kannan, 2018). It shows neuroanatomical, behavioral, and pathophysiological similarities with patients living with AD and has therefore been used in neurophysiological and neuropharmacological research (Saleem & Kannan, 2018). Transgenic Zebrafish models expressing tau proteins have played a role in understanding tauopathies and tau-dependent neuronal loss (Paquet et al., 2010).

Fruit Flies

Drosophila melanogaster has many advantages for the study of human neurological disorders. Its nervous system is arranged similarly to that of humans and functions through a neurotransmitter system. It also has a well-structured BBB identical to the mammalian barrier but with different permeability (DeSalvo et al., 2014). Another major benefit of *D. melanogaster* is its ease of genomic manipulation. The most widely used *D. melanogaster* model for neurodegeneration is the GAL4-UAS (upstream activation sequences) model, in which a GAL4 driver binding to UAS sequences represents a bipartite system that activates gene expression (del Valle Rodríguez et al., 2011).

For reseearch purposes, gain-of-function models are created using the GAL4-UAS system. Such a model involves transgenic flies expressing human DNA sequences to identify genes, proteins, or molecular pathways involved in a particular disease (Deal & Yamamoto, 2018). Alternatively, a loss-of-function model uses techniques to create a knockout model for some genes. The RNAi (RNA interference) model knocks out gene expression at an RNA level in a tissue-specific manner. Genes with neuroprotective functions, like Presenilin and Nicastrin subunits of γ-secretase, have been identified through RNAi (Casas-Tinto et al., 2011).

Mice and Rats

Mice and rats are among the most valuable models for studying human diseases. Humans and murine animals share about 85% of their genes (Uhl & Warner, 2015). Because human and murine nervous systems are significantly similar, these rodents provide good models for studying NDDs (Uhl & Warner, 2015).

Aging is the most critical factor for NDDs. Various models of aging have been created by promoting or inducing aging. Natural aging models are generated by raising mice to 18 to 24 months of age. Upon aging, mice develop several pathological features similar to NDD manifestations, such as neuronal degeneration and behavioral and memory impairment. Among the limitations of this model are the long time investment and the co-occurrence of other age-related diseases that may lead to the animal's death, making it challenging to maintain consistency within the group of mice (Dawson et al., 2018).

One example of an induced aging model is the D-galactose-induced model, in which D-galactose is injected into mice to accelerate aging. This process generates characteristics similar to those seen in elderly mice, such as learning and memory decline, a reduced number of neurons, and decreased superoxide dismutase activity in the brain (Kosel et al., 2020).

The senescence-accelerated model (SAMP8) is a spontaneous animal model of accelerated aging. After 6 months of life, these models start to express accelerated aging and exhibit many brain aging features such as cognitive decline, Aβ plaque deposition, aberrant hyperphosphorylation of tau-like neurofibrillary tangles, and oxidative stress, among other changes (Kosel et al., 2020; Liu et al., 2020).

The Aβ-induced AD model is made by injecting Aβ fragments into the mice's hippocampal CA1 region or lateral ventricle. This results in Aβ plaque deposition in the hippocampus and other brain regions. Cognitive decline and pathophysiological manifestations similar to those seen in AD, such as Aβ deposition in the brain and the proliferation of stellate glial cells around amyloid plaques, are observed in this model (Lanz et al., 2003).

Lastly, in genome-editing mouse models, AD-related genes are introduced into the mouse genome to exhibit AD pathological features. The most frequently used models are APP/PS1, 5xFAD, Tg2576, and 3xTg-AD. The main issue with these models is that most of them exhibit amyloid plaque formation, but only a few involve the pathological aggregation of tau proteins (Billings et al., 2005; Jackson Laboratory, n.d.).

Non-Human Primates

Non-human primates represent the best animal model for studying NDDs. Genetically and physiologically, non-human primates are the closest to humans. Spontaneous models can show manifestations similar to AD or other subtypes of dementia, with cognitive decline, Aβ plaque-like structures, and neuropathy (Li et al., 2019). The major drawback of these models is their cost, as they must be cared for until they reach an advanced age (5 to 24 years). Artificially induced models can be created by injecting the lysate of an AD brain into an animal's brain or by inducing neuronal damage through formaldehyde, inhibition of cholinergic receptors, or intracerebroventricular injection of streptozocin (Dawson et al., 2018). Developing these models is usually expensive and can cause significant harm to the animal, raising ethical concerns.

ANIMAL MODELS OF ADDICTION

Addiction is a chronic, relapsing condition characterized by compulsive illicit drug intake, continued use despite harmful consequences, and long-lasting changes in the brain. Understanding the complexities of addiction in humans is challenging due to ethical and practical limitations. Therefore, animal models (Table 1.3) play a crucial role in unraveling the neurobiological underpinnings of addiction and in developing potential treatments.

Face Validity of Animal Models in Addiction Research

The appetite for drugs is a trait shared between humans and animals. When given unrestricted access to drugs such as heroin, cocaine, or alcohol, animal models often overdose fatally or reach intoxication doses. This behavior closely mimics human drug-seeking behavior, suggesting high face validity. Face validity usually stems from anthropomorphic interpretations of animal behaviors. However, a more genuine face validity could be inferred if the behavioral traits have evolutionarily developed. For instance, fear responses, such as freezing behavior, transmitted identically during evolution in several species, including humans and rodents, provide genuine examples of face validity (Spanagel, 2017).

TABLE 1.3
Advantages and Limits of Animal Models of Addiction

Animal Model	Advantages	Limits
Fixed-ratio model	• Simple and consistent. • Allows rapid acquisition of behavior. • Allows easy measurement of the reinforcement rate.	• Provides a limited insight into motivation (does not assess an increase in motivation over time). • Does not capture impulsive and compulsive drug-seeking behaviors. • Does not mimic the dynamic and variable nature of real-world drug-seeking behaviors.
Progressive-ratio model	• Represents a simple task model. • Can be implemented several times throughout training. • Assesses the motivation to obtain the drug. • Provides insights into the reinforcing properties of substances. • Has good face validity.	• Lacks contextual or social factors. • Animals may reach a limit of responding due to fatigue or frustration, which can be confounding factors.
Non-operant self-administration model	• Features a more straightforward design compared to the operant models. • Mimics more naturalistic drug-taking behaviors.	• Lacks motivation assessment. • Offers limited behavioral insights. • Contains several confounding factors.
Conditioned place preference model	• Effective for evaluating the rewarding effects of drugs without requiring self-administration. • Provides insight into learning processes. • Has a simple experimental design.	• Lacks motivation assessment. • Results are context-dependent, limiting the external validity of the findings.
Intracranial self-stimulation model	• Allows direct assessment of reward pathways. • Gives a real-time measurement of animal behaviors in response to drug administration. • Highly sensitive to changes in the reinforcing properties of drugs.	• Requires invasive surgery. • Does not mimic drug-seeking behaviors in a naturalistic setting.
Spontaneous withdrawal model	• Offers a naturalistic representation of withdrawal. • Studies the long-term effects of withdrawal and how withdrawal symptoms evolve without external intervention. • Allows the comparison of different drug withdrawal profiles.	• Shows variability in withdrawal severity symptoms between animals. • Withdrawal behaviors are complex to measure. • Raises ethical concerns due to animal suffering.
Precipitated withdrawal model	• Allows rapid induction of withdrawal with better control over timing and symptom onset. • Has good consistency and reproducibility. • Precisely measures symptoms and behaviors.	• Does not ideally mimic the natural withdrawal process. • Withdrawal symptoms could be exaggerated because of the use of the antagonist. • Focuses on acute withdrawal rather than long-term adaptations. • Raises ethical concerns due to animal suffering.

(*Continued*)

TABLE 1.3 (CONTINUED)

Animal Model	Advantages	Limits
Reinstatement model	• Closely mirrors the relapse process seen in humans living with SUDs. • Allows the study of specific triggers (drug, cue, and stress).	• While it mimics some aspects of relapse, it does not reproduce the complexities of human relapse under real-world conditions. • Emphasizes extinction rather than abstinence. • Lacks long-term relapse insight.
Protracted abstinence model	• Models long-term relapse vulnerability. • Addresses addiction as a chronic and relapsing condition. • More accurately reflects the human experience of addiction compared to short-term models.	• Requires an extended time frame to study protracted abstinence. • Biased by the individual variability in relapse susceptibility over long abstinence periods. • Humans' social, environmental, and psychological relapse factors are difficult to replicate in animal models.
Incubation of craving model	• Mimics the tendency of craving for drugs to increase over time, which is observed in humans. • Allows the study of temporal dynamics of craving and relapse periods. • Allows the study of cue-triggered relapses. • Has good face validity.	• Focuses on cue-triggered craving only without considering the other relapse factors. • Does not study spontaneous relapse. • Does not reproduce the subjective experience of craving seen in humans.

Given that addictive behaviors in humans are complex and culturally shaped, how valid could animal models be for human studies?

One argument suggests that drug and alcohol consumption in mammals and other species has been a consistent behavior throughout evolutionary history, given that the evolution of animals and human predecessors took place in an environment that included drugs. According to this notion, the consumption of alcohol from fermenting sugar-rich plant products or psychoactive chemicals from a variety of plants on an occasional or even long-term basis may have developed over millions of years. Remarkably, studies show that long before humans developed the ability to ferment food, hominids evolved to metabolize alcohol. This adaptation coincided with the shift to a terrestrial lifestyle, as fruit taken from the forest floor had significantly higher concentrations of fermenting yeast and alcohol than fruit from trees. This change most likely signified the first real exposure and adaptation to large doses of alcohol. These results provide credence to the hypothesis that our typical behavioral repertoire includes alcohol and drug consumption behaviors, which have been evolutionarily shaped over millions of years. Consequently, these evolutionary roots confirm the genuine face validity of animal models of drug self-administration (Spanagel, 2017).

Types of Animal Models in Addiction Research

Self-Administration Models

Self-administration models have long been considered the gold standard for studying drug addiction. Rats and mice are the animal models of choice for addiction studies and are trained to perform a task, for example, pressing a lever to receive a drug. The amount of work an animal is willing to perform to receive the drug reflects the drug's reinforcing properties and potential for dependence. This setup closely mimics human drug-taking behavior and offers a powerful tool to investigate the reinforcing properties of drugs and related behaviors in a controlled setting (Spanagel, 2017; Kuhn et al., 2019; Sanchis-Segura & Spanagel, 2006; Panlilio & Goldberg, 2007).

Operant Self-Administration Models

A popular experimental paradigm in addiction research for examining the mechanisms behind drug-seeking behavior and the reinforcing effects of substances is the operant drug self-administration model. In this model, animals—typically rats or non-human primates—are trained to perform a specific behavior, such as pressing a lever or sticking their snout into a hole, to receive a drug reward. This setup replicates human voluntary drug consumption and enables researchers to explore a range of features related to addiction, such as the effects of different drugs, the role of specific brain regions, and the efficacy of potential therapies.

Animals are gradually trained to perform the operant task by rewarding them with the drug or another reinforcer during initial sessions (Spanagel, 2017; Kuhn et al., 2019; Sanchis-Segura & Spanagel, 2006). The research employs different conditioning schedules:

a. **Fixed-ratio schedules:** The animal is given a drug dose after a predetermined number of responses (e.g., under a fixed-ratio 1 schedule, a small and previously defined drug dose is delivered after a single press of a lever). Studying the drug's reinforcing efficacy is made easier with this model.
b. **Progressive-ratio schedules:** The rodent must perform more responses to obtain the drug. The progressive-ratio breakpoint is defined as the maximal number of presses resulting in the final administration of the drug. This model provides insight into the motivational aspects of drug-seeking behavior.

Non-Operant Self-Administration Models

An alternative to the operant conditioning paradigm for studying drug-taking behavior is the non-operant drug self-administration model, which does not require animals to complete a task to obtain a drug. This model provides a more straightforward method for investigating voluntary drug usage, which might help researchers comprehend the fundamental characteristics of drug reinforcement and addiction.

This setup is based on choice paradigms: the rats choose between two levers, one providing a drug and the other a non-drug reward. This procedure helps in understanding the preferences and decision-making processes involved in addiction.

Conditioned Place Preference (CPP)

This model is used to assess the rewarding or aversive effects of drugs. Animals are repeatedly exposed to a drug in one specific environment and saline (non-drug) in another. When given free access to both compartments, rodents that develop a preference for the drug-paired environment are considered to have a stronger motivation to seek the drug's rewarding properties. CPP models are beneficial for studying the associative learning aspects of addiction (Tzschentke, 2007).

Intracranial Self-Stimulation (ICSS)

Rodents in ICSS models self-administer electrical stimulation to specific brain regions involved in reward circuits. When drugs enhance the rewarding effect of the stimulation, they are considered to have a potential for abuse. This model helps to understand how drugs interact with brain reward circuits (Spanagel, 2017; Kuhn et al., 2019; Sanchis-Segura & Spanagel, 2006).

Withdrawal and Relapse Models

Researchers induce drug dependency in animals, observe withdrawal symptoms, and search for relapse behaviors to understand the negative reinforcement and relapse elements of addiction (Spanagel, 2017; Kuhn et al., 2019; Sanchis-Segura & Spanagel, 2006).

a. **Withdrawal models**
   - *Spontaneous withdrawal*: After inducing drug dependence in animals, the drug is abruptly discontinued.
   - *Precipitated withdrawal*: After inducing drug dependence in animals, withdrawal symptoms are triggered using antagonists.

b. **Relapse models**
   - *Reinstatement model*: Animals that have self-administered drugs for a while are trained to stop desiring drugs, a process known as extinction training. Subsequently, relapse triggers such as stress, exposure to drug-associated cues, or a modest dose of the drug (drug priming) might cause relapse in the animal. This model facilitates research into the mechanisms that underlie relapse and the testing of behavioral and pharmacological therapies that may be used to prevent relapses.
   - *Protracted abstinence model*: After a period of drug self-administration, animals are forced to abstain from drug consumption for a prolonged duration (weeks to months). Next, relapse-like behaviors are evaluated. This model helps researchers better understand the long-term alterations in brain function and behavior linked to relapse risk by simulating the extended period of abstinence experienced by individuals with SUDs.
   - *Incubation of craving model*: Animals are examined for drug-seeking behaviors at various stages of abstinence after drug self-administration and extinction training. This model helps explain the time-dependent character of cravings and the neurobiological alterations that occur over extended periods of abstinence.

CONVERGENCE POINTS BETWEEN ADDICTION AND NEURODEGENERATION

There are many similarities between addiction and neurodegenerative pathophysiology.

Neurocircuitry and Neurotransmission

Dopamine neuronal loss is a common feature of many NDDs. Experiments on rodents show that dopamine neurons in the ventral tegmental area (VTA) are compromised from the very early stages of many NDDs. In mouse models, VTA dopamine neuronal degeneration is visible as early as 3 months of age and is accompanied by marked local inflammation. As the mice age, cell loss becomes more severe. Reduced dopamine outflow in the hippocampus and nucleus accumbens (NAc) due to dopamine neuronal degeneration is temporally correlated with deficits in hippocampal neuronal excitability, synaptic plasticity, memory, and reward function, with the latter representing the connection between NDDs and SUDs (Krashia et al., 2019).

Dysregulation of glutamate transmission is also a common feature of both disorders. Glutamate is a ubiquitous neurotransmitter that plays a role in many brain functions, such as neuroplasticity and cognitive functions.

An alteration in cholinergic transmission has been reported in dementia, mainly in AD. Cholinergic transmission also plays a central role in activating the reward system. Nicotine binding to the α7 and α4β2 subtypes of the nicotinic acetylcholine receptors (nAChRs) in the VTA promotes the initiation of addictive behavior by stimulating dopamine release in the NAc. Afterward, they impact the GABAergic, dopaminergic, and glutamatergic releases, thereby playing a central role in the initiation of addictive behavior. In parallel, impaired cholinergic transmission has been described in both AD and PD. A depletion of the α7 subtype of the AChRs was found in the brains of patients with AD and was shown to correlate with Aβ plaque deposition and cognitive impairment. In contrast, a depletion of dopamine characterizes PD; nevertheless, presynaptic nAChR agonists may ameliorate dyskinetic symptoms in PD by promoting the endogenous release of dopamine (LaFerla et al., 2007). Animal models of alcohol dementia also provide another argument in favor of the cholinergic hypothesis. In particular, rats given chronic ethanol treatment may serve as a suitable animal model to develop treatment strategies for dementia disorders such as AD and alcoholic dementia, which are linked to the degeneration of the Nucleus basalis of Meynert (Ittner & Götz, 2011).

Neuroinflammation

A chronic neuroinflammatory state of the brain characterizes SUDs. Numerous drugs increase the brain's susceptibility to inflammation and the ensuing neuropathology. It has been shown that through the activation of microglia and other glia-mediated synaptic remodeling, drug-induced dysregulation of neuroimmune signaling may impair neuronal function, exacerbate neurodegeneration, and enhance neurotoxicity. Proliferation and modifications of the morphology and function of microglia and astrocytes are among the hallmarks of the neuroimmune response to drugs of abuse. Drug-induced central nervous system injury results in microglia activation by stimulating chemokine and cytokine receptors or by peripheral signals that may arise from disruption to the BBB. Microglia activation triggers several downstream processes, such as reactive oxygen and nitrogen species generation, which cause neuronal damage, cell migration to the site of injury and phagocytosis, and the production of pro-inflammatory factors like interleukin (IL)-1β, IL-6, and tumor necrosis factor-α (TNF-α). Like microglia, astrocytes also release pro-inflammatory cytokines in reaction to stressors such as exposure to drug abuse or tissue damage. Therefore, excess neurotransmitters (dopamine and glutamate, for example) released from drug use may bind to glial cell receptors and intensify inflammatory signaling by releasing more cytokines and chemokines, which may contribute to the positive feedback loop that stimulates inflammation (Liddelow & Barres, 2017).

Along the same lines, NDDs are accompanied by chronic inflammation of the central nervous system beginning from the very early stages of the disease. Neuroinflammation and microglia activation are significant events leading to neurodegeneration. Microglia can either promote nerve growth or induce neurotoxicity depending on their activation state. When hyperactivated, microglia contribute to tissue toxicity. Specifically, the activation of M1 subtype microglia via Toll-like receptors and the gamma interferon signaling pathway triggers the secretion of pro-inflammatory cytokines such as IL-1, IL-6, IL-1β, TNF-α, and NF-κB, along with various chemokines. Astrocytes also play a role in the neurodegenerative process. Astrocytes are activated during neuroinflammation and release a range of cytokines and chemokines. Furthermore, astrocytes play a role in regulating the BBB, and during neuroinflammation, their activation might alter its permeability, allowing immune cells and neurotoxic substances to enter the brain parenchyma (Rajendran & Paolicelli, 2018).

Thus, a neuroinflammatory environment characterized by microglial and astrocyte activation, along with the release of inflammatory markers, is a common feature in both disorders, leading to subsequent neuronal damage in each condition.

Synaptic Plasticity

In SUDs, animal model studies show that drugs can change synaptic plasticity in many brain areas, remodeling the brain's reward circuitry and the mesolimbic dopamine system. Drug-induced widespread adaptations of the glutamatergic system in specific brain regions, such as the VTA, the NAc, and the prefrontal cortex, have been noted due to a disruption of the AMPA/NMDA receptor ratio in these regions. Long-term neuronal potentiation (LTP) and long-term depression (LTD) follow this glutamatergic adaptation, leading to particular synaptic plasticity with the corollary of behavioral disorders, cognitive disorders, and the strengthening of the addiction circuit (Nelson et al., 2011).

Similarly, strengthening (via LTP) and weakening (via LTD) of excitatory synapses in the hippocampus have significant functions in learning and memory processes. The impaired synaptic plasticity observed in both conditions plays a major role in the cognitive deficits observed in some NDDs and SUDs. Cellular and animal models show impaired NMDA receptors in both conditions, leading to synaptic and neuroplastic alterations (Cramer et al., 2012).

Other neuroplastic alterations are observed in both disorders. For example, glycogen synthase kinase-3β (GSK3β) is a molecular hub linking numerous signaling pathways and regulating many cellular processes, including cytoskeletal rearrangement, cell migration, apoptosis, proliferation, and neuroplasticity. Disruption of GSK-3β activity may result in the development of synaptic plasticity dysfunction, leading to the development of both NDDs and SUDs (Selkoe, 2019; Karran et al., 2011). Growth factors, such as the brain-derived neurotrophic factor (BDNF), also appear to play a fundamental role in both disorders. BDNF is implicated in reward and motivation systems in addiction (Gomez-Isla & Frosch, 2017). In NDDs, impaired levels of growth factors, including BDNF, glial cell line-derived neurotrophic factor (GDNF), and nerve growth factor (NGF), have been identified (Selkoe, 2019).

Genetic and Epigenetic Factors

The aging process is characterized by a widespread reduction in DNA methylation within the brain. Reduced synaptic density, especially in the cortex and hippocampus, correlates with age-related brain DNA demethylation and is hypothesized to promote cognitive impairment in older adults. Recent research suggests that environmental enrichment can enhance adult neurogenesis and synaptic plasticity in the hippocampus while mitigating age-related changes in DNA methylation. Most of the differential methylation resulting from environmental enrichment occurs at methyl-CpG binding protein 2 (MeCP2) binding sites, and human studies of age-related cognitive decline and neurodegeneration have shown a significant representation of the genes affected by these changes. These results imply, again, that proactive involvement with the environment can modify brain function over time through DNA methylation, potentially guarding against the detrimental effects of aging (Wang & Mandelkow, 2016).

The epigenetic clock reflects the biological aging of individuals according to their DNA methylation at CpG sites. Individual longevity and general health have been predicted using discrepancies between epigenetic and chronological ages. However, only a limited number of studies have examined how drug addiction affects clock CG methylation. The DNA epigenetic age of heroin users was found to be younger than that of matched drug-naïve controls, based on an analysis of DNA changes in neurons isolated from the postmortem orbitofrontal cortex of individuals with opioid use disorder. The tet methylcytosine dioxygenase 3 (TET3) locus showed differential methylation with heroin use, indicating that drug-induced control of DNA epigenetic machinery may affect the brain's DNA epigenetic age. In the future, it will be interesting to investigate the impact of drug consumption on epigenetic age and its possible correlation with addiction vulnerability (Wang & Mandelkow, 2016).

Cognitive Impairment

Addiction and dementia are both associated with impairments in behavioral flexibility. Examples of these deficits include difficulties in stopping drug-seeking behaviors, rigidity in thinking, and problem-solving difficulties (Walker & Jucker, 2015). Chronic drug use also causes deficits in executive function and memory that are comparable to those seen in dementia. Maze tests, object recognition tasks, and other cognitive assays are frequently used in animal models to investigate these issues. In this regard, chronic alcoholism has been found to be associated with cognitive deficits in frontal cortical-dependent tasks (Walker & Jucker, 2015).

Clinically, beyond memory impairment, patients with chronic alcoholism show deficits in fluency, cognitive flexibility, and perseverative responding (Walker & Jucker, 2015). Both anterograde and retrograde memory functions are impaired in alcohol use disorder, particularly in the associated Wernicke–Korsakov syndrome. Recent studies also indicate that alcohol may be a critical factor in brain aging and cognitive decline, depending on the dose, duration of ethanol exposure, and withdrawal episodes. Conversely, some research suggests that low-to-moderate alcohol exposure throughout aging could have a protective effect on cognitive functions (Shankar et al., 2008).

Animal models have played a role in discovering the etiological factors behind the cognitive decline associated with drug misuse. Cognitive dysfunction appears to be primarily caused by white matter loss in the brain. Nonetheless, specific frontal brain regions have also shown evidence of neuronal loss. The frontal lobes regulate the complex cognitive abilities known as "executive functions," including working memory, sequential ordering, and reversal learning. Chronic alcohol users tend to perform poorly on these types of tasks. Additionally, recurrent withdrawal symptoms from binge drinking appear to contribute to hippocampal dysfunction, including impaired neurogenesis and memory (Walker & Jucker, 2015).

Pharmacological Interventions

Cholinesterase inhibitors have primarily been used in the treatment of AD. Such drugs, including donepezil, have also been shown to attenuate nicotine and heroin-seeking behaviors in rats. These compounds are also being investigated for their potential effects on addiction-related cognitive deficits (Ferreira & Klein, 2011). Concurrently, anti-inflammatory agents are currently being investigated in dementia and addiction based on the neuroinflammatory component identified in both disorders (Selkoe, 2019; Musiek & Holtzman, 2016).

In the realm of drugs, cannabidiol (CBD) is emerging as a novel therapeutic alternative for the treatment of dementia. A study conducted on AD mouse models showed that the hippocampal volume tended to increase proportionally with the dose of CBD (Saido & Leissring, 2012). Other studies have also documented the neuroprotective and anti-inflammatory properties of CBD, suggesting it could represent an early treatment for the neurodegenerative processes observed in AD (Boche et al., 2013). Finally, a study using knockout mice as an animal model of Lafora disease showed that CBD-enriched extract partially reduced the cognitive symptoms of the disorder (Bennett & Robbins, 2018), reinforcing the potential role of CBD in the treatment of cognitive deficits in NDDs.

Amphetamines have also begun to establish themselves as potential alternatives in the treatment of NDDs. Recent studies show that amphetamines play a role in the normalization of norepinephrine depletion in AD. Additionally, amphetamines have been found to enhance dendritic spine growth and, more significantly, demonstrate preclinical efficacy in ameliorating memory deficits in aged rats by potentiating norepinephrine neurotransmission at β-adrenergic receptors (Wilson & Djakiew, 2019). In AD mouse models, methamphetamine improved cognitive and executive functions and positively affected neural activity and plasticity in the medial prefrontal cortex and hippocampus (Wolfe, 2006).

CONCLUSION

The similarities and overlaps between animal models of addiction and dementia underscore the common neural pathways and processes underlying both conditions. Researchers can develop more comprehensive and effective therapeutic strategies for NDDs and SUDs by integrating insights from various preclinical models. The future will undoubtedly bring more accurate, human-relevant models that incorporate cutting-edge technologies while addressing the ethical challenges of animal research. Cross-disciplinary research could uncover new intervention targets, benefiting those affected by these severe disorders and paving the way for truly personalized treatments.

REFERENCES

Alexander, A. G., Marfil, V., & Li, C. (2014). Use of *Caenorhabditis elegans* as a model to study Alzheimer's disease and other neurodegenerative diseases. *Frontiers in Genetics, 5*, 279. https://doi.org/10.3389/fgene.2014.00279

Bennett, R. E., & Robbins, E. M. (2018). Tau and neuronal function in health and disease. *Brain Research, 1691*(Pt A), 1–13. https://doi.org/10.1016/j.brainres.2018.05.042

Bilinski, T., Bylak, A., & Zadrag-Tecza, R. (2017). The budding yeast *Saccharomyces cerevisiae* as a model organism: Possible implications for gerontological studies. *Biogerontology, 18*(5), 631–640. https://doi.org/10.1007/s10522-017-9694-2

Billings, L. M., Oddo, S., Green, K. N., McGaugh, J. L., & LaFerla, F. M. (2005). Intraneuronal Abeta causes the onset of early Alzheimer's disease-related cognitive deficits in transgenic mice. *Neuron, 45*(5), 675–688. https://doi.org/10.1016/j.neuron.2005.01.040

Boche, D., Perry, V. H., & Nicoll, J. A. (2013). Review: Activation patterns of microglia and their identification in the human brain. *Neuropathology and Applied Neurobiology, 39*(1), 3–18. https://doi.org/10.1111/nan.12011

Botstein, D., Chervitz, S. A., & Cherry, J. M. (1997). Yeast as a model organism. *Science, 277*(5330), 1259–1260. https://doi.org/10.1126/science.277.5330.1259

Casas-Tinto, S., Zhang, Y., Sanchez-Garcia, J., Gomez-Velazquez, M., Rincon-Limas, D. E., & Fernandez-Funez, P. (2011). The ER stress factor XBP1s prevents amyloid-beta neurotoxicity. *Human Molecular Genetics, 20*(11), 2144–2160. https://doi.org/10.1093/hmg/ddr101

Cramer, P. E., Cirrito, J. R., Wesson, D. W., Lee, C. Y., Karlo, J. C., Zinn, A. E., ... & Landreth, G. E. (2012). ApoE-directed therapeutics rapidly clear β-amyloid and reverse deficits in AD mouse models. *Science, 335*(6075), 1503–1506. https://doi.org/10.1126/science.1217697

Dawson, T. M., Golde, T. E., & Lagier-Tourenne, C. (2018). Animal models of neurodegenerative diseases. *Nature Neuroscience, 21*(10), 1370–1379. https://doi.org/10.1038/s41593-018-0236-8

Deal, S. L., & Yamamoto, S. (2018). Unraveling novel mechanisms of neurodegeneration through a large-scale forward genetic screen in Drosophila. *Frontiers in Genetics, 9*, 700. https://doi.org/10.3389/fgene.2018.00700

del Valle Rodríguez, A., Didiano, D., & Desplan, C. (2011). Power tools for gene expression and clonal analysis in Drosophila. *Nature Methods, 9*(1), 47–55. https://doi.org/10.1038/nmeth.1800

Delenclos, M., Burgess, J. D., Lamprokostopoulou, A., Outeiro, T. F., Vekrellis, K., & McLean, P. J. (2019). Cellular models of alpha-synuclein toxicity and aggregation. *Journal of Neurochemistry, 150*(5), 566–576. https://doi.org/10.1111/jnc.14819

DeSalvo, M. K., Hindle, S. J., Rusan, Z. M., Orng, S., Eddison, M., Halliwill, K., ... & Bainton, R. J. (2014). The Drosophila surface glia transcriptome: Evolutionary conserved blood-brain barrier processes. *Frontiers in Neuroscience, 8*, 346. https://doi.org/10.3389/fnins.2014.00346

Domínguez-Oliva, A., Hernández-Ávalos, I., Martínez-Burnes, J., Olmos-Hernández, A., Verduzco-Mendoza, A., & Mota-Rojas, D. (2023). The importance of animal models in biomedical research: Current insights and applications. *Animals, 13*(7), 1223. https://doi.org/10.3390/ani13071223

Ferreira, S. T., & Klein, W. L. (2011). The Aβ oligomer hypothesis for synapse failure and memory loss in Alzheimer's disease. *Neurobiology of Learning and Memory, 96*(4), 529–543. https://doi.org/10.1016/j.nlm.2011.08.003

Gomez-Isla, T., & Frosch, M. P. (2017). Tau, amyloid, and the pathobiology of Alzheimer's disease. *Brain Pathology, 27*(5), 612–616. https://doi.org/10.1111/bpa.12523

Hobert, O. (2010). Neurogenesis in the nematode *Caenorhabditis elegans*. *WormBook*, 1–24. https://doi.org/10.1895/wormbook.1.12.2

Ittner, L. M., & Götz, J. (2011). Amyloid-β and tau—A toxic pas de deux in Alzheimer's disease. *Nature Reviews Neuroscience, 12*(2), 65–72. https://doi.org/10.1038/nrn2967

Jackson Laboratory. (n.d.). B6.Cg-Tg(PDGFB-APPSwInd)20Lms/2Mmjax. The Jackson Laboratory. Retrieved from https://www.jax.org/strain/005864

Kalueff, A. V., Stewart, A. M., & Gerlai, R. (2014). Zebrafish as an emerging model for studying complex brain disorders. *Trends in Pharmacological Sciences, 35*(2), 63–75. https://doi.org/10.1016/j.tips.2013.12.002

Karran, E., Mercken, M., & De Strooper, B. (2011). The amyloid cascade hypothesis for Alzheimer's disease: An appraisal for the development of therapeutics. *Nature Reviews Drug Discovery, 10*(9), 698–712. https://doi.org/10.1038/nrd3505

Kim, Y., Park, Y., Hwang, J., & Kwack, K. (2018). Comparative genomic analysis of the human and nematode *Caenorhabditis elegans* uncovers potential reproductive genes and disease associations in humans. *Physiological Genomics, 50*(11), 1002–1014. https://doi.org/10.1152/physiolgenomics.00048.2018

Korhonen, P., Malm, T., & White, A. R. (2018). 3D human brain cell models: New frontiers in disease understanding and drug discovery for neurodegenerative diseases. *Neurochemistry International, 120*, 191–199. https://doi.org/10.1016/j.neuint.2018.08.011

Kosel, F., Pelley, J., & Franklin, T. B. (2020). Behavioral and psychological symptoms of dementia in mouse models of Alzheimer's disease-related pathology. *Neuroscience & Biobehavioral Reviews, 112*, 634–647. https://doi.org/10.1016/j.neubiorev.2020.02.017

Krashia, P., Nobili, A., & D'Amelio, M. (2019). Unifying hypothesis of dopamine neuron loss in neurodegenerative diseases: Focusing on Alzheimer's disease. *Frontiers in Molecular Neuroscience, 12*, 123. https://doi.org/10.3389/fnmol.2019.00123

Kuhn, B. N., Kalivas, P. W., & Bobadilla, A. C. (2019). Understanding addiction using animal models. *Frontiers in Behavioral Neuroscience, 13*, 262. https://doi.org/10.3389/fnbeh.2019.00262

LaFerla, F. M., Green, K. N., & Oddo, S. (2007). Intracellular amyloid-beta in Alzheimer's disease. *Nature Reviews Neuroscience, 8*(7), 499–509. https://doi.org/10.1038/nrn2168

Lai, C. H., Chou, C. Y., Ch'ang, L. Y., Liu, C. S., & Lin, W. (2000). Identification of novel human genes evolutionarily conserved in *Caenorhabditis elegans* by comparative proteomics. *Genome Research, 10*(5), 703–713. https://doi.org/10.1101/gr.10.5.703

Langhans, S. A. (2018). Three-dimensional in vitro cell culture models in drug discovery and drug repositioning. *Frontiers in Pharmacology, 9*, 6. https://doi.org/10.3389/fphar.2018.00006

Lanz, T. A., Carter, D. B., & Merchant, K. M. (2003). Dendritic spine loss in the hippocampus of young PDAPP and Tg2576 mice and its prevention by the ApoE2 genotype. *Neurobiology of Disease, 13*(3), 246–253. https://doi.org/10.1016/s0969-9961(03)00055-7

Li, H. W., Zhang, L., & Qin, C. (2019). Current state of research on non-human primate models of Alzheimer's disease. *Animal Models and Experimental Medicine, 2*(4), 227–238. https://doi.org/10.1002/ame2.12087

Li, J., & Le, W. (2013). Modeling neurodegenerative diseases in *Caenorhabditis elegans*. *Experimental Neurology, 250*, 94–103. https://doi.org/10.1016/j.expneurol.2013.09.012

Liddelow, S. A., & Barres, B. A. (2017). Reactive astrocytes: Production, function, and therapeutic potential. *Immunity, 46*(6), 957–967. https://doi.org/10.1016/j.immuni.2017.06.006

Link, C. D. (2006). *C. elegans* models of age-associated neurodegenerative diseases: Lessons from transgenic worm models of Alzheimer's disease. *Experimental Gerontology, 41*(10), 1007–1013. https://doi.org/10.1016/j.exger.2006.06.059

Liu, B., Liu, J., & Shi, J. S. (2020). SAMP8 mice as a model of age-related cognition decline with underlying mechanisms in Alzheimer's disease. *Journal of Alzheimer's Disease, 75*(2), 385–395. https://doi.org/10.3233/jad-190784

Liu, W., Li, L., Ye, H., Chen, H., Shen, W., & Zhong, Y. (2017). From *Saccharomyces cerevisiae* to human: The important gene co-expression modules. *Biomedical Reports, 7*(2), 153–158. https://doi.org/10.3892/br.2017.942

Medda, X., Mertens, L., Versweyveld, S., Diels, A., Barnham, L., Bretteville, A., ... & Verheijen, M. (2016). Development of a scalable, high-throughput-compatible assay to detect tau aggregates using iPSC-derived cortical neurons maintained in a three-dimensional culture format. *Journal of Biomolecular Screening, 21*(8), 804–815. https://doi.org/10.1177/1087057116655408

Musiek, E. S., & Holtzman, D. M. (2016). Mechanisms linking circadian clocks, sleep, and neurodegeneration. *Science, 354*(6315), 1004–1008. https://doi.org/10.1126/science.aah4968

National Institute on Drug Abuse. (n.d.). Drugs, brains, and behavior: The science of addiction. National Institute on Drug Abuse. Retrieved August 26, 2024, from https://nida.nih.gov/publications/drugs-brains-behavior-science-addiction/drug-misuse-addiction

Nelson, P. T., Head, E., & Schmitt, F. A. (2011). Alzheimer's disease is not "brain aging": Neuropathological, genetic, and epidemiological human studies. *Acta Neuropathologica, 121*(5), 571–587. https://doi.org/10.1007/s00401-011-0826-y

Panlilio, L. V., & Goldberg, S. R. (2007). Self-administration of drugs in animals and humans as a model and an investigative tool. *Addiction, 102*(12), 1863–1870. https://doi.org/10.1111/j.1360-0443.2007.02011.x

Paquet, D., Schmid, B., & Haass, C. (2010). Transgenic zebrafish as a novel animal model to study tauopathies and other neurodegenerative disorders in vivo. *Neurodegenerative Diseases, 7*(1–3), 99–102. https://doi.org/10.1159/000288725

Penney, J., Ralvenius, W. T., & Tsai, L. H. (2020). Modeling Alzheimer's disease with iPSC-derived brain cells. *Molecular Psychiatry, 25*, 148–167. https://doi.org/10.1038/s41380-019-0468-3

Rajendran, L., & Paolicelli, R. C. (2018). Microglia-mediated synapse loss in Alzheimer's disease. *Journal of Neuroscience, 38*(12), 2911–2919. https://doi.org/10.1523/JNEUROSCI.1136-17.2018

Saido, T. C., & Leissring, M. A. (2012). Proteolytic degradation of amyloid β-protein. *Cold Spring Harbor Perspectives in Medicine, 2*(6), a006379. https://doi.org/10.1101/cshperspect.a006379

Saleem, S., & Kannan, R. R. (2018). Zebrafish: An emerging real-time model system to study Alzheimer's disease and neurospecific drug discovery. *Cell Death Discovery, 4*, 45. https://doi.org/10.1038/s41420-018-0109-5

Sanchis-Segura, C., & Spanagel, R. (2006). Review: Behavioural assessment of drug reinforcement and addictive features in rodents: An overview of current protocols. *Current Protocols in Neuroscience, 35*(1), 9–10. https://doi.org/10.1002/0471142301.ns0910s35

Selkoe, D. J. (2019). Alzheimer disease and aducanumab: Adjusting our approach. *Nature Reviews Neurology, 15*(7), 365–366. https://doi.org/10.1038/s41582-019-0220-7

Shankar, G. M., Li, S., Mehta, T. H., Garcia-Munoz, A., Shepardson, N. E., Smith, I., ... & Selkoe, D. J. (2008). Amyloid-β protein dimers isolated directly from Alzheimer's brains impair synaptic plasticity and memory. *Nature Medicine, 14*(8), 837–842. https://doi.org/10.1038/nm1782

Slanzi, A., Iannoto, G., Rossi, B., Zenaro, E., & Constantin, G. (2020). In vitro models of neurodegenerative diseases. *Frontiers in Cell and Developmental Biology, 8*, 328. https://doi.org/10.3389/fcell.2020.00328

Spanagel, R. (2017). Animal models of addiction. *Dialogues in Clinical Neuroscience, 19*(3), 247–258. https://doi.org/10.31887/DCNS.2017.19.3/rspanagel

Tello, J. A., Williams, H. E., Eppler, R. M., Steinhilb, M. L., & Khanna, M. (2022). Animal models of neurodegenerative disease: Recent advances in fly highlight innovative approaches to drug discovery. *Frontiers in Molecular Neuroscience, 15*, 883358. https://doi.org/10.3389/fnmol.2022.883358

Tzschentke, T. M. (2007). Measuring reward with the conditioned place preference (CPP) paradigm: Update of the last decade. *Addiction Biology, 12*(3–4), 227–262. https://doi.org/10.1111/j.1369-1600.2007.00070.x

Uhl, E. W., & Warner, N. J. (2015). Mouse models as predictors of human responses: Evolutionary medicine. *Current Pathobiology Reports, 3*(3), 219–223. https://doi.org/10.1007/s40139-015-0075-5

Walker, L. C., & Jucker, M. (2015). Neurodegenerative diseases: Expanding the prion concept. *Annual Review of Neuroscience, 38*, 87–103. https://doi.org/10.1146/annurev-neuro-071714-033828

Wang, Y., & Mandelkow, E. (2016). Tau in physiology and pathology. *Nature Reviews Neuroscience, 17*(1), 5–21. https://doi.org/10.1038/nrn.2015.1

Wilson, C. A., & Djakiew, D. (2019). Inflammatory cytokines and neurodegeneration: A role for the NLRP3 inflammasome in Alzheimer's disease. *Frontiers in Cellular Neuroscience, 13*, 122. https://doi.org/10.3389/fncel.2019.00122

Winderickx, J., Delay, C., De Vos, A., Klinger, H., Pellens, K., Vanhelmont, T., ... & Wésolowski-Louvel, M. (2008). Protein folding diseases and neurodegeneration: Lessons learned from yeast. *Biochimica et Biophysica Acta (BBA)—Molecular Basis of Disease, 1783*(7), 1381–1395. https://doi.org/10.1016/j.bbadis.2008.03.003

Wolfe, M. S. (2006). The γ-secretase complex: Membrane-embedded proteolytic ensemble for Alzheimer's disease and notch signaling. *Current Issues in Molecular Biology, 8*(2), 69–84. https://doi.org/10.21775/cimb.008.069

World Health Organization. (n.d.). Aging and health. World Health Organization. Retrieved from https://www.who.int/news-room/fact-sheets/detail/ageing-and-health

World Health Organization. (n.d.). Dementia. World Health Organization. Retrieved from https://www.who.int/news-room/fact-sheets/detail/dementia#:~:text=Key%20facts,injuries%20that%20affect%20the%20brain

2 Alzheimer's Disease and Drug Addiction

A Global and Sub-Saharan African Perspective

Joy Louise Gumikiriza-Onoria, Dennis Buwembo, Rita Nassanga, and Kamada Lwere

INTRODUCTION

Alzheimer's disease (AD) and drug addiction are significant public health challenges that affect populations worldwide, transcending cultural and national boundaries. This chapter explores the impact and interconnections of these two conditions globally, explicitly focusing on low- and middle-income countries (LMICs), such as Sub-Saharan Africa (SSA).

Overview of Alzheimer's Disease and Drug Addiction

AD, the most common form of dementia, is characterized by progressive neurodegenerative changes that lead to severe cognitive decline and the loss of daily living abilities (Alzheimer's Disease Facts and Figures, 2023). Symptoms such as memory loss, problem-solving difficulties, and behavioral changes gradually result in an inability to perform simple tasks independently. The global prevalence of AD is rising, with over 70 million people currently affected, a number expected to triple by 2050 due to aging populations (World Health Organization, 2022).

Drug addiction, or substance use disorder (SUD), is a complex condition manifested by compulsive drug use despite harmful consequences. It involves structural and functional changes in the brain that lead to behavioral, perceptual, and physiological alterations (APA, 2013; Subramaniam et al., 2023). The global burden of drug addiction is immense, impacting healthcare systems, economies, and families. According to the 2022 World Drug Report released by the United Nations Office on Drugs and Crime (UNODC), in 2020 an estimated 284 million people worldwide aged 15–64 had used a drug within the last 12 months (Marinelli et al., 2023).

The Neurobiological Intersection of Alzheimer's Disease and Drug Addiction

Recent research suggests a significant neurobiological intersection between AD and drug addiction. Both conditions involve alterations in neurotransmitter systems, particularly those related to dopamine signaling, which is crucial in the brain's reward and learning pathways (Ekhtiari et al., 2021). SUDs can exacerbate neurodegenerative processes, potentially accelerating cognitive decline through mechanisms such as chronic neuroinflammation, oxidative stress, and synaptic dysfunction (Volkow et al., 2014; Collins-Praino & Corrigan, 2017).

Chronic use of substances like alcohol, methamphetamine, and opioids has been shown to induce neurotoxic effects that mimic or exacerbate Alzheimer's pathology, including β-amyloid (Aβ) plaque accumulation and tau tangle formation (Kinney et al., 2018; Zhang et al., 2023). These shared pathological mechanisms underscore the importance of addressing substance use in populations at risk

DOI: 10.1201/9781003478485-2

for or already suffering from AD, especially in regions such as SSA, where both conditions are increasing.

SUDs complicate the management of AD, particularly in terms of treatment adherence. Individuals with SUDs are often less likely to adhere to prescribed treatments due to cognitive and behavioral impairments associated with addiction (Levran et al., 2012). This non-adherence can lead to faster disease progression, poorer health outcomes, and increased caregiver burden, further complicating public health responses in resource-limited settings (Dhana et al., 2021).

The Importance of Global and Local Perspectives

Understanding AD and SUD in a global context is crucial for developing effective public health strategies, particularly in LMICs where resources are limited and healthcare systems may struggle with the growing burden of these conditions. SSA faces unique challenges, including limited access to healthcare, high rates of infectious diseases that complicate treatment strategies, and socioeconomic factors exacerbating health disparities (World Health Organization, Regional Office for Africa, 2022).

In SSA, the prevalence of Alzheimer's is increasing, but the condition is underdiagnosed and under-reported due to a lack of awareness and diagnostic facilities (World Health Organization, 2021; Akinyemi et al., 2023). Similarly, drug addiction remains a critical public health issue, with rising usage of both legal and illegal substances hindering treatment and prevention efforts (Lo et al., 2020).

This chapter will delve deeper into these issues by examining the socioeconomic, cultural, and healthcare factors that influence the prevalence and management of AD and drug addiction in SSA. By comparing LMICs with global trends, we aim to highlight unique regional challenges and potential strategies that can be adapted from global successes.

GLOBAL PERSPECTIVES ON ALZHEIMER'S DISEASE

Prevalence, Socioeconomic Impact, and Evolution of Alzheimer's Disease

AD is a pervasive global health issue, with approximately 55 million people living with dementia worldwide, 60–70% of whom are affected by AD (Alzheimer's Disease International, 2023). The economic and social impact of AD is profound, with the global cost of dementia care exceeding US$1 trillion annually—a figure projected to double by 2030 if effective treatments are not developed (World Health Organization, 2022).

Table 2.1 below provides an overview of the estimated global prevalence rates of AD across various regions, highlighting the prevalence percentages and relevant factors influencing diagnosis and care.

TABLE 2.1
Global Prevalence Rates of Alzheimer's Disease (*World AlzheimerReport*, 2022)

Region	Estimated Prevalence (%)	Comments
North America	1.5	High diagnosis rates due to advanced healthcare systems.
Western Europe	1.4	Extensive elderly care options, similar to North America.
Sub-Saharan Africa	0.5	Underdiagnosed due to lack of facilities and awareness.
East Asia	1.0	Rising prevalence due to an aging population.
Latin America	0.8	Increasing awareness and diagnostic rates.
Middle East and North Africa	0.7	Cultural factors influencing reporting and care approaches.

Socioeconomic Factors and Global Variations in Alzheimer's Diagnosis and Treatment

In high-income countries, access to advanced diagnostic technologies like neuroimaging and biomarker testing has led to higher diagnosis rates and comprehensive treatment plans, including pharmacological interventions and caregiver support (Dafre & Wasnik, 2023). However, the situation is markedly different in LMICs, particularly in SSA, where diagnostic rates are significantly lower due to insufficient resources, trained healthcare professionals, and diagnostic tools.

The stigma associated with neurological and mental health conditions further complicates care, often leading to delays in diagnosis and limiting treatment options to symptom management rather than comprehensive disease management (Sexton et al., 2021). In SSA, several factors, in addition to limited access to diagnostic and healthcare facilities and low awareness of AD, exacerbate these challenges, making it difficult to manage AD effectively. Various diagnostic/treatment and care pathways have been reported, but the choice of pathway taken depends on perceptions towards AD. Non-use of modern healthcare systems is associated with the perception of AD as a natural part of the aging process. In contrast, the use of both modern healthcare systems and traditional care is linked to seeking treatment for other comorbidities, attributing AD to spirits or witchcraft, previous positive experiences with similar services, and concerns related to stigma (Mwenda et al., 2022).

Evolution of Alzheimer's Disease Understanding and Regional Insights

Since Dr Alois Alzheimer first identified the pathological features of Alzheimer's in 1906, our understanding of the disorder has significantly evolved. Initially considered a normal part of aging, Alzheimer's was later redefined as a significant cause of dementia thanks to mid-20th-century biomedical advancements (Beach, 1987; Jack et al., 2018; Yang et al., 2016).

The diagnostic criteria for AD have been refined over time, incorporating biomarkers and imaging tests to identify Aβ plaques and tau protein tangles. However, applying these advanced diagnostic tools remains challenging in SSA due to resource limitations (Carroll, 2019). Treatment approaches have also evolved, from symptom management to targeting underlying disease processes, such as immunotherapies to reduce Aβ plaques and tau proteins (Vogt et al., 2023; Zhang et al., 2024).

SSA's participation in international clinical trials offers hope for tailored approaches to Alzheimer's treatment, considering the region's unique genetic, environmental, and cultural factors (Akinyemi et al., 2023). This regional focus is crucial for addressing the specific needs and challenges that LMICs face in managing AD.

Imaging for Alzheimer's Disease in Sub-Saharan Africa

The limited availability and accessibility of advanced neuroimaging modalities significantly challenge the diagnosis of AD in SSA. Essential imaging techniques such as computed tomography (CT), magnetic resonance imaging (MRI), functional MRI (fMRI), MRI morphometry, and positron emission tomography-computed tomography (PET-CT) are crucial for the early detection and management of AD. However, their utilization in SSA is constrained by several factors, including resource limitations, inadequate infrastructure, and a shortage of trained healthcare professionals.

CT imaging is the most commonly used modality in SSA due to its lower cost, faster scan times, and broader availability compared to other advanced imaging techniques. It is a practical tool for ruling out other reversible causes of dementia-like symptoms, such as brain tumors or hemorrhages (Mackenzie et al., 2014). However, CT's utility in diagnosing Alzheimer's is limited because it lacks the sensitivity needed to detect the subtle and early neurodegenerative changes associated with AD, such as hippocampal atrophy, a disease hallmark (Pasi et al., 2011).

MRI offers superior sensitivity and specificity in detecting structural brain changes indicative of AD, particularly hippocampal atrophy (Schuff et al., 2009). Despite its advantages, the adoption of MRI in SSA remains limited due to the high costs of MRI machines, maintenance challenges, and a critical shortage of trained radiologists and technicians who can interpret these images (Akinyemi et al., 2023). Additionally, the availability of MRI is concentrated in a few urban centers, further restricting access for many populations in rural areas.

fMRI provides valuable insights into brain activity by measuring changes in blood flow, potentially enhancing our understanding of the neural mechanisms underlying Alzheimer's and related conditions, such as drug addiction (Buckner, 2012). However, implementing fMRI in SSA is almost non-existent due to the substantial financial investments required, the complexity of the operation, and the need for highly specialized personnel (Mogaji et al., 2022).

Similarly, MRI morphometry, which allows for the quantitative analysis of brain structures, is an invaluable tool for early AD detection. Yet, its application in SSA is rare, primarily due to the lack of advanced imaging software and the technical expertise necessary to perform these analyses (Sexton et al., 2021).

PET-CT, particularly amyloid PET, represents one of the most advanced imaging modalities available for early AD diagnosis, as it allows for the direct visualization of Aβ plaques in the brain, a key pathological feature of AD (Villemagne et al., 2017). Nevertheless, the availability of PET-CT in SSA is minimal, with access confined to a few specialized centers in countries like South Africa and Kenya. The high cost of PET-CT scans, the need for specialized facilities, and the logistical challenges associated with producing and distributing radiopharmaceuticals hinder its widespread use in the region (Dafre & Wasnik, 2023).

To overcome these challenges, SSA urgently needs regional collaborations that can pool resources and expertise, investments in imaging infrastructure, and the development of innovative healthcare delivery models tailored to its unique needs. Additionally, training programs aimed at building local capacity in neuroimaging are crucial for enhancing the diagnostic and research capabilities in the region. These efforts could significantly improve the early diagnosis and management of AD and associated conditions like drug addiction in SSA, thereby contributing to global research and clinical practice.

DRUG ADDICTION: A GLOBAL HEALTH CHALLENGE

Overview of Drug Addiction as a Global Health Issue

Drug addiction, or SUD, is a major global health challenge affecting millions of people across all regions and societies (Table 2.2). According to the World Health Organization, approximately 35 million individuals worldwide suffer from SUDs, with only one in seven receiving treatment (UNODC, 2023). The consequences of drug addiction extend beyond the individual, affecting families, communities, and national healthcare systems, leading to an increased burden of disease, crime, and societal costs.

Substance Use in Older Adults

Substance use among older adults is a growing concern, particularly given the aging global population. According to recent studies, alcohol, prescription opioids and benzodiazepines are among

TABLE 2.2
The Top Five Most Misused Drugs Globally (UNODC, 2023)

Rank	Drug	Global Prevalence	Comments
1	Alcohol	18%	Legal substance; a significant cause of morbidity and mortality.
2	Tobacco	15%	Highly addictive; leads to respiratory and cardiovascular diseases.
3	Cannabis	3.8%	Increasing legalization; concerns remain about mental health effects.
4	Opioids	0.9%	High addiction potential; significant public health burden.
5	Amphetamines	0.7%	Includes methamphetamine; associated with severe health issues.

the most commonly used substances in individuals aged ≥ 65 years (Kuerbis, 2020; Lin et al., 2023). The Substance Abuse and Mental Health Services Administration (SAMHSA) estimates that approximately 5.7 million adults over the age of 50 will require treatment for SUDs by 2030 (SAMHSA, 2024). Age-related physiological changes exacerbate substance use complications in this age group. For example, alcohol misuse in older adults increases the risk of falls, liver damage, and cognitive decline. Prescription drug misuse, particularly opioids and benzodiazepines, can lead to respiratory depression, sedation, and an increased risk of overdose due to slower metabolism and increased drug sensitivity (Markota et al., 2016). While alcohol remains the most commonly misused substance in this demographic, opioid misuse has sharply increased in recent years. Older adults are particularly vulnerable to opioid-related complications, including respiratory failure and adverse drug interactions, owing to the frequent use of multiple medications. This group is also more likely to experience chronic pain, which can lead to higher rates of prescription opioid use and potential dependency.

Commonly Misused Drugs among Older People and Their Impact on Health

Substance dependence among older adults is a growing concern, particularly with the increasing use of both legal and illegal substances among individuals in this vulnerable population. The substances most commonly misused by older individuals include alcohol, tobacco, opioids, stimulants, and cannabis. The health impacts of these substances are often more severe in older adults than their younger counterparts owing to age-related changes in metabolism, preexisting health conditions, and the increased likelihood of drug interactions. Below is a summary of commonly misused substances and their effects on the health of older people.

Opioids (e.g., Heroin)

- *Respiratory depression and overdose risk*: Opioids, including prescription pain relievers and illicit drugs such as heroin, are hazardous for older adults because of their sedative properties. Opioid use can lead to respiratory depression, wherein breathing slows or stops altogether. This is the leading cause of death from opioid overdose and can be exacerbated by slower metabolism and decreased organ function, both common in older people (National Institute on Drug Abuse, 2023). Additionally, opioids are highly addictive, and their long-term use often leads to tolerance, requiring higher doses to achieve the same pain relief, which increases the risk of overdose.
- *Infectious disease risk*: In older adults who use opioids intravenously, the risk of infectious diseases such as HIV and hepatitis C increases significantly, secondary to unsafe injection practices, such as needle sharing, which facilitate the transmission of bloodborne pathogens (CDC, 2024). Chronic infections, including hepatitis C, can lead to liver damage, cirrhosis, and liver cancer, compounding the health challenges faced by older adults.

Stimulants (e.g., Cocaine and Methamphetamine)

- *Cardiovascular problems*: The use of cocaine and methamphetamine is associated with severe cardiovascular risks, which are magnified in older adults. Stimulants increase heart rate, blood pressure, and the risk of arrhythmias, heart attacks, and strokes. These substances cause vasoconstriction, reduce blood flow to the heart, and increase the likelihood of acute cardiovascular events (Coles et al., 2022). Older adults are particularly vulnerable to these effects because of the prevalence of preexisting heart conditions and age-related vascular changes in this population.
- *Neurological damage and mental health issues*: The long-term use of stimulants can lead to significant neurological damage (Shukla & Vincent, 2020). Cocaine and methamphetamine overstimulate the dopamine system in the brain, which can lead to permanent changes in brain function, including cognitive deficits and memory loss. Stimulant misuse

is also associated with mental health issues such as anxiety, paranoia, and psychosis. In older adults, these psychiatric problems can exacerbate preexisting conditions such as depression or anxiety and increase the risk of dementia (Pariente et al., 2018).

Depressants (e.g., Alcohol)

- *Liver disease and cognitive impairment*: Alcohol is one of the most commonly misused substances in older adults. Chronic alcohol intake leads to a higher incidence of liver diseases, such as cirrhosis and alcoholic hepatitis. The ability of the liver to process alcohol diminishes with age, causing a higher blood alcohol concentration and increased toxicity. Chronic alcohol use in older adults is also strongly associated with cognitive impairment, including memory loss, impaired judgment, and an increased risk of developing dementia (National Institute on Drug Abuse, 2023).
- *Interaction with prescription medications*: Older adults are more likely to be taking multiple prescription medications, many of which interact with alcohol. Benzodiazepines and opioids are commonly prescribed for anxiety, sleep disorders, and pain management, and their depressant effects are significantly amplified when combined with alcohol. This can lead to severe sedation, respiratory depression, increased risk of falls, and overdose, which are particularly concerning among older adults (Taipale et al., 2018).
- *Social isolation*: Alcohol misuse can also contribute to social isolation, a significant issue among older adults. As alcohol dependence increases, social interactions and participation in community activities often decrease, leading to loneliness and further deterioration of mental health. This isolation can fuel the cycle of increased alcohol consumption as a means of coping with loneliness and depression.

Cannabis

- *Use for pain management*: Cannabis use is on the rise among older adults, often for pain management and particularly in the context of chronic conditions such as arthritis. Although cannabis can provide relief from pain and improve sleep, its long-term use in older adults presents risks that need to be carefully managed. These include cognitive impairments, such as short-term memory loss and concentration difficulties, which might worsen preexisting cognitive decline (Coles et al., 2022).
- *Mental health risks*: For older adults with preexisting neurological and mental health conditions, such as dementia, anxiety, or depression, cannabis use can exacerbate these problems. Some studies have linked long-term cannabis use to an increased risk of psychosis and other serious mental health issues, particularly in individuals with a genetic predisposition to mental illnesses. While cannabidiol (CBD) is being investigated for its neuroprotective properties in neurodegenerative conditions like AD, the psychoactive component of cannabis, tetrahydrocannabinol (THC), can have adverse effects on mental health in older populations (Pariente et al., 2018).

Demographic and Socioeconomic Influences on Drug Use in Low- and Middle-Income Countries

In LMICs, drug addiction is strongly influenced by demographic and socioeconomic factors. War zones and refugee populations are particularly vulnerable to substance misuse due to immense stress, trauma, and instability. Refugee camps and conflict areas often have limited access to mental health and addiction treatment services, exacerbating the risk of drug use as a coping mechanism. For example, opioid use has been documented among refugees in conflict zones in the Middle East, where access to healthcare is severely restricted, and the psychological toll of displacement is immense (Laban et al., 2007).

Cultural and social dynamics also play a significant role in substance use in various LMICs. For instance, cannabis cultivation and use are deeply rooted in cultural practices in many Arab countries, often normalized despite legal restrictions. In Morocco, cannabis ("kif") has been traditionally cultivated for centuries and remains widely consumed despite efforts to curtail its production and consumption (Chouvy, 2016).

Similarly, khat, a plant chewed for its stimulant effects, is commonly used in Uganda, Kenya, Somalia, and Ethiopia. The use of khat is socially accepted and embedded in cultural practices, but its widespread use has raised concerns due to its addictive properties and associated health risks, including mental health disorders and economic burdens (Al Shubbar, 2024; Hassen et al., 2021; Odenwald et al., 2007). In Uganda, khat use is prevalent in regions like Karamoja, where it contributes to significant social and health problems (Beckerleg, 2010).

Drug Addiction in Sub-Saharan Africa

In SSA, drug addiction is complicated by socioeconomic factors such as poverty, unemployment, and limited access to education, which exacerbate substance dependence issues. The region faces a growing problem with traditional substances like alcohol and khat, alongside an increase in synthetic drugs and opioids. Health systems in many SSA countries are under-resourced and ill-equipped to deal with the complex needs of individuals with addiction, resulting in inadequate treatment and support services (Nyaga et al., 2021).

SUDs are intertwined with sociopolitical instability in Uganda and other East African nations. Prolonged conflicts in Northern Uganda, for example, have led to an increase in alcohol and drug use as coping mechanisms among affected populations. Former child soldiers and displaced persons in the region are particularly vulnerable to substance dependence, with alcohol and khat misuse being prominent issues (Bhui & Warfa, 2007; Makoha & Denov, 2024; Veale & Stavrou, 2007). These substances are often consumed to mitigate psychological distress caused by war and displacement, leading to long-term addiction problems that challenge post-conflict recovery and reintegration efforts.

THE INTERSECTION OF ALZHEIMER'S DISEASE AND DRUG ADDICTION

Globally, evidence suggests a significant overlap in the neurobiological pathways affected by AD and SUDs. Dysfunction in dopamine signaling, critical to the brain's reward system and a hallmark of both Alzheimer's pathology and drug addiction, highlights this connection (Volkow et al., 2014).

In SSA, this intricate relationship and potential comorbidity present an added burden. Countries such as Nigeria, South Africa, and Kenya are witnessing a rise in substance misuse among older adults. For instance, in South Africa, the use of methamphetamine ("tik") and its impact on cognitive decline has been a critical area for research, given the high abuse rates in regions like the Western Cape (Plüddemann et al., 2010). The interaction between other local drug use patterns, such as the widespread consumption of khat and cannabis, and cognitive health necessitates region-specific studies to understand the full impact of these substances on neurodegenerative processes (Ekhtiari et al., 2021).

Potential Mechanisms Linking Alzheimer's Disease and Substance Use Disorders

The potential mechanisms linking AD and SUD are diverse and complex, involving neurobiological and genetic pathways. Substance misuse can exacerbate AD pathology, accelerating cognitive decline and neurodegeneration. Specific substances may increase or decrease the risk of developing AD. This section explores how neuroinflammation, neuroplasticity alterations, genetic and epigenetic factors, and clinical studies shape our understanding of the connections between AD and SUD.

Neuroinflammation

Neuroinflammation plays a central role in the pathology of AD and chronic substance use. Chronic alcohol, methamphetamine, and opioid use induce neuroinflammatory responses that mirror those found in AD, where activated microglia release cytokines that lead to neuronal damage. Microglial activation, a hallmark of neuroinflammation, has also been observed in AD, as Aβ plaques and tau tangles accumulate (Collins-Praino & Corrigan, 2017).

When it comes to alcohol, alcohol-feeding studies in AD mouse models reinforce the concept that extreme alcohol intake promotes AD while low doses have protective properties (Chandrashekar et al., 2023). Long-term alcohol misuse has been shown to increase Aβ plaque deposition in the brain, worsening inflammation and accelerating neurodegeneration (Ramos et al., 2022). Venkataraman and colleagues discuss how alcohol worsens the cognitive burden of AD, contributes to its genesis, and allows its sustenance via neuroinflammation (Venkataraman et al., 2017). This is mediated by direct effects, mainly through the modulation of Aβ plaque synthesis and clearance, as well as the stimulation of Toll-like receptors, glycogen synthase kinase 3-β, protein kinase-B (Akt)/mammalian target of rapamycin (mTOR) pathway, cyclin-dependent kinase-5, cyclic adenosine monophosphate (cAMP) response element-binding protein phosphorylation pathway, and insulin-like growth factor type-1 receptor. In addition to these brain-centric pathways, indirect processes include alcohol-mediated liver and small bowel injury, through alterations in the peripheral-to-central Aβ plaque homeostasis, withdrawal, kindling effect, and increased risk of traumatic brain injury (Chandrashekar et al., 2023; Venkataraman et al., 2017).

Shukla and Vincent, 2020Similarly, methamphetamine misuse induces oxidative stress and neuroinflammation that damage neurons, and studies have suggested that methamphetamine may increase the risk of AD (Shukla and Vincent, 2020). In their extensive review, Shukla and Vincent elaborate on preclinical and clinical data, discussing the analogy between the profiles of neuronal loss and atrophy in methamphetamine misuse and AD. The authors discuss evidence linking methamphetamine misuse with a higher probability of developing premature AD through various pathways, including neuroinflammation, anatomical links, and genetic factors. Alternatively, the authors highlight some studies exploring the potential therapeutic effects of methamphetamine under specific conditions, which may help enhance cognitive function in early-stage AD models (Shukla & Vincent, 2020). This dichotomy highlights the complex effects of substance use on AD progression.

Neuroplasticity Alterations

Substance use also disrupts neuroplasticity, the brain's ability to form and reorganize synaptic connections, which is crucial for learning and memory. In preclinical mouse models of AD, alcohol was found to increase the onset and magnitude of AD-like neural and behavioral pathology (Hoffman et al., 2019). Along the same lines, chronic use of cocaine, methamphetamine, and opioids was found to alter synaptic plasticity, contributing to the cognitive decline observed in both AD and SUDs (Levran et al., 2012).

Methamphetamine, in particular, has profound effects on the brain's dopamine system, damaging neurons and disrupting memory pathways. The ability of methamphetamine to alter synaptic plasticity in regions such as the hippocampus is particularly concerning (Wu et al., 2024), as the hippocampus is critical for memory and is one of the first areas affected in AD (Rao et al., 2022). The neurotoxic effects of methamphetamine may accelerate AD progression by impairing the neural circuits required for cognitive function. This makes it a substance to avoid at all costs, particularly in populations at risk for AD (Shukla & Vincent, 2020).

Genetic and Epigenetic Factors

Genetic and epigenetic factors present another common ground between SUDs and AD. The presence of the APOE ε4 allele, a genetic variant that increases the risk of AD, is associated with a higher likelihood of substance misuse, particularly alcohol. The interaction of this allele with chronic substance use can accelerate Aβ plaque accumulation, thus worsening cognitive decline (Calderón et al., 2023). Additionally, variations in the dopamine transporter gene (DAT1) have been implicated in both conditions, potentially linking dopaminergic dysregulation to addiction and AD development. Epigenetic changes resulting from chronic substance use, such as DNA methylation and histone modification, can further contribute to the pathological features of AD (Emrani et al., 2020). For instance, chronic alcohol use induces epigenetic modifications that affect gene expression in the brain, leading to cognitive impairment and exacerbating AD pathology (Taipale et al., 2018).

Roussotte and colleagues provide a comprehensive review of neuroimaging and genetic risk in AD. They highlight an overlap between genes that code for neurotransmitter transporters and receptors involved in the etiology of both AD and substance misuse. These include SLC1 glutamate transporters and delta opioid receptors encoded by the OPRD1 gene. Other relevant genes are the COL25A1 gene, which encodes collagen XXV alpha 1, also known as the collagen-like Alzheimer's Aβ plaque component precursor, and the brain-derived neurotrophic factor (BDNF) gene. In this regard, gene therapy targeting receptors and transporters involved in the etiology or pathophysiology of AD and drug addiction is being investigated, with experimental studies specifically targeting excitatory amino acid transporters, human sigma-1 receptor, and delta opioid receptors (Roussotte et al. 2014).

CLINICAL STUDIES: SUBSTANCE USE AND RISK OF ALZHEIMER'S DISEASE

Several clinical studies have provided insights into the impact of different substances on AD risk. A 2022 study by Ramos et al. investigated substance use-related cognitive decline in families with autosomal dominant AD. The study highlighted that chronic misuse of substances, such as methamphetamine and alcohol, accelerated the onset of cognitive symptoms in genetically predisposed individuals (Ramos et al., 2022). In the realm of stimulants, one study noted that amphetamine users exhibited an increase in several inflammatory markers, such as α1-anti-chymotrypsin, pigment epithelium-derived factor, and macrophage inflammatory protein-4, potentially predisposing them to an accelerated risk for AD (Alrafiah et al., 2022).

Findings on opioid use suggest a different narrative. Research conducted by Taipale et al. (2018) found no significant association between opioid use and increased risk of AD. This finding is significant as it hints that, despite the well-documented neuroinflammatory effects of opioids, they may not exacerbate AD pathology in the same way as other substances. These results have important implications for pain management in older adults at risk of AD, where opioid use may be necessary. Further studies on this particular topic are needed.

Regarding benzodiazepines, findings on their use and the development of AD are inconclusive. A 2018 study by Pariente et al. found that long-term benzodiazepine use was associated with a higher risk of AD, particularly when prescribed for extended periods to manage anxiety or sleep disorders. The results of this study highlighted the need for cautious use of benzodiazepines in older adults, especially those already at risk for AD (Pariente et al., 2018). A more recent review noted controversy regarding the use of benzodiazepines and their risk for the development or progression of AD (Al-Kuraishy et al., 2023a). While benzodiazepines may exert a protective effect against AD development by reducing neuroinflammation and tau phosphorylation, chronic use of intermediate-acting benzodiazepines, such as lorazepam, has been found to be associated with amnesic effects due to the reduction of synaptic plasticity and impairment of recognition memory (Al-Kuraishy et al., 2023b).

Lastly, a systematic review of 21 case-control studies and eight cohort studies found conflicting evidence on the direction of the association between smoking tobacco and AD (Almeida et al., 2002). A more recent meta-analysis of 27 studies found a significant association between smoking and AD, with the effect being stable over time (Niu et al., 2018).

Cannabinoids as an Emerging Treatment Avenue for Alzheimer's Disease

As research advances, cannabinoid-based therapies have emerged as potential treatment options for AD (Sánchez Montero et al., 2021) and its associated neuropsychiatric symptoms, such as agitation (Outen et al., 2021). The cannabinoid effect is mediated via competitive inhibition of the acetylcholinesterase enzyme and competitive antagonism at cannabinoid CB_1 and CB_2 receptors (Sánchez Montero et al., 2021). Preclinical studies also show that cannabinoids possess neuroprotective properties that may reduce Aβ plaque accumulation, decrease neuroinflammation, and modulate neurotransmitter systems in AD (Coles et al., 2022; Farkhondeh et al., 2020). Specifically, CBD was found to reverse and prevent the development of cognitive deficits in AD mouse models, while low-dose THC improved cognition in aging mice. Together, these findings suggest that a CBD-THC combination might be more efficacious compared to cannabinoid isolate treatment strategies (Coles et al., 2022).

Clinically, a recent meta-analysis of six double-blind, placebo-controlled studies assessing the effects of cannabinoids on agitation in patients with AD was inconclusive, although significant effects on agitation were noted in those with greater cognitive impairment (Ruthirakuhan et al., 2019). An older literature review echoed the same results. Findings from six studies showed significant benefits from synthetic cannabinoids (e.g., dronabinol and nabilone) on agitation and aggression. However, definitive conclusions were limited by small sample sizes, lack of placebo, and short trial durations (Liu et al., 2015). In their comprehensive review of preclinical and clinical evidence, Abate et al. summarize the potential and limitations of cannabinoids in the treatment of AD (Abate et al., 2021). Given the findings to date, further preclinical studies and rigorous prospective clinical trials are recommended to determine the safety and efficacy of cannabinoids for the treatment of AD and its associated neuropsychiatric symptoms.

CONCLUSION

The co-occurrence of AD and SUDs represents a multifaceted public health challenge, intertwining neurological, behavioral, and societal dimensions. This chapter has examined the neurobiological intersections between these conditions, emphasizing shared mechanisms such as disruptions in dopamine signaling, neuroinflammation, and neuroplasticity. Chronic substance misuse, including alcohol and methamphetamine, accelerates neurodegenerative processes, contributing to earlier onset and faster progression of cognitive decline in AD. Despite these critical connections, research on the comorbidity of AD and SUDs remains limited, leaving significant gaps in understanding their implications for diagnosis, treatment, and prevention, particularly in resource-limited settings.

In SSA, the dual burden of AD and SUDs is exacerbated by systemic barriers such as underdeveloped healthcare infrastructure, a scarcity of trained professionals, and pervasive stigma. Misconceptions, including the belief that dementia is a natural part of aging or that addiction reflects moral failure, further delay diagnosis and care. Poverty, unemployment, and other socioeconomic factors also fuel substance misuse and hinder access to effective interventions.

Addressing these challenges in SSA, and by extension in LMICs, requires region-specific strategies. Strengthening research is paramount to elucidating the cognitive effects of locally prevalent substances and the pathways linking SUDs and AD. Integrating dementia and addiction care into primary healthcare systems can improve early and simultaneous management. Public awareness campaigns, led by community and religious leaders, can combat stigma and misconceptions, fostering help-seeking behaviors. Policies must emphasize harm reduction and rehabilitation while addressing socioeconomic determinants through expanded education and employment initiatives.

In LMICs, similar challenges prevail, and healthcare systems must be adapted to manage the dual burden of AD and SUDs effectively. Local solutions must ensure that interventions are accessible and culturally appropriate, while international collaborations can help bring proven strategies from high-income countries to the region's unique contexts.

Further research into the intertwining aspects of AD and SUDs is essential for meaningful progress. Understanding how these conditions mutually exacerbate each other will guide the development of targeted, innovative treatments. Exploring novel treatment avenues, such as the potential role of cannabinoids in treating AD, along with neuroprotective therapies and personalized care approaches, holds promise for significantly improving patient outcomes and quality of life.

REFERENCES

Abate, G., Uberti, D., & Tambaro, S. (2021). Potential and limits of cannabinoids in Alzheimer's disease therapy. *Biology, 10*(6), 542. https://doi.org/10.3390/biology10060542

Akinyemi, R. O., Owolabi, M. O., Okubadejo, N., Ogunniyi, A., Kalaria, R. N., & The African Dementia Consortium. (2023). Challenges and opportunities in Alzheimer's disease research in Africa: Insights from the African Dementia Consortium. *Lancet Neurology, 22*(1), 28–29. https://doi.org/10.1016/S1474-4422(22)00475-6

Al Shubbar M. D. (2024). Understanding khat: Its sociocultural and health implications in Saudi Arabia. *Cureus, 16*(3), e56657. https://doi.org/10.7759/cureus.56657

Al-Kuraishy, H. M., Al-Gareeb, A. I., Saad, H. M., & Batiha, G. E. (2023a). Benzodiazepines in Alzheimer's disease: Beneficial or detrimental effects. *Inflammopharmacology, 31*(1), 221–230. https://doi.org/10.1007/s10787-022-01099-4

Al-Kuraishy, H. M., Al-Gareeb, A. I., Alsayegh, A. A., Abusudah, W. F., Almohmadi, N. H., Eldahshan, O. A., Ahmed, E. A., & Batiha, G. E. (2023b). Insights on benzodiazepines' potential in Alzheimer's disease. *Life Sciences, 320*, 121532. https://doi.org/10.1016/j.lfs.2023.121532

Almeida, O. P., Hulse, G. K., Lawrence, D., & Flicker, L. (2002). Smoking as a risk factor for Alzheimer's disease: Contrasting evidence from a systematic review of case-control and cohort studies. *Addiction (Abingdon, England), 97*(1), 15–28. https://doi.org/10.1046/j.1360-0443.2002.00016.x

Alrafiah, A. R., Abu-Illah, M. M., Magadmi, R. M., Aqeel, A., Najmi, A., & Jaddoh, S. (2022). Increased inflammatory markers at AMPH-addicts are related to neurodegenerative conditions: Alzheimer's disease. *Applied Sciences, 12*(11), 5536. https://doi.org/10.3390/app12115536

Alzheimer's Disease Facts and Figures (2023). Alzheimer's & dementia. *Journal of the Alzheimer's Association, 19*(4), 1598–1695.

Alzheimer's Disease International (ADI). (2023). World Alzheimer report 2022: Life after diagnosis: Navigating treatment, care, and support. *Alzheimer's & Dementia, 19*(4), 1598–1695. https://doi.org/10.1002/alz.13016

American Psychiatric Association. (2013). *Diagnostic and statistical manual of mental disorders* (5th ed.). https://doi.org/10.1176/appi.books.9780890425596

Barry, K. L., & Blow, F. C. (2016). Drinking over the lifespan: Focus on older adults. *Alcohol Research: Current Reviews, 38*(1), 115–120.

Beach, T. G. (1987). The history of Alzheimer's disease: Three debates. *Journal of the History of Medicine and Allied Sciences, 42*(3), 327–349. https://doi.org/10.1093/jhmas/42.3.327

Beckerleg, S. (2010). 'Idle and disorderly' khat users in Western Uganda. *Drugs: Education, Prevention and Policy, 17*(4), 303–314.

Bhui, K., & Warfa, N. (2007). Drug consumption in conflict zones in Somalia. *PLoS Medicine, 4*(12), e354.

Buckner, R. L. (2012). The serendipitous discovery of the default network. *NeuroImage, 62*(2), 1137–1145. https://doi.org/10.1016/j.neuroimage.2012.01.020

Calderón-Garcidueñas, L., Hernández-Luna, J., Aiello-Mora, M., Brito-Aguilar, R., Evelson, P. A., Villarreal-Ríos, R., & Mukherjee, P. S. (2023). APOE peripheral and brain impact: APOE4 carriers accelerate their Alzheimer continuum and have a high risk of suicide in PM2.5 polluted cities. *Biomolecules, 13*(6), 927. https://doi.org/10.3390/biom13060927

Carroll, W. M. (2019). The global burden of neurological disorders. *Lancet Neurology, 18*(5), 418–419. https://doi.org/10.1016/S1474-4422(19)30029-8

CDC. (2024). Data Summary: Vulnerable Areas for Infectious Diseases in Persons Who Inject Drugs. Retrieved from https://www.cdc.gov/persons-who-inject-drugs/vulnerable/index.html

Chandrashekar, D. V., Steinberg, R. A., Han, D., & Sumbria, R. K. (2023). Alcohol as a modifiable risk factor for Alzheimer's disease-evidence from experimental studies. *International Journal of Molecular Sciences, 24*(11), 9492. https://doi.org/10.3390/ijms24119492

Chouvy, P. A. (2016). The supply of Moroccan kif and hashish in Europe. *Journal of Drug Issues, 46*(1), 61–81. https://doi.org/10.1177/0022042615610745

Coles, M., Steiner-Lim, G. Z., & Karl, T. (2022). Therapeutic properties of multi-cannabinoid treatment strategies for Alzheimer's disease. *Frontiers in Neuroscience, 16*, 962922. https://doi.org/10.3389/fnins.2022.962922

Collins-Praino, L. E., & Corrigan, F. (2017). The impact of neuroinflammation in Alzheimer's disease: The role of TLR4. *Journal of Alzheimer's Disease, 60*(1), 1–13. https://doi.org/10.3233/JAD-170123

Crews, F. T., Sarkar, D. K., Qin, L., Zou, J., Boyadjieva, N., & Vetreno, R. P. (2015). Neuroimmune function and the consequences of alcohol exposure. *Alcohol Research: Current Reviews, 37*(2), 331–351. https://doi.org/10.15288/arcr.2015.37.2.331

Dafre, R., & Wasnik, P. (2023). Current diagnostic and treatment methods of Alzheimer's disease: A narrative review. *Cureus, 15*(9), e45649. https://doi.org/10.7759/cureus.45649

Dhana, K., Aggarwal, N. T., Rajan, K. B., Barnes, L. L., Evans, D. A., & Morris, M. C. (2021). Impact of the apolipoprotein E ε4 allele on the relationship between healthy lifestyle and cognitive decline: A population-based study. *American Journal of Epidemiology, 190*(7), 1225–1233. https://doi.org/10.1093/aje/kwab033

Ekhtiari, H., Zare-Bidoky, M., & Verdejo-Garcia, A. (2021). Neurocognitive disorders in substance use disorders. In N. el-Guebaly, G. Carrà, M. Galanter, & A. M. Baldacchino (Eds.), *Textbook of addiction treatment* (pp. 81–91). Springer. https://doi.org/10.1007/978-3-030-36391-8_81

Emrani, S., Arain, H. A., DeMarshall, C., & Nuriel, T. (2020). APOE4 is associated with cognitive and pathological heterogeneity in patients with Alzheimer's disease: A systematic review. *Alzheimer's Research & Therapy, 12*, 141. https://doi.org/10.1186/s13195-020-00740-0

Farkhondeh, T., et al. (2020). Impact of cannabis-based medicine on Alzheimer's disease by focusing on the Amyloid β-modifications: A systematic study. *CNS & Neurological Disorders—Drug Targets, 19*(5), 334–343. https://doi.org/10.2174/1871527319666200708130745

Hassen, M. T., Soboka, M., Widmann, M., Keller, L., Zeller, A. C., Büchele, N., Barnewitz, E., Yitayih, Y., Schiller, S., Senger, J., Adorjan, K., & Odenwald, M. (2021). Khat use patterns, associated features, and psychological problems in a khat-treatment-seeking student sample of Jimma University, Southwestern Ethiopia. *Frontiers in Public Health, 9*, 645980. https://doi-org.ezproxy.aub.edu.lb/10.3389/fpubh.2021.645980

Hoffman, J. L., Faccidomo, S., Kim, M., Taylor, S. M., Agoglia, A. E., May, A. M., Smith, E. N., Wong, L. C., & Hodge, C. W. (2019). Alcohol drinking exacerbates neural and behavioral pathology in the 3xTg-AD mouse model of Alzheimer's disease. *International Review of Neurobiology, 148*, 169–230. https://doi.org/10.1016/bs.irn.2019.10.017

Jack, C. R., Bennett, D. A., Blennow, K., Carrillo, M. C., Dunn, B., Haeberlein, S. B., ... & Silverberg, N. (2018). NIA-AA Research Framework: Toward a biological definition of Alzheimer's disease. *Lancet Neurology, 17*(5), 389–406. https://doi.org/10.1016/S1474-4422(18)30062-1

Kinney, J. W., Bemiller, S. M., Murtishaw, A. S., Leisgang, A. M., Salazar, A. M., & Lamb, B. T. (2018). Inflammation as a central mechanism in Alzheimer's disease. *Alzheimer's & Dementia: Translational Research & Clinical Interventions, 4*, 575–590. https://doi.org/10.1016/j.trci.2018.06.014

Kuerbis, A. (2020). Substance use among older adults: An update on prevalence, etiology, assessment, and intervention. *Gerontology, 66*(3), 249–258. https://doi.org/10.1159/000504363.

Laban, C. J., Komproe, I. H., Gernaat, H. B., & De Jong, J. T. (2007). The impact of a long asylum procedure on quality of life, disability, and physical health in Iraqi asylum seekers in the Netherlands. *Social Psychiatry and Psychiatric Epidemiology, 43*(7), 507–515. https://doi.org/10.1007/s00127-008-0333-1

Levran, O., Yuferov, V., & Kreek, M. J. (2012). The genetics of the opioid system and specific drug addictions. *Human Genetics, 131*(6), 823–842. https://doi.org/10.1007/s00439-012-1172-4

Lin, J., Arnovitz, M., Kotbi, N., & Francois, D. (2023). Substance use disorders in the geriatric population: A review and synthesis of the literature of a growing problem in a growing population. *Current Treatment Options in Psychiatry, 10*(3), 313–332. https://doi.org/10.1007/s40501-023-00291-9

Liu, C. S., Chau, S. A., Ruthirakuhan, M., Lanctôt, K. L., & Herrmann, N. (2015). Cannabinoids for the treatment of agitation and aggression in Alzheimer's disease. *CNS Drugs, 29*(8), 615–623. https://doi.org/10.1007/s40263-015-0270-y

Lo, T. W., Yeung, J. W. K., & Tam, C. H. L. (2020). Substance abuse and public health: A multilevel perspective and multiple responses. *International Journal of Environmental Research and Public Health, 17*(7), 2610. https://doi.org/10.3390/ijerph17072610

Mackenzie, I. R., Neumann, M., Bigio, E. H., Cairns, N. J., Alafuzoff, I., Kril, J., & Halliday, G. (2014). Nomenclature and nosology for neuropathologic subtypes of frontotemporal lobar degeneration: An update. *Acta Neuropathologica, 127*(1), 1–14. https://doi.org/10.1007/s00401-013-1228-9

Makoha, G., & Denov, M. (2024). War, forced displacement, and alcohol abuse: Experiences and perceptions of war-affected South Sudanese refugee youth living in Bidibidi refugee settlement in northern Uganda. *Frontiers in Public Health, 12*, 1232504.

Marinelli, S., Basile, G., Zaami, S., & Manfredini, R. (2023). Sex- and gender-specific drug abuse dynamics: the need for tailored therapeutic approaches. *Journal of Personalized Medicine, 13*(6), 965. https://doi.org/10.3390/jpm13060965

Markota, M., Rummans, T. A., Bostwick, J. M., & Lapid, M. I. (2016). Benzodiazepine use in older adults: Dangers, management, and alternative therapies. *Mayo Clinic proceedings, 91*(11), 1632–1639. https://doi.org/10.1016/j.mayocp.2016.07.024

Mogaji, H. O., Ataguba, J. E., & Anumudu, C. (2022). Addressing the gaps in neuroimaging research in Sub-Saharan Africa. *Journal of Neuroscience Methods, 378*, 109527. https://doi.org/10.1016/j.jneumeth.2022.109527

Mwenda, P., Lawlor, B., Kroll, T., & De Brun, A. (2022). A systematic review and narrative synthesis of experiences with caring for older people living with dementia in Sub-Saharan Africa. *BMC Geriatrics, 22*, 961. https://doi.org/10.1186/s12877-022-03668-2

National Institute on Drug Abuse (NIDA). (2023). NIDA.NIH.GOV. https://www.unodc.org/unodc/en/data-and-analysis/world-drug-report-2023.html

Niu, H., Qu, Y., Li, Z., Wang, R., Li, L., Li, M., Lv, X., Gao, C., Song, Y., & Li, B. (2018). Smoking and risk for Alzheimer disease: A meta-analysis based on both case-control and cohort study. *Journal of Nervous and Mental Disease, 206*(9), 680–685. https://doi.org/10.1097/NMD.0000000000000859

Nyaga, J., Mwaura, P., Mutundu, K., Njeru, D., Juma, G., & Were, T. (2021). Socio-economic and health consequences of drugs and substance use in Gachie: A peri-urban town on the outskirts of Nairobi. *African Journal of Alcohol and Drug Abuse (AJADA), 6*(1), 67–80. https://ajada.nacada.go.ke/index.php/ajada/article/view/39

Odenwald, M., Hinkel, H., Schauer, E., Neuner, F., Schauer, M., Elbert, T. R., & Rockstroh, B. (2007). The consumption of khat and other drugs in Somali combatants: A cross-sectional study. *PLoS Medicine, 4*(12), e341. https://doi.org/10.1371/journal.pmed.0040341

Outen, J. D., Burhanullah, M. H., Vandrey, R., Amjad, H., Harper, D. G., Patrick, R. E., May, R. L., Agronin, M. E., Forester, B. P., & Rosenberg, P. B. (2021). Cannabinoids for agitation in Alzheimer's disease. *American Journal of Geriatric Psychiatry: Official Journal of the American Association for Geriatric Psychiatry, 29*(12), 1253–1263. https://doi.org/10.1016/j.jagp.2021.01.015

Pariente, A., et al. (2018). Benzodiazepines as a risk factor for Alzheimer's disease. *Acta Psychiatrica Scandinavica*. https://doi.org/10.1111/acps.12921

Pasi, M., Poggesi, A., & Pantoni, L. (2011). The use of CT in dementia. *International Psychogeriatrics*, 23 Suppl 2, S6–S12.

Plüddemann, A., Flisher, A. J., McKetin, R., Parry, C. D., & Lombard, C. J. (2010). The emerging threat of methamphetamine use in South Africa. *South African Medical Journal, 95*(6), 403–405. https://doi.org/10.1097/ADM.0b013e3181b8964a

Rao, Y. L., Ganaraja, B., Murlimanju, B. V., Joy, T., Krishnamurthy, A., & Agrawal, A. (2022). Hippocampus and its involvement in Alzheimer's disease: A review. 3 *Biotech, 12*(2), 55. https://doi.org/10.1007/s13205-022-03123-4

Ramos, C., Quiroz, Y. T., Baena, A., Arboleda-Velasquez, J. F., Lopera, F., & Reiman, E. M. (2022). Substance use-related cognitive decline in families with autosomal dominant Alzheimer's disease: A cohort study. *Alzheimer's & Dementia, 18*(1), 1423–1439.

Roussotte, F. F., Daianu, M., Jahanshad, N., Leonardo, C. D., & Thompson, P. M. (2014). Neuroimaging and genetic risk for Alzheimer's disease and addiction-related degenerative brain disorders. *Brain Imaging and Behavior, 8*(2), 217–233. https://doi.org/10.1007/s11682-013-9263-y

Ruthirakuhan, M., Lanctôt, K. L., Vieira, D., & Herrmann, N. (2019). Natural and synthetic cannabinoids for agitation and aggression in Alzheimer's disease: A meta-analysis. *Journal of Clinical Psychiatry, 80*(2), 18r12617. https://doi.org/10.4088/JCP.18r12617

Sánchez Montero, J. M., Agis-Torres, A., Solano, D., Söllhuber, M., Fernandez, M., Villaro, W., Gómez-Cañas, M., García-Arencibia, M., Fernández-Ruiz, J., Egea, J., Martín, M. I., & Girón, R. (2021). Analogues of cannabinoids as multitarget drugs in the treatment of Alzheimer's disease. *European Journal of Pharmacology, 895*, 173875. https://doi.org/10.1016/j.ejphar.2021.173875

Schuff, N., Woerner, N., Boreta, L., Kornak, J., Flores, R., Kwon, D., & Weiner, M. W. (2009). MRI of hippocampal volume loss in early Alzheimer's disease in relation to ApoE genotype and biomarkers. *Brain, 132*(4), 1062–1077. https://doi.org/10.1093/brain/awp007

Sexton, C., Snyder, H. M., Chandrasekaran, L., Worley, S., & Carrillo, M. C. (2021). Expanding representation of low- and middle-income countries in global dementia research: Commentary from the Alzheimer's association. *Frontiers in Neurology, 12*, 633777. https://doi.org/10.3389/fneur.2021.633777

Shukla, M., & Vincent, B. (2020). The multifaceted impact of methamphetamine on Alzheimer's disease: From a triggering role to a possible therapeutic use. *Ageing Research Reviews, 60*, 101062. https://doi.org/10.1016/j.arr.2020.101062

Subramaniam, G. A., Nolan, L., Huntley, K., Corbin, M., Crenshaw, K., Mandell, T., & Linton, J. (2023). National institute on drug abuse: Dissemination of scientific knowledge to improve adolescent health. *Psychiatric Clinics of North America, 46*(4), 789–799. https://doi.org/10.1016/j.psc.2023.03.009

Substance Abuse and Mental Health Services Administration. (2024). *Annual National Survey on Drug use and health*. US Department of Health and Human Services. https://www.samhsa.gov

Taipale, H., Hamina, A., Lampela, P., Tanskanen, A., Tiihonen, J., Karttunen, N., Tolppanen, Anna-Maija, & Hartikainen, S. (2018). Is Alzheimer's disease associated with previous opioid use? *Pain Medicine, 19*(11), 2115–2121. https://doi.org/10.1093/pm/pnx210.

United Nations Office on Drugs and Crime. (2023). *World drug report 2023*. United Nations.

Venkataraman, A., Kalk, N., Sewell, G., Ritchie, C. W., & Lingford-Hughes, A. (2017). Alcohol and Alzheimer's disease-does alcohol dependence contribute to beta-amyloid deposition, neuroinflammation and neurodegeneration in Alzheimer's disease? *Alcohol and Alcoholism (Oxford, Oxfordshire), 52*(2), 151–158. https://doi.org/10.1093/alcalc/agw092

Villemagne, V. L., Burnham, S., Bourgeat, P., Brown, B., Ellis, K. A., Salvado, O., & Rowe, C. C. (2017). Amyloid β deposition, neurodegeneration, and cognitive decline in sporadic Alzheimer's disease: A prospective cohort study. *Lancet Neurology, 12*(4), 357–367. https://doi.org/10.1016/S1474-4422(13)70255-3

Veale, A., & Stavrou, A. (2007). Former Lord's Resistance Army child soldier abductees: Explorations of identity in reintegration and reconciliation. *Peace and Conflict: Journal of Peace Psychology, 13*(3), 273.

Vogt, A. S., Jennings, G. T., Mohsen, M. O., Vogel, M., & Bachmann, M. F. (2023). Alzheimer's disease: A brief history of immunotherapies targeting amyloid β. *International Journal of Molecular Sciences, 24*(4), 3895.

Volkow, N. D., Koob, G. F., & McLellan, A. T. (2014). Neurobiologic advances from the brain disease model of addiction. *New England Journal of Medicine, 374*(4), 363–371. https://doi.org/10.1056/NEJMra1511480

World Alzheimer Report 2022 (2022). Alzheimer's Disease International. https://www .alzint .org /resource / world -alzheimer -report -2022.

World Health Organization. (2017, December 7). Dementia: Number of people affected to triple in next 30 years. https://www.who.int/news/item/07-12-2017 dementia number-of-people-affected-to-triple-in-next-30-years

Wu, L., Liu, X., Jiang, Q., Li, M., Liang, M., Wang, S., Wang, R., Su, L., Ni, T., Dong, N., Zhu, L., Guan, F., Zhu, J., Zhang, W., Wu, M., Chen, Y., Chen, T., & Wang, B. (2024). Methamphetamine-induced impairment of memory and fleeting neuroinflammation: Profiling mRNA changes in mouse hippocampus following short-term and long-term exposure. *Neuropharmacology*, 261, 110175. https://doi-org.ezproxy.aub.edu.lb/10.1016/j.neuropharm.2024.110175

Yang, H. D., Kim, D. H., Lee, S. B., & Young, L. D. (2016). History of Alzheimer's disease. *Dementia and Neurocognitive Disorders 15*(4), 115–121.

Zhang, J., Zhang, Y., Wang, J. et al. (2024). Recent advances in Alzheimer's disease: mechanisms, clinical trials and new drug development strategies. *Signal Transduction and Targeted Therapy 9*, 211.

Zhang, W., Zhou, J., Su, H., Zhang, X., Song, W., Wang, Z., Tang, C., Uludağ, K., Zhao, M., Xiong, Z. Q., Zhai, R., & Jiang, H. (2023). Repeated methamphetamine exposure decreases plasma brain-derived neurotrophic factor levels in rhesus monkeys. *General Psychiatry, 36*(5), e101127. https://doi.org/10.1136/gpsych-2023-101127

3 Frontotemporal Dementia and Drug Addiction

Victor Olufolahan Lasebikan, Victor Makanjuola, Morufat Abosede Alabi, Tiwatayo Olufolahan Lasebikan, Oluremi Oladipo Oladele, and Oyeyemi Afolabi

INTRODUCTION

Addictive disorders are a critical public health issue, particularly in older adults. With increasing age, addictive disorders become more prevalent, an issue of significant concern for both clinical practice and research (Lasebikan & Gureje, 2015). Older adults are more likely to have several health comorbidities and may be taking various prescription medications, increasing their risk for drug–drug interactions (Roux-Marson et al., 2020). The relationship between addictive disorders and frontotemporal dementia (FTD) is of growing research interest due to overlapping symptoms such as behavioral disinhibition and hyperorality (Kalapatapu et al., 2016). Individuals with the behavioral subtype of FTD (bvFTD) frequently report substance use, highlighting the need for integrated treatment approaches (de Paula França Resende et al., 2022).

This chapter delves into the intricate overlap between addictive disorders and FTD. It explores the history and epidemiology of FTD, examines its neuropathological underpinnings and how it intersects with addiction, and discusses innovative therapeutic strategies, including the use of cannabinoids and psychedelics. Additionally, it addresses the role of psychological interventions in managing FTD, providing a comprehensive overview of current and emerging treatment modalities.

FRONTOTEMPORAL DEMENTIA

FTD is a cluster of diseases presenting with behavioral changes, language impairments, and executive dysfunction, often accompanied by motor symptoms (Bott et al., 2014). The primary subtypes of FTD include bvFTD, the "non-fluent/agrammatic variant of primary progressive aphasia" (nfvPPA), and the "semantic variant of primary progressive aphasia" (svPPA) (Bott et al., 2014). There is also the "logopenic variant of primary progressive aphasia" (lvPPA). These subtypes can evolve, sometimes leading to motor features in keeping with amyotrophic lateral sclerosis, parkinsonism, progressive supranuclear palsy, or corticobasal syndrome (Bocchetta et al., 2021). Understanding the full spectrum of FTD and its potential progression is crucial for developing effective management and treatment strategies.

HISTORICAL PERSPECTIVE

A neurologist, Arnold Pick (1851–1924), was the first to describe FTD in a patient with left temporal lobe atrophy and progressive deterioration of language (Olney et al., 2017).

Atrophic changes in the frontotemporal lobes characterize Pick's disease without affecting the posterior lobes. Neuropathological changes in Pick's disease include ballooned cells and cortical-subcortical gliosis (Itoh et al., 1997). Alternatively, Alzheimer's disease (AD) is marked by more diffuse cerebral atrophy and histological findings such as neurofibrillary tangles and senile plaques (Scheltens et al., 2021).

 DOI: 10.1201/9781003478485-3

Unlike in Pick's disease, where there are no symptoms of agnosia, apraxia, and problems with spatial orientation, patients with AD typically exhibit these symptoms (Thibodeau & Miller, 2013).

Epidemiology of Frontotemporal Dementia

FTD is common in individuals below 65 (Seo et al., 2018). Usually, the age at onset of FTD is between 45 and 65, although cases outside this age range have been reported (Snowden et al., 2002). FTD is equally prevalent in men and women (Hogan et al., 2016); bvFTD is the most common type (60%), with the rest being the language variants (Onyike & Diehl-Schmid, 2013). Age and genetic factors are the only certainties recognized as risk factors (Warren et al., 2013). The pathology is characterized by the mutation of an autosomal dominant gene in the "tau protein," "chromosome 9 open reading frame 72" in the microtubule, or "progranulin," observed in about a third of FTD cases (Warren et al., 2013).

Behavioral Variant Frontotemporal Dementia

bvFTD typically presents with behavioral and personality changes, emotional problems, and loss of executive control. Thus, many individuals with bvFTD exhibit disinhibition, compulsions, hyperorality, changes in dietary habits, lack of empathy, perseveration, apathy, and impairment in executive function (Rascovsky et al., 2011).

Neuropathological Considerations and Causal Link

Anatomically, the behavioral symptoms of bvFTD are due to dysfunction in the medial and orbital frontal cortices, as well as frontoinsular and anterior cingulate cortices, notably in the right hemisphere. Alternatively, the dorsolateral prefrontal cortex and the bilateral parietal lobes are involved in the loss of executive control (Olney et al., 2017). Although FTD is traditionally considered a cortical dementia, the insula, anterior cingulate, and subcortical structures are also involved (Bocchetta et al., 2021). For example, the dopamine reward pathway, related to the limbic system, is crucial in regulating a given stimulus's motivational and emotional aspects (Boccardi et al., 2005). Dysfunction of this pathway in bvFTD is responsible for abnormal responses to rewards such as food, sex, and addictive behaviors (Reddy et al., 2019). Supporting this, Macfarlane et al. (2015) found that primary reward-seeking was associated with atrophy of the right ventral putamen and pallidum in FTD. The differential roles of brain structures in FTD and addiction are described below and presented in Table 3.1.

a. *Basal ganglia*: The corpus striatum, an essential nucleus of the basal ganglia, comprises the dorsal and ventral regions (Bocchetta et al., 2021). Degeneration of the striatum presents with multiple symptoms of FTD secondary to its widespread parallel projections to other brain regions. These symptoms include disinhibition, perseveration, binge eating, poor memory recall, akinesia and parkinsonism, and abnormal reward (Bertoux et al., 2018). Affected patients lose the ability to determine short and long-term consequences of behaviors, the correct interpretation of risks, and the sense of immediate versus delayed gratification (Basar et al., 2010). This explains why substance misuse is a common presentation in both bvFTD and svPPA (Perry et al., 2014).

   Degeneration of the pallidum, a basal ganglia nucleus, may present with motor perseveration, applause sign, disinhibition (Schönecker et al., 2019), poor performance on the "Go/No-Go" task of inhibitory control (O'Callaghan et al., 2013), and with substance misuse secondary to an increase in reward-seeking behaviors (Perry et al., 2014).

   The substantia nigra, another basal ganglia nucleus, is connected to the globus pallidus and is involved in motor, reward, and executive functions (Bocchetta et al., 2021). Degeneration of the substantia nigra is associated with a reduced level of dopamine,

TABLE 3.1
Differential Roles of Brain Structures in FTD

Brain Structure	Function	Role in FTD
Orbital frontal cortex (Schoenbaum & Shaham, 2008)	Decision-making, social behavior	Degeneration leads to disinhibition and socially inappropriate behavior.
Medial prefrontal cortex (Goldstein & Volkow, 2011)	Emotion regulation, motivation	Atrophy leads to apathy.
Anterior cingulate cortex (Zhao et al., 2020)	Error detection, emotional regulation	Atrophy leads to loss of executive control and apathy.
Dorsolateral prefrontal cortex (Alizadehgoradel et al., 2020)	Executive functions, such as planning and reasoning	Atrophy leads to loss of executive control.
Basal ganglia (Sivils et al., 2021)	Motor control, behavior modulation	Degeneration leads to symptoms like disinhibition, binge eating, and poor memory.
Hippocampus (Kutlu & Gould, 2016)	Memory formation, learning, emotion	Atrophy leads to memory deficits and problems with emotion modulation.
Amygdala (Cheng & Liu, 2020)	Emotion processing, motivation	Atrophy leads to disinhibition, poor social behavior, and lack of insight.

typically associated with parkinsonism, commonly seen in bvFTD, while pathology in the subthalamic nucleus may present with disinhibition (Rowe, 2019).

The basal ganglia have also been implicated in the craving, preoccupation, and anticipation stages of addiction (Koob & Volkow, 2016), as well as in compulsive drug seeking (Koob, 2021).

b. *Medial temporal lobe*: This structure plays an essential role in the development of FTD (Landin-Romero et al., 2017). For example, the hippocampus of individuals with bvFTD is significantly smaller than that of individuals without FTD (Landin-Romero et al., 2017). Given that the hippocampus is a hub of the brain network of memory, memory deficits and problems with emotion modulation and evaluation of facial emotion may be present in persons with bvFTD (Anderson et al., 2000). There is clear evidence that chronic use of cannabis (Daniju et al., 2020), tobacco, alcohol (Gazdzinski et al., 2008), and cocaine (Jacobsen et al., 2001) causes hippocampal damage.

c. *Amygdala*: A part of the limbic system, the amygdala is involved in emotion, learning, motivation, and the cognitive functions of attention, perception, and memory (Rolls, 2023). There are reports of at least a 50% reduction in the volume of the amygdala in FTD (Mann & South, 1993), with the basolateral nucleus affected early in the disease process (Brettschneider et al., 2014). The core behavioral features of FTD, such as disinhibition, poor social behavior, and lack of insight, are similar to the characteristics seen in amygdala dysfunction. The amygdala mediates the reward and punishment process, and atrophy of the subnuclei is responsible for several behavioral and neuropsychiatric features (Sollberger et al., 2014), while atrophy of the medial nuclei is likely to lead to abnormal reward and emotional processing (Seeley et al., 2007).

d. *Basal forebrain*: The basal forebrain comprises cholinergic nuclei that link the cortex and limbic system. They are critical for memory, learning, and attention (Sarter et al., 2003). Neuropathological evidence shows that the volume of the forebrain is reduced in bvFTD, svPPA, nfvPPA, and all PPA variants when compared with healthy older adults. This has implications for diverse symptoms of FTD, including impairment in social cognition and attachment, abnormal prosocial sentiments, cholinergic dysfunction, and behavioral symptomatology (Silverman et al., 2020).

Similarly, there is research evidence involving the neural elements of the basal forebrain in the process of addiction. This anatomical site is indeed involved in the positive reinforcing effects of addictive drugs (Koob, 2000), and maladaptive changes in the neural circuits of the basal forebrain contribute to the reward and aversion observed in substance use disorders (Chen, 2022).

e. *Thalamus*: The thalamus is a relay station comprised of several nuclei connected to other brain areas (Ide et al., 2015). FTD is usually associated with thalamic atrophy and volume loss. The heterogeneous nature of thalamic nuclei gives rise to diverse symptoms and signs in FTD, such as disrupted pain processing, perceptual disorders, impaired affective and reward processing, and psychotic manifestations (Ducharme et al., 2017). In the same vein, research evidence suggests that the thalamus plays a vital role in drug addiction, particularly regarding the structural and functional changes it undergoes during drug-seeking behavior (Huang et al., 2018). Indeed, Millan et al. (2017) describe the paraventricular thalamic nucleus as a gateway to drug addiction due to its role in motivation. Similarly, Zhou and Zhu (2019) highlight the paraventricular nuclei as a critical hub within neural circuits implicated in drug addiction. Drug exposure, related cues, and the act of substance use all activate these neurons.
f. *Hypothalamus*: This brain structure plays a vital role in satiety, hunger control, and reward and is at the center of the pathogenesis of addictive behaviors. This role is supported by a wide range of endocrine and molecular mechanisms linking the hypothalamus to addiction (Hildebrandt & Greif, 2013). For example, the neurotensin-containing pathways in the hypothalamus mediate some of the reward properties of psychoactive substances (Ferraro et al., 2016). The lateral hypothalamus also produces the melanin-concentrating hormone (MCH). With its network with other parts of the central nervous system, this neuropeptide is implicated in motivation and drug use by directly influencing the dopaminergic, glutamatergic, and GABAergic systems (Morganstern et al., 2020).
g. *Brainstem*: The brainstem comprises the midbrain, pons, and medulla oblongata. The nuclei of the brainstem supply noradrenergic, dopaminergic, and serotonergic fibers to the forebrain, and its dysfunction in FTD is associated with significant cognitive and behavioral changes, including aggression, compulsivity, and hyperorality (Hughes et al., 2015).

   Alternatively, evidence suggests the existence of an addiction circuit between the cortex and the brainstem. This circuit, for instance, regulates the transition from moderate to compulsive drinking (Siciliano et al., 2019). The use of cannabis or cocaine may also be associated with diffuse brainstem dysfunction, disrupting the transmission of auditory stimuli (Weich et al., 2012). This finding was corroborated by Garg et al. (2015), who found that alcohol, tobacco, and opioid users had alterations in brainstem auditory evoked potentials, implying the neurotoxic effects of the drugs on the functional integrity of the sensory tract in the brainstem area.
h. *Cerebellum*: The cerebellum is essential in cognitive processing, emotional processing, and movement coordination (Bocchetta et al., 2021). There is increasing evidence indicating that the cerebellum plays a significant role in addiction, in the control of motivation, salience, and drug reward. This activity is mediated by the glutamatergic and endocannabinoid systems (Wang & Lan, 2023). The cerebellum significantly contributes to the plasticity mechanisms associated with drug use, mainly due to the widespread presence of receptors, such as opioid receptors, that play a prominent role in synaptic activity (Ranjbar et al., 2021).

ALCOHOL, DRUGS, AND BEHAVIORAL ADDICTIONS: A PRESENTATION OF FRONTOTEMPORAL DEMENTIA

Disinhibition, impulsivity, and executive dysfunction are common symptoms of drug addiction (Cadet & Bisagno, 2013), and are also core features observed in FTD (Sabodash et al., 2013). Therefore, and given their similar neurostructural underpinnings, it is theoretically plausible that individuals with FTD may be prone to drug addiction. Literature shows that drug addiction may occur before, during, or following the onset of FTD (Kalapatapu et al., 2016). Consequently, addictive behavior can be a prominent clinical manifestation of FTD.

Alcohol and Frontotemporal Dementia

bvFTD commonly presents with hyperorality and compulsive behavior (Rascovsky et al., 2011). In a review of case histories by Perry et al. (2014), 20.4% of the sample presented with new or increased alcohol consumption. To corroborate this, a comparative study by Mendez et al. (1997) found that compulsive behaviors as presenting symptoms were significantly more common in FTD compared with AD. A cross-sectional retrospective study showed a significantly higher proportion of patients with bvFTD having late-onset "alcohol abuse" compared with AD but not svPPA, indicating that late-onset "alcohol abuse" could be a presenting symptom of bvFTD (de Paula França Resende et al., 2022). In support of this, a relevant criterion for diagnosing bvFTD reads "binge eating, increased consumption of alcohol or cigarette" (Rascovsky et al., 2011). Unfortunately, when an older adult develops alcohol misuse, increased daily consumption, or binge drinking for the first time, a neurocognitive disorder is not usually the first consideration (de Paula França Resende et al., 2022).

Tobacco and Frontotemporal Dementia

There is a dearth of research on modifiable risk factors for FTD. Nevertheless, the few available studies suggest that smoking is a modifiable risk factor for FTD (Kalkonde et al., 2012), regardless of the variant (Soppela et al., 2022). In this regard, Atkins et al. (2012) found that patients with FTD, especially those with early-onset FTD, were three times more likely to be current smokers than those with AD. A case report revealed a female patient with FTD who presented with nicotine dependence (Reddy et al., 2019), and there are reports that chain smoking leads to degeneration of the frontotemporal lobes (Isaacs et al., 2011). It should also be noted that smoking may be a prodromal symptom of FTD (Rasmussen et al., 2018).

When it comes to memory, although tobacco is a nicotinic cholinergic receptor agonist, its chronic exposure causes desensitization of the receptors, leading to cognitive impairment (Sun et al., 2013). This occurs by activating the phosphodiesterase-5 signaling pathway, leading to the cleaving of cyclic guanosine monophosphate and cyclic adenosine monophosphate, which activate the pathways contributing to memory impairment (Peixoto et al., 2015).

Cannabis and Frontotemporal Dementia

Several research findings have shown that cannabinoid products ameliorate FTD behaviors and initial symptoms (Costa et al., 2022). Despite these results, a case report showed that early and continued use of cannabinoids from early adulthood was associated with atrophy of the hippocampus and cognitive impairment in middle age, making cannabis smoking a potential risk factor for early-onset FTD, especially in genetically vulnerable individuals (Moshfeghinia et al., 2023). A possible underlying pathway is that many illicit drug users, particularly those who smoke cannabis, are at high risk of consuming cannabis contaminated with pollutants, such as heavy metals, microbes, and pesticides, in large and unsafe quantities (Dryburgh et al., 2018), which may increase their risks of developing FTD (Adani et al., 2020).

In addition, the work of Amen et al. (2017) showed that chronic cannabis users demonstrated persistent hypoperfusion in the orbitofrontal cortex, posterior cingulate, medial temporal lobe, hippocampus, and parahippocampus, as well as reduced gray matter volumes in these areas (Battistella et al., 2014). Chronic cannabis use is also characterized by dopamine deficits, which resemble the picture observed in FTD (Huber et al., 2022).

TABLE 3.2
Case Reports of Addictive Disorders and FTD

Addictive Disorder	Case Report
Nicotine dependence and FTD	Reddy et al. (2019)
Gambling disorder and FTD	Manes et al. (2010)
	Ozel-Kizil et al. (2013)
	Nakaaki et al. (2007)
	Lo Coco and Nacci (2004)
Cannabis use and FTD	Moshfeghinia et al. (2023)

Gambling Disorder and Frontotemporal Dementia

Gambling disorder is a behavioral addiction and the only addictive disorder diagnosable in the latest *Diagnostic and Statistical Manual of Mental Disorders* (DSM-5-TR). According to the manual, gambling disorder is a persistent and problematic behavior that causes significant impairment or distress, occurring over one year (American Psychiatric Association, 2022). The ICD-11 also has "disorders due to addictive behaviors," in which there are two behavioral addictions, gambling disorder and gaming disorder (WHO, 2019).

Increasing research evidence indicates that gambling disorder is associated with frontal lobe dysfunction and has also been linked to FTD. Indeed, there are four case reports of FTD presenting with pathological gambling (Lo Coco & Nacci, 2004; Manes et al., 2010; Nakaaki et al., 2007; Ozel-Kizil et al., 2013). Neuropsychological evidence shows that the orbitofrontal cortex, the first part of the brain to atrophy in FTD (Viskontas et al., 2007), is critical in the pathophysiology of pathological gambling (Fuentes et al., 2006). Along the same lines, bvFTD is characterized by various personality and behavioral changes, including disinhibition and compulsive behavior (Mendez et al., 1997). The case reports of addictive disorders and FTD are presented in Table 3.2.

THERAPEUTIC INTERVENTIONS

Innovative Treatments

Recent evidence suggests that psychedelics (serotonergic hallucinogens) enhance cortical neuronal growth by stimulating serotonergic 5-HT_{2A} receptors, and their use could be applicable in the treatment of neuropsychiatric disorders, such as mood disorders and drug addiction, where cortical atrophy is present (Carhart-Harris & Goodwin, 2017). Following a single administration, their therapeutic response can last for months, which may hypothetically improve symptoms of FTD, as this condition has been consistently linked to a loss of 5-HT_{2A} receptor density (Franceschi et al., 2005). Psychedelics also activate α-amino-3-hydroxy-5-methyl-4-isoxazolepropionic acid (AMPA) receptors, tropomyosin receptor kinase B (TrkB) receptors, and mammalian target of rapamycin (mTOR) signaling, thereby prolonging their therapeutic effects and making them unique therapeutic agents for neurodegenerative disorders, such as FTD (Oña & Bouso, 2021).

Furthermore, psychedelics stimulate mitochondrial biogenesis, have potent anti-inflammatory effects by activating peripheral immune cells' 5-HT_{2A} receptors, and are considered psychoplastogens due to their ability to cross the blood–brain barrier and induce plasticity (Moda-Sava et al., 2019). Ketamine is an example of a psychoplastogen that increases the dendritic spine and synaptic density in the prefrontal cortex (Holmes et al., 2022). It holds high promise in the treatment of addiction (Krupitsky et al., 2002), depression (Ionescu et al., 2016), and neurodegenerative disorders (Oña & Bouso, 2021).

Antipsychotics and Mood Stabilizers

FTD has no specific pharmacological treatment, although off-label medications are available (Bei et al., 2010). However, the use of psychopharmacological agents, including acetylcholinesterase inhibitors, antidepressants, atypical antipsychotics, and N-methyl-D-aspartate (NMDA) glutamate receptor antagonists, is characterized by adverse effects, thereby limiting their use (Wylie et al., 2013). Individuals with FTD are highly susceptible to extrapyramidal symptoms induced by antipsychotics (Kerssens & Pijnenburg, 2008).

Non-Pharmacological Interventions

Typically, non-pharmacological interventions are regarded as the first-line treatment for FTD unless they are unsuccessful or there is a significant risk to the patient (Sadowsky & Galvin, 2012). The aim of using non-pharmacological strategies is to prevent and ameliorate problematic behaviors and reduce caregiver distress (Gitlin et al., 2012).

Environmental Approaches

Patients with FTD have difficulty accurately interpreting and responding to subtle emotional cues (Kortte & Rogalski, 2013). In this regard, FTD-specific environmental manipulations such as standard lighting, moderate sound, having a small number of people around, changes in mealtime, removing mirrors, and music therapy can reduce behavioral symptoms (Livingston et al., 2005; Raglio et al., 2008; Trahan et al., 2014).

Behavioral Approaches

Cognitive behavioral therapy can potentially ameliorate mood and behavioral symptoms in FTD (Sadowsky & Galvin, 2012). Behavioral management techniques such as reintroducing old hobbies and favorite games, dietary or monetary rewards for desired behaviors, and "routinizing therapy," in which stimulus-bound and stereotypical behaviors are replaced with appropriate behaviors, are some of the techniques reported to be beneficial (Ikeda et al., 1995).

Caregiver Approaches

Among caregivers of individuals with FTD, psychoeducation on disease progression and home safety, such as preventing aspiration, choking, injuries, and acute intoxication from any psychoactive agents, is crucial. Additionally, problem-solving, provision of support groups, and acceptance of the disease have helped reduce the burden of caring (Alberici et al., 2012). Social issues, such as financial concerns and the ability to continue driving, also require attention as they remain significant sources of distress for patients with FTD and their families.

CONCLUSION

Addictive disorders are receiving increasing research attention due to widespread evidence indicating their prevalence in older adults, a population where there is an increased likelihood of medical comorbidities and neurodegenerative disorders. Of particular interest is the overlap between the symptoms of FTD and addictive disorders, as both conditions affect similar brain structures. While cannabinoids show limited evidence for therapeutic effects on FTD, psychedelics may hold promise as novel pharmacological interventions for neurodegenerative disorders such as FTD. Additionally, psychological interventions should be prioritized when treating the behavioral symptoms of FTD, as the use of psychotropic medications in this population is often associated with adverse events.

This review presents suggestions and recommendations. First, there is a need for more research on the neuropathological commonalities between FTD and addictive disorders. A better understanding of the biological basis of this comorbidity may lead to increased focus on potential biological interventions. Furthermore, additional research is necessary to explore innovative treatments for FTD.

REFERENCES

Adani, G., Filippini, T., Garuti, C., Malavolti, M., Vinceti, G., Zamboni, G., Tondelli, M., Galli, C., Costa, M., Vinceti, M., & Chiari, A. (2020). Environmental risk factors for early-onset Alzheimer's dementia and frontotemporal dementia: A case-control study in Northern Italy. *Int J Environ Res Public Health*, *17*(21). https://doi.org/10.3390/ijerph17217941

Alberici, A., Cottini, E., Cosseddu, M., Borroni, B., & Padovani, A. (2012). Suicide risk in frontotemporal lobe degeneration: To be considered, to be prevented. *Alzheimer Dis Assoc Disord*, *26*(2), 194–196. https://doi.org/10.1097/WAD.0b013e3182223254

Alizadehgoradel, J., Nejati, V., Sadeghi Movahed, F., Imani, S., Taherifard, M., Mosayebi-Samani, M., Vicario, C. M., Nitsche, M. A., & Salehinejad, M. A. (2020). Repeated stimulation of the dorsolateral-prefrontal cortex improves executive dysfunctions and craving in drug addiction: A randomized, double-blind, parallel-group study. *Brain Stimul*, *13*(3), 582–593. https://doi.org/10.1016/j.brs.2019.12.028

Amen, D. G., Darmal, B., Raji, C. A., Bao, W., Jorandby, L., Meysami, S., & Raghavendra, C. S. (2017). Discriminative properties of hippocampal hypoperfusion in marijuana users compared to healthy controls: Implications for marijuana administration in Alzheimer's dementia. *J Alzheimers Dis*, *56*(1), 261–273. https://doi.org/10.3233/jad–160833

American Psychiatric Association. (2022). *Diagnostic and Statistical Manual of Mental Disorders* (5th ed., text rev.). https://doi.org/10.1176/appi.books.9780890425787.

Anderson, A. K., Spencer, D. D., Fulbright, R. K., & Phelps, E. A. (2000). Contribution of the anteromedial temporal lobes to the evaluation of facial emotion. *Neuropsychology*, *14*(4), 526–536. https://doi.org/10.1037//0894-4105.14.4.526

Atkins, E. R., Bulsara, M. K., & Panegyres, P. K. (2012). Cerebrovascular risk factors in early-onset dementia. *J Neurol Neurosurg Psychiatry*, *83*(6), 666–667. https://doi.org/10.1136/jnnp.2009.202846

Basar, K., Sesia, T., Groenewegen, H., Steinbusch, H. W., Visser-Vandewalle, V., & Temel, Y. (2010). Nucleus accumbens and impulsivity. *Prog Neurobiol*, *92*(4), 533–557. https://doi.org/10.1016/j.pneurobio.2010.08.007

Battistella, G., Fornari, E., Annoni, J. M., Chtioui, H., Dao, K., Fabritius, M., Favrat, B., Mall, J. F., Maeder, P., & Giroud, C. (2014). Long-term effects of cannabis on brain structure. *Neuropsychopharmacology*, *39*(9), 2041–2048. https://doi.org/10.1038/npp.2014.67

Bei, H., Ross, L., Neuhaus, J., Knopman, D., Kramer, J., Boeve, B., Caselli, R. J., Graff-Radford, N., Mendez, M. F., Miller, B. L., & Boxer, A. L. (2010). Off-label medication use in frontotemporal dementia. *Am J Alzheimers Dis Other Demen*, *25*(2), 128–133. https://doi.org/10.1177/1533317509356692

Bertoux, M., Flanagan, E. C., Hobbs, M., Ruiz-Tagle, A., Delgado, C., Miranda, M., Ibáñez, A., Slachevsky, A., & Hornberger, M. (2018). Structural anatomical investigation of long-term memory deficit in behavioral frontotemporal dementia. *J Alzheimers Dis*, *62*(4), 1887–1900. https://doi.org/10.3233/jad-170771

Boccardi, M., Sabattoli, F., Laakso, M. P., Testa, C., Rossi, R., Beltramello, A., Soininen, H., & Frisoni, G. B. (2005). Frontotemporal dementia as a neural system disease. *Neurobiol Aging*, *26*(1), 37–44. https://doi.org/10.1016/j.neurobiolaging.2004.02.019

Bocchetta, M., Malpetti, M., Todd, E. G., Rowe, J. B., & Rohrer, J. D. (2021). Looking beneath the surface: The importance of subcortical structures in frontotemporal dementia. *Brain Commun*, *3*(3), fcab158. https://doi.org/10.1093/braincomms/fcab158

Bott, N. T., Radke, A., Stephens, M. L., & Kramer, J. H. (2014). Frontotemporal dementia: Diagnosis, deficits and management. *Neurodegener Dis Manag*, *4*(6), 439–454. https://doi.org/10.2217/nmt.14.34

Brettschneider, J., Del Tredici, K., Irwin, D. J., Grossman, M., Robinson, J. L., Toledo, J. B., Fang, L., Van Deerlin, V. M., Ludolph, A. C., Lee, V. M., Braak, H., & Trojanowski, J. Q. (2014). Sequential distribution of pTDP-43 pathology in behavioral variant frontotemporal dementia (bvFTD). *Acta Neuropathol*, *127*(3), 423–439. https://doi.org/10.1007/s00401-013-1238-y

Cadet, J. L., & Bisagno, V. (2013). The primacy of cognition in the manifestations of substance use disorders. *Front Neurol*, *4*, 189. https://doi.org/10.3389/fneur.2013.00189

Carhart-Harris, R. L., & Goodwin, G. M. (2017). The therapeutic potential of psychedelic drugs: Past, present, and future. *Neuropsychopharmacology*, *42*(11), 2105–2113. https://doi.org/10.1038/npp.2017.84

Chen, W. (2022). Neural circuits provide insights into reward and aversion. *Front Neural Circuits*, *16*, 1002485. https://doi.org/10.3389/fncir.2022.1002485

Cheng, H., & Liu, J. (2020). Alterations in amygdala connectivity in internet addiction disorder. *Scientific Reports*, *10*(1), 2370. https://doi.org/10.1038/s41598-020-59195-w

Costa, A. C., Joaquim, H. P. G., Pedrazzi, J. F. C., Pain, A. O., Duque, G., & Aprahamian, I. (2022). Cannabinoids in late life Parkinson's disease and dementia: Biological pathways and clinical challenges. *Brain Sci*, *12*(12). https://doi.org/10.3390/brainsci12121596

Daniju, Y., Bossong, M. G., Brandt, K., & Allen, P. (2020). Do the effects of cannabis on the hippocampus and striatum increase risk for psychosis? *Neurosci Biobehav Rev*, *112*, 324–335. https://doi.org/10.1016/j.neubiorev.2020.02.010

de Paula França Resende, E., Ketelle, R., Karydas, A., Allen, I., Grinberg, L. T., Spina, S., Seeley, W. W., Perry, D. C., Miller, B., & Naasan, G. (2022). Late-onset alcohol abuse as a presenting symptom of neurodegenerative diseases. *J Alzheimers Dis*, *86*(3), 1073–1080. https://doi.org/10.3233/jad-215369

Dryburgh, L. M., Bolan, N. S., Grof, C. P. L., Galettis, P., Schneider, J., Lucas, C. J., & Martin, J. H. (2018). Cannabis contaminants: Sources, distribution, human toxicity and pharmacologic effects. *Br J Clin Pharmacol*, *84*(11), 2468–2476. https://doi.org/10.1111/bcp.13695

Ducharme, S., Bajestan, S., Dickerson, B. C., & Voon, V. (2017). Psychiatric presentations of C9orf72 mutation: What are the diagnostic implications for clinicians? *J Neuropsychiatry Clin Neurosci*, *29*(3), 195–205. https://doi.org/10.1176/appi.neuropsych.16090168

Ferraro, L., Tiozzo Fasiolo, L., Beggiato, S., Borelli, A. C., Pomierny-Chamiolo, L., Frankowska, M., Antonelli, T., Tomasini, M. C., Fuxe, K., & Filip, M. (2016). Neurotensin: A role in substance use disorder? *J Psychopharmacol*, *30*(2), 112–127. https://doi.org/10.1177/0269881115622240

Franceschi, M., Anchisi, D., Pelati, O., Zuffi, M., Matarrese, M., Moresco, R. M., Fazio, F., & Perani, D. (2005). Glucose metabolism and serotonin receptors in the frontotemporal lobe degeneration. *Ann Neurol*, *57*(2), 216–225. https://doi.org/10.1002/ana.20365

Fuentes, D., Tavares, H., Artes, R., & Gorenstein, C. (2006). Self-reported and neuropsychological measures of impulsivity in pathological gambling. *J Int Neuropsychol Soc*, *12*(6), 907–912. https://doi.org/10.1017/s1355617706061091

Garg, S., Sharma, R., Mittal, S., & Thapar, S. (2015). Alterations in brain-stem auditory evoked potentials among drug addicts. *A Cross-Sectional Study*, *20*(3), 253–258. https://doi.org/10.17712/nsj.2015.3.20150105 %

Gazdzinski, S., Durazzo, T. C., Yeh, P. H., Hardin, D., Banys, P., & Meyerhoff, D. J. (2008). Chronic cigarette smoking modulates injury and short-term recovery of the medial temporal lobe in alcoholics. *Psychiatry Res*, *162*(2), 133–145. https://doi.org/10.1016/j.pscychresns.2007.04.003

Gitlin, L. N., Kales, H. C., & Lyketsos, C. G. (2012). Nonpharmacologic management of behavioral symptoms in dementia. *JAMA*, *308*(19), 2020–2029. https://doi.org/10.1001/jama.2012.36918

Goldstein, R. Z., & Volkow, N. D. (2011). Dysfunction of the prefrontal cortex in addiction: Neuroimaging findings and clinical implications. *Nat Rev Neurosci*, *12*(11), 652–669. https://doi.org/10.1038/nrn3119

Hildebrandt, T., & Greif, R. (2013). Stress and addiction. *Psychoneuroendocrinology*, *38*(9), 1923–1927. https://doi.org/10.1016/j.psyneuen.2013.06.017

Hogan, D. B., Jetté, N., Fiest, K. M., Roberts, J. I., Pearson, D., Smith, E. E., Roach, P., Kirk, A., Pringsheim, T., & Maxwell, C. J. (2016). The prevalence and incidence of frontotemporal dementia: A systematic review. *Can J Neurol Sci*, *43 Suppl 1*, S96–s109. https://doi.org/10.1017/cjn.2016.25

Holmes, S. E., Finnema, S. J., Naganawa, M., DellaGioia, N., Holden, D., Fowles, K., Davis, M., Ropchan, J., Emory, P., Ye, Y., Nabulsi, N., Matuskey, D., Angarita, G. A., Pietrzak, R. H., Duman, R. S., Sanacora, G., Krystal, J. H., Carson, R. E., & Esterlis, I. (2022). Imaging the effect of ketamine on synaptic density (SV2A) in the living brain. *Mol Psychiatry*, *27*(4), 2273–2281. https://doi.org/10.1038/s41380-022-01465-2

Huang, A. S., Mitchell, J. A., Haber, S. N., Alia-Klein, N., & Goldstein, R. Z. (2018). The thalamus in drug addiction: From rodents to humans. *Philos Trans R Soc Lond B Biol Sci*, *373*(1742). https://doi.org/10.1098/rstb.2017.0028

Huber, N., Korhonen, S., Hoffmann, D., Leskelä, S., Rostalski, H., Remes, A. M., Honkakoski, P., Solje, E., & Haapasalo, A. (2022). Deficient neurotransmitter systems and synaptic function in frontotemporal lobar degeneration—Insights into disease mechanisms and current therapeutic approaches. *Mol Psychiatry*, *27*(3), 1300–1309. https://doi.org/10.1038/s41380-021-01384-8

Hughes, L. E., Rittman, T., Regenthal, R., Robbins, T. W., & Rowe, J. B. (2015). Improving response inhibition systems in frontotemporal dementia with citalopram. *Brain*, *138*(Pt 7), 1961–1975. https://doi.org/10.1093/brain/awv133

Ide, S., Kakeda, S., & Korogi, Y. (2015). [Anatomy of the Thalamus]. *Brain Nerve*, *67*(12), 1459–1469. https://doi.org/10.11477/mf.1416200323

Ikeda, M., Tanabe, H., Horino, T., Komori, K., Hirao, K., Yamada, N., Hashimoto, M., Kazui, H., & Mori, T. (1995). [Care for patients with Pick's disease—By using their preserved procedural memory]. *Seishin Shinkeigaku Zasshi*, *97*(3), 179–192.

Ionescu, D. F., Swee, M. B., Pavone, K. J., Taylor, N., Akeju, O., Baer, L., Nyer, M., Cassano, P., Mischoulon, D., Alpert, J. E., Brown, E. N., Nock, M. K., Fava, M., & Cusin, C. (2016). Rapid and sustained reductions in current suicidal ideation following repeated doses of intravenous ketamine: Secondary analysis of an open-label study. *J Clin Psychiatry*, *77*(6), e719–725. https://doi.org/10.4088/JCP.15m10056

Isaacs, A. M., Johannsen, P., Holm, I., & Nielsen, J. E. (2011). Frontotemporal dementia caused by CHMP2B mutations. *Curr Alzheimer Res*, *8*(3), 246–251. https://doi.org/10.2174/156720511795563764

Itoh, Y., Inoue, M., Amano, N., & Yagashita, S. (1997). Three dimensional analysis of abnormal filaments of Pick's disease by scanning electron microscopy. *Neuropathol Appl Neurobiol*, *23*(4), 326–330.

Jacobsen, L. K., Giedd, J. N., Kreek, M. J., Gottschalk, C., & Kosten, T. R. (2001). Quantitative medial temporal lobe brain morphology and hypothalamic-pituitary-adrenal axis function in cocaine dependence: a preliminary report. *Drug Alcohol Depend*, *62*(1), 49–56. https://doi.org/10.1016/s0376-8716(00)00159-9

Kalapatapu, R. K., Delucchi, K. L., Wang, S., Harbison, J. D., Nelson, E. E., & Kramer, J. H. (2016). Substance use history in behavioral-variant frontotemporal dementia versus primary progressive aphasia. *J Addict Dis*, *35*(1), 36–41. https://doi.org/10.1080/10550887.2015.1102026

Kalkonde, Y. V., Jawaid, A., Qureshi, S. U., Shirani, P., Wheaton, M., Pinto-Patarroyo, G. P., & Schulz, P. E. (2012). Medical and environmental risk factors associated with frontotemporal dementia: A case-control study in a veteran population. *Alzheimers Dement*, *8*(3), 204–210. https://doi.org/10.1016/j.jalz.2011.03.011

Kerssens, C. J., & Pijnenburg, Y. A. (2008). Vulnerability to neuroleptic side effects in frontotemporal dementia. *Eur J Neurol*, *15*(2), 111–112. https://doi.org/10.1111/j.1468-1331.2007.02035.x

Koob, G. F. (2000). Neurobiology of addiction. Toward the development of new therapies. *Ann N Y Acad Sci*, *909*, 170–185. https://doi.org/10.1111/j.1749-6632.2000.tb06682.x

Koob, G. F. (2021). Drug addiction: Hyperkatifeia/negative reinforcement as a framework for medications development. *Pharmacol Rev*, *73*(1), 163–201. https://doi.org/10.1124/pharmrev.120.000083

Koob, G. F., & Volkow, N. D. (2016). Neurobiology of addiction: A neurocircuitry analysis. *Lancet Psychiatry*, *3*(8), 760–773. https://doi.org/10.1016/s2215-0366(16)00104-8

Kortte, K. B., & Rogalski, E. J. (2013). Behavioural interventions for enhancing life participation in behavioural variant frontotemporal dementia and primary progressive aphasia. *Int Rev Psychiatry*, *25*(2), 237–245. https://doi.org/10.3109/09540261.2012.751017

Krupitsky, E., Burakov, A., Romanova, T., Dunaevsky, I., Strassman, R., & Grinenko, A. (2002). Ketamine psychotherapy for heroin addiction: immediate effects and two-year follow-up. *J Subst Abuse Treat*, *23*(4), 273–283. https://doi.org/10.1016/s0740-5472(02)00275-1

Kutlu, M. G., & Gould, T. J. (2016). Effects of drugs of abuse on hippocampal plasticity and hippocampus-dependent learning and memory: Contributions to development and maintenance of addiction. *Learn Mem*, *23*(10), 515–533. https://doi.org/10.1101/lm.042192.116

Landin-Romero, R., Kumfor, F., Leyton, C. E., Irish, M., Hodges, J. R., & Piguet, O. (2017). Disease-specific patterns of cortical and subcortical degeneration in a longitudinal study of Alzheimer's disease and behavioural-variant frontotemporal dementia. *Neuroimage*, *151*, 72–80. https://doi.org/10.1016/j.neuroimage.2016.03.032

Lasebikan, V. O., & Gureje, O. (2015). Lifetime and 7-day alcohol consumption in the elderly, prevalence and correlates: Reports from the Ibadan Study of Aging. *Afr J Med Med Sci*, *44*(1), 33–41.

Livingston, G., Johnston, K., Katona, C., Paton, J., & Lyketsos, C. G. (2005). Systematic review of psychological approaches to the management of neuropsychiatric symptoms of dementia. *Am J Psychiatry*, *162*(11), 1996–2021. https://doi.org/10.1176/appi.ajp.162.11.1996

Lo Coco, D., & Nacci, P. (2004). Frontotemporal dementia presenting with pathological gambling. *J Neuropsychiatry Clin Neurosci*, *16*(1), 117–118. https://doi.org/10.1176/jnp.16.1.117

Macfarlane, M. D., Jakabek, D., Walterfang, M., Vestberg, S., Velakoulis, D., Wilkes, F. A., Nilsson, C., van Westen, D., Looi, J. C., & Santillo, A. F. (2015). Striatal atrophy in the behavioural variant of frontotemporal dementia: Correlation with diagnosis, negative symptoms and disease severity. *PLoS One*, *10*(6), e0129692. https://doi.org/10.1371/journal.pone.0129692

Manes, F. F., Torralva, T., Roca, M., Gleichgerrcht, E., Bekinschtein, T. A., & Hodges, J. R. (2010). Frontotemporal dementia presenting as pathological gambling. *Nat Rev Neurol*, *6*(6), 347–352. https://doi.org/10.1038/nrneurol.2010.34

Mann, D. M., & South, P. W. (1993). The topographic distribution of brain atrophy in frontal lobe dementia. *Acta Neuropathol*, *85*(3), 334–340. https://doi.org/10.1007/bf00227731

Mendez, M. F., Perryman, K. M., Miller, B. L., Swartz, J. R., & Cummings, J. L. (1997). Compulsive behaviors as presenting symptoms of frontotemporal dementia. *J Geriatr Psychiatry Neurol*, *10*(4), 154–157. https://doi.org/10.1177/089198879701000405

Millan, E. Z., Ong, Z., & McNally, G. P. (2017). Paraventricular thalamus: Gateway to feeding, appetitive motivation, and drug addiction. *Prog Brain Res*, *235*, 113–137. https://doi.org/10.1016/bs.pbr.2017.07.006

Moda-Sava, R. N., Murdock, M. H., Parekh, P. K., Fetcho, R. N., Huang, B. S., Huynh, T. N., Witztum, J., Shaver, D. C., Rosenthal, D. L., Alway, E. J., Lopez, K., Meng, Y., Nellissen, L., Grosenick, L., Milner, T. A., Deisseroth, K., Bito, H., Kasai, H., & Liston, C. (2019). Sustained rescue of prefrontal circuit dysfunction by antidepressant-induced spine formation. *Science*, *364*(6436). https://doi.org/10.1126/science.aat8078

Morganstern, I., Gulati, G., & Leibowitz, S. F. (2020). Role of melanin-concentrating hormone in drug use disorders. *Brain Res*, *1741*, 146872. https://doi.org/10.1016/j.brainres.2020.146872

Moshfeghinia, R., Oji, B., Hosseinzadeh, M., Pourfridoni, M., & Ahmadi, J. (2023). Early onset frontotemporal dementia following cannabis abuse: A case report. *BMC Psychiatry*, *23*(1), 484. https://doi.org/10.1186/s12888-023-04956-w

Nakaaki, S., Murata, Y., Sato, J., Shinagawa, Y., Hongo, J., Tatsumi, H., Mimura, M., Furukawa, T. A. (2007). Impairment of decision-making cognition in a case of frontotemporal lobar degeneration (FTLD) presenting with pathologic gambling and hoarding as the initial symptoms. *Cogn Behav Neurol*, *20*(2), 121–125. https://doi.org/10.1097/WNN.0b013e31804c6ff7

O'Callaghan, C., Naismith, S. L., Hodges, J. R., Lewis, S. J., & Hornberger, M. (2013). Fronto-striatal atrophy correlates of inhibitory dysfunction in Parkinson's disease versus behavioural variant frontotemporal dementia. *Cortex*, *49*(7), 1833–1843. https://doi.org/10.1016/j.cortex.2012.12.003

Olney, N. T., Spina, S., & Miller, B. L. (2017). Frontotemporal dementia. *Neurol Clin*, *35*(2), 339–374. https://doi.org/10.1016/j.ncl.2017.01.008

Oña, G., & Bouso, J. C. (2021). Therapeutic potential of natural psychoactive drugs for central nervous system disorders: A perspective from polypharmacology. *Curr Med Chem*, *28*(1), 53–68. https://doi.org/10.2174/0929867326666191212103330

Onyike, C. U., & Diehl-Schmid, J. (2013). The epidemiology of frontotemporal dementia. *Int Rev Psychiatry*, *25*(2), 130–137. https://doi.org/10.3109/09540261.2013.776523

Ozel-Kizil, E., Sakarya, A., Arica, B., & Haran, S. (2013). A case of frontotemporal dementia with amyotrophic lateral sclerosis presenting with pathological gambling. *J Clin Neurol*, *9*(2), 133–137. https://doi.org/10.3988/jcn.2013.9.2.133

Peixoto, C. A., Nunes, A. K., & Garcia-Osta, A. (2015). Phosphodiesterase-5 inhibitors: Action on the signaling pathways of neuroinflammation, neurodegeneration, and cognition. *Mediators Inflamm*, *2015*, 940207. https://doi.org/10.1155/2015/940207

Perry, D. C., Sturm, V. E., Seeley, W. W., Miller, B. L., Kramer, J. H., & Rosen, H. J. (2014). Anatomical correlates of reward-seeking behaviours in behavioural variant frontotemporal dementia. *Brain*, *137*(Pt 6), 1621–1626. https://doi.org/10.1093/brain/awu075

Raglio, A., Bellelli, G., Traficante, D., Gianotti, M., Ubezio, M. C., Villani, D., & Trabucchi, M. (2008). Efficacy of music therapy in the treatment of behavioral and psychiatric symptoms of dementia. *Alzheimer Dis Assoc Disord*, *22*(2), 158–162. https://doi.org/10.1097/WAD.0b013e3181630b6f

Ranjbar, H., Soti, M., Banazadeh, M., Saleki, K., Kohlmeier, K. A., & Shabani, M. (2021). Addiction and the cerebellum with a focus on actions of opioid receptors. *Neurosci Biobehav Rev*, *131*, 229–247. https://doi.org/10.1016/j.neubiorev.2021.09.021

Rascovsky, K., Hodges, J. R., Knopman, D., Mendez, M. F., Kramer, J. H., Neuhaus, J., van Swieten, J. C., Seelaar, H., Dopper, E. G., Onyike, C. U., Hillis, A. E., Josephs, K. A., Boeve, B. F., et al. (2011). Sensitivity of revised diagnostic criteria for the behavioural variant of frontotemporal dementia. *Brain*, *134*(Pt 9), 2456–2477. https://doi.org/10.1093/brain/awr179

Rasmussen, H., Stordal, E., & Rosness, T. A. (2018). Risk factors for frontotemporal dementia. *Tidsskr Nor Laegeforen*, *138*(14). https://doi.org/10.4045/tidsskr.17.0763 (Risikofaktorer for frontotemporal demens.)

Reddy, P. V., Anuroop, L., Shetageri, V., Kumar, R. K., & Gopalakrishnan, G. (2019). A case of frontotemporal dementia presenting as nicotine dependence and carbohydrate craving. *Indian J Psychol Med*, *41*(4), 391–393. https://doi.org/10.4103/ijpsym.Ijpsym_390_18

Rolls, E. T. (2023). Emotion, motivation, decision-making, the orbitofrontal cortex, anterior cingulate cortex, and the amygdala. *Brain Struct Funct*, *228*(5), 1201–1257. https://doi.org/10.1007/s00429-023-02644-9

Roux-Marson, C., Baranski, J. B., Fafin, C., Exterman, G., Vigneau, C., Couchoud, C., Moranne, O., & Investigators, P. (2020). Medication burden and inappropriate prescription risk among elderly with advanced chronic kidney disease. *BMC Geriatr*, *20*(1), 87. https://doi.org/10.1186/s12877-020-1485-4

Rowe, J. B. (2019). Parkinsonism in frontotemporal dementias. *Int Rev Neurobiol*, *149*, 249–275. https://doi.org/10.1016/bs.irn.2019.10.012

Sabodash, V., Mendez, M. F., Fong, S., & Hsiao, J. J. (2013). Suicidal behavior in dementia: A special risk in semantic dementia. *Am J Alzheimers Dis Other Demen*, *28*(6), 592–599. https://doi.org/10.1177/1533317513494447

Sadowsky, C. H., & Galvin, J. E. (2012). Guidelines for the management of cognitive and behavioral problems in dementia. *J Am Board Fam Med*, *25*(3), 350–366. https://doi.org/10.3122/jabfm.2012.03.100183

Sarter, M., Bruno, J. P., & Givens, B. (2003). Attentional functions of cortical cholinergic inputs: What does it mean for learning and memory? *Neurobiol Learn Mem*, *80*(3), 245–256. https://doi.org/10.1016/s1074-7427(03)00070-4

Scheltens, P., De Strooper, B., Kivipelto, M., Holstege, H., Chételat, G., Teunissen, C. E., Cummings, J., & van der Flier, W. M. (2021). Alzheimer's disease. *Lancet*, *397*(10284), 1577–1590. https://doi.org/10.1016/s0140-6736(20)32205-4

Schoenbaum, G., & Shaham, Y. (2008). The role of orbitofrontal cortex in drug addiction: A review of preclinical studies. *Biol Psychiatry*, *63*(3), 256–262. https://doi.org/10.1016/j.biopsych.2007.06.003

Schönecker, S., Hell, F., Bötzel, K., Wlasich, E., Ackl, N., Süßmair, C., … & Danek, A. (2019). The applause sign in frontotemporal lobar degeneration and related conditions. *J Neurol*, *266*(2), 330–338. https://doi.org/10.1007/s00415-018-9134-y

Seeley, W. W., Menon, V., Schatzberg, A. F., Keller, J., Glover, G. H., Kenna, H., … & Greicius, M. D. (2007). Dissociable intrinsic connectivity networks for salience processing and executive control. *J Neurosci*, *27*(9), 2349–2356. https://doi.org/10.1523/jneurosci.5587-06.2007

Seo, S. W., Thibodeau, M. P., Perry, D. C., Hua, A., Sidhu, M., Sible, I., Vargas, J. N. S., Gaus, S. E., Rabinovici, G. D., Rankin, K. D., Boxer, A. L., Kramer, J. H., Rosen, H. J., Gorno-Tempini, M. L., Grinberg, L. T., Huang, E. J., DeArmond, S. J., Trojanowski, J. Q., Miller, B. L., & Seeley, W. W. (2018). Early vs late age at onset frontotemporal dementia and frontotemporal lobar degeneration. *Neurology*, *90*(12), e1047–e1056. https://doi.org/10.1212/wnl.0000000000005163

Siciliano, C. A., Noamany, H., Chang, C. J., Brown, A. R., Chen, X., Leible, D., Lee, J. J., Wang, J., Vernon, A. N., Vander Weele, C. M., Kimchi, E. Y., Heiman, M., & Tye, K. M. (2019). A cortical-brainstem circuit predicts and governs compulsive alcohol drinking. *Science*, *366*(6468), 1008–1012. https://doi.org/10.1126/science.aay1186

Silverman, H. E., Gazes, Y., Barker, M. S., Manoochehri, M., Goldman, J. S., Wassermann, E. M., Tierney, M. C., Cosentino, S., Grafman, J., & Huey, E. D. (2020). Frontal pole hypometabolism linked to reduced prosocial sexual behaviors in frontotemporal dementia and corticobasal syndrome. *J Alzheimers Dis*, *77*(2), 821–830. https://doi.org/10.3233/jad-200346

Sivils, A., Wang, J. Q., & Chu, X.-P. (2021). Striatonigrostriatal spirals in addiction [Review]. *Front Neural Circuits, 15*. https://doi.org/10.3389/fncir.2021.803501

Snowden, J. S., Neary, D., & Mann, D. M. (2002). Frontotemporal dementia. *Br J Psychiatry*, *180*, 140–143. https://doi.org/10.1192/bjp.180.2.140

Sollberger, M., Rosen, H. J., Shany-Ur, T., Ullah, J., Stanley, C. M., Laluz, V., Weiner, M. W., Wilson, S. M., Miller, B. L., & Rankin, K. P. (2014). Neural substrates of socioemotional self-awareness in neurodegenerative disease. *Brain Behav*, *4*(2), 201–214. https://doi.org/10.1002/brb3.211

Soppela, H., Katisko, K., Gadola, Y., Krüger, J., Hartikainen, P., Alberici, A., Benussi, A., Koivisto, A., Haapasalo, A., Remes, A. M., Borroni, B., & Solje, E. (2022). Modifiable potential risk factors in familial and sporadic frontotemporal dementia. *Ann Clin Transl Neurol*, *9*(8), 1195–1205. https://doi.org/10.1002/acn3.51619

Sun, Z., Smyth, K., Garcia, K., Mattson, E., Li, L., & Xiao, Z. (2013). Nicotine inhibits memory CTL programming. *PLoS One*, *8*(7), e68183. https://doi.org/10.1371/journal.pone.0068183

Thibodeau, M. P., & Miller, B. L. (2013). 'Limits and current knowledge of Pick's disease: Its differential diagnosis'. A translation of the 1957 Delay, Brion, Escourolle article. *Neurocase*, *19*(5), 417–422. https://doi.org/10.1080/13554794.2012.667133

Trahan, M. A., Kuo, J., Carlson, M. C., & Gitlin, L. N. (2014). A systematic review of strategies to foster activity engagement in persons with dementia. *Health Educ Behav*, *41*(1 Suppl), 70s–83s. https://doi.org/10.1177/1090198114531782

Viskontas, I. V., Possin, K. L., & Miller, B. L. (2007). Symptoms of frontotemporal dementia provide insights into orbitofrontal cortex function and social behavior. *Ann N Y Acad Sci*, *1121*, 528–545. https://doi.org/10.1196/annals.1401.025

Wang, Y.-b., & Lan, Y. (2023). The role of the cerebellum in drug reward: A review. *22*(6). https://doi.org/10.31083/j.jin2206147

Warren, J. D., Rohrer, J. D., & Rossor, M. N. (2013). Clinical review. Frontotemporal dementia. *BMJ*, *347*, f4827. https://doi.org/10.1136/bmj.f4827

Weich, T. M., Tochetto, T. M., & Seligman, L. (2012). Brain stem evoked response audiometry of former drug users. *Brazilian Journal of Otorhinolaryngology*, *78*(5), 90–96. https://doi.org/https://doi.org/10.5935/1808-8694.20120014

World Health Organization. (2019). *Risk reduction of cognitive decline and dementia: WHO guidelines.*

Wylie, M. A., Shnall, A., Onyike, C. U., & Huey, E. D. (2013). Management of frontotemporal dementia in mental health and multidisciplinary settings. *Int Rev Psychiatry*, *25*(2), 230–236. https://doi.org/10.3109/09540261.2013.776949

Zhao, Y., Sallie, S. N., Cui, H., Zeng, N., Du, J., Yuan, T., Li, D., De Ridder, D., & Zhang, C. (2020). Anterior cingulate cortex in addiction: New insights for neuromodulation. *Neuromodulation*. https://doi.org/10.1111/ner.13291

Zhou, K., & Zhu, Y. (2019). The paraventricular thalamic nucleus: A key hub of neural circuits underlying drug addiction. *Pharmacol Res*, *142*, 70–76. https://doi.org/10.1016/j.phrs.2019.02.014

4 Huntington's Disease, Amyotrophic Lateral Sclerosis, and Addiction

Yara Ghanem, Mounir Fakhouri, Karl Khalil, and Mario Eid

INTRODUCTION

The huntingtin gene (HTT), located on chromosome 4, is present in all individuals and codes for the huntingtin protein (Htt). One part of HTT typically consists of a trinucleotide repeat of cytosine-adenine-guanine (CAG). However, when the number of CAG repeats exceeds 36, it becomes pathological, leading to Huntington's disease (HD) (Bates et al., 2015). HD is transmitted in an autosomal dominant manner, meaning that inheriting just one HTT mutant allele is enough to cause the disorder. In addition, since its penetrance is very high, individuals with the mutated allele almost always develop the disease (Bates et al., 2015). Patients with HD typically carry the HTT allele with CAG repeats ranging from 36 to 55, while in cases of juvenile-onset HD, the number of CAG repeats is usually greater than 60. Anticipation is a common genetic phenomenon observed in the paternal inheritance of HD. This occurs due to the instability of CAG repeats during spermatogenesis, resulting in a trend where affected offspring develop the condition at a younger age than their affected parent (Dayalu & Albin, 2015). Because the symptoms usually start around the age of 40, HD does not interfere with reproduction. The worldwide incidence and prevalence of the disorder are estimated to be 0.48 cases per 100,000 and 5–10 cases per 100,000, respectively (Medina et al., 2022). It is important to note that prevalence varies geographically: it is highest in people of European ancestry and lower in people of Asian and African descent (Medina et al., 2022). Over the past decades, there seems to be a slight increase in the prevalence of HD, potentially due to greater access to molecular testing, increased awareness of HD's symptoms and presentation, and advancements in treating associated conditions, enabling patients to live longer (Medina et al., 2022).

Amyotrophic lateral sclerosis (ALS) is a degenerative motor neuron disease that tends to be fatal (Rowland, 1994). It is a rare disease requiring specialized expertise for diagnosis and ongoing multidisciplinary care (Wolfson et al., 2023). The age of symptom onset is between 54 and 67. However, there is usually a diagnosis delay of around one year (Chiò et al., 2013). The incidence rate of ALS varies between 0.26 and 1.40 per 100,000 years, and the prevalence rate ranges between 4.90 and 8.86 per 100,000 people. Studies of incidence and prevalence rates highlight the lowest rates in South America and the highest rates in Europe (Wolfson et al., 2023). The male-to-female case ratio is around 1.42 (Camacho-Soto, 2022).

Recently, several studies have emerged highlighting addiction as a risk factor for developing HD and ALS. Substance misuse worsens psychiatric symptoms in HD and reduces the age of motor symptom onset due to its neurodegenerative effects (Byars et al., 2012; Ehret et al., 2007; Guo et al., 2023; Schultz et al., 2017). Alternatively, neuronal damage in HD affects the cortico-striatal circuits implicated in addiction, which can contribute to the development of behavioral and chemical addictions (Milton et al., 2020; Waldvogel et al., 2015; Xu et al., 2021). In ALS, research has established drug misuse as a potential risk factor among young adults (Azhari et al., 2015). While some studies highlighted the association between cocaine use and early-onset ALS (Argyriou et al., 2011), others concluded a protective role of opioids in patients with the disease (D'Ovidio et al., 2016).

DOI: 10.1201/9781003478485-4

This chapter provides a comprehensive overview of the epidemiological, genetic, and clinical dimensions of HD and ALS, with a focus on their intersections with addiction. It explores the role of substance misuse as a risk factor for earlier symptom onset and disease progression. The discussion extends to related topics, such as behavioral addictions in HD and the use of opioids for pain management in ALS. Furthermore, the chapter highlights advancements in the management of HD and ALS, including emerging therapeutic strategies and the potential role of cannabinoids in symptom alleviation.

HUNTINGTON'S DISEASE

Signs and Symptoms

The clinical manifestations of HD usually become noticeable between the ages of 30 and 50. HD presents with motor, psychiatric, and cognitive problems. Other features include unintended weight loss and sleep disturbances. In about 50% of cases, psychiatric problems can be the first manifestation, while motor symptoms become increasingly noticeable and more common as the condition progresses (Jensen et al., 2018).

Psychiatric Symptoms

Early symptoms, such as anxiety, depression, and irritability, are common; depression is particularly difficult to diagnose because it can overlap with other symptoms, such as apathy, weight loss, and decreased activity (van Duijn et al., 2007). While apathy typically worsens as the illness progresses, studies have shown that anxiety and depression can strike at any stage, affecting up to 76% of patients. Up to 20% of HD mutation carriers experience suicidal ideations, particularly prominent during the periods after diagnosis and the loss of independence (Di Maio et al., 1993). From mood swings and interpersonal conflicts to physical aggression, irritability is frequently one of the earliest signs. Aggressive and obsessive-compulsive behaviors are also noteworthy, as they often lead to severe emotional distress. Delusions and hallucinations are common late-stage psychotic symptoms that often coexist with cognitive decline and may resemble schizophrenia. Another prevalent problem is sexual dysfunction, where early hypersexuality can occasionally give way to later hyposexuality (van Duijn et al., 2007).

Motor Symptoms

Chorea, characterized by brief, involuntary, and often erratic movements, is the most noticeable motor symptom in the early stages of HD (McCusker et al., 2013). It typically starts in the fingers, toes, or facial muscles and eventually affects the trunk and limbs. Over time, hypokinetic features such as bradykinesia (e.g., slowed movements), dystonia (e.g., permanent aberrant postures), and rigidity frequently join or replace hyperkinetic symptoms (Walker, 2007). Dystonia, which can cause aberrant twisting or repetitive movements and postures, can appear at any stage but worsens as the disease progresses (Walker, 2007).

As motor control deteriorates, walking, standing, and balance problems arise, resulting in an unstable stride and frequent falls. Dysarthria (e.g., speech difficulties) and dysphagia (e.g., difficulty swallowing) also become prevalent as involuntary movements affect muscles in the throat and mouth, increasing the risk of choking and aspiration pneumonia (Walker, 2007). In later phases, chorea may progress to a stiffer, akinetic state, similar to parkinsonism, in which motions become extremely sluggish or absent (Walker, 2007).

Cognitive Problems

The predominant cognitive impairment in HD spans executive processes necessary for planning, organizing, multitasking, and adapting to new conditions. Patients usually struggle to initiate actions, stop improper responses, and alter their mental methods, which causes substantial hurdles in managing daily life activities and making judgments. As the disease progresses, executive dysfunctions become accompanied by memory loss. At first, patients may struggle to focus on tasks and process novel information, while short- and long-term memory impairments emerge over time. Memory problems might include difficulties with episodic memory (e.g., remembering personal events), procedural memory (e.g., skills and procedures), and working memory (e.g., keeping and manipulating information). Semantic memory, or the knowledge of facts, often remains intact despite these limitations (van Duijn et al., 2007; Walker, 2007).

Unlike cortical dementias such as Alzheimer's disease, which severely impair language and motor planning, HD is characterized by subcortical dementia involving inadequate memory retrieval rather than complete memory loss. Language dysfunction and apraxia (e.g., difficulty with motor tasks) are less common, and patients may have psychomotor slowness (e.g., delayed reactions and slower processing speed) (van Duijn et al., 2007; Walker, 2007).

Juvenile Huntington's Disease

Juvenile HD (JHD) is an uncommon type of HD that appears before age 20 and is usually associated with a CAG repeat length of more than 55 (McAllister et al., 2021). Unlike adult-onset HD, JHD frequently begins with behavioral abnormalities and learning difficulties noted at school. Seizures are common, and patients may have significant cognitive decline, cerebellar abnormalities, and delays in motor, speech, and language development. Motor symptoms commonly seen in JHD include hypokinesia, bradykinesia, and dystonia, with chorea being less common. Almost 75% of cases involve an afflicted father, indicating a potential paternal transmission pattern (McAllister et al., 2021).

Mortality

The leading causes of death in HD include (van der Burg et al., 2009):

- *Respiratory insufficiency and aspiration pneumonia*: Due to decreased motor control and swallowing difficulties, patients are at significant risk of developing aspiration pneumonia, a primary cause of death in HD.
- *Heart disease*: Cardiovascular difficulties are substantial, accounting for nearly a quarter of mortalities.
- *Suicide*: Approximately 7.3% of people with HD die by suicide, with up to 27% attempting it.
- *Other risks*: Other causes of death include choking due to swallowing difficulty, physical injuries from falls, and malnutrition caused by reduced motor capabilities.

In summary, the evolution of HD encompasses a steady deterioration in health over 15 to 20 years, with severe problems resulting in high fatality rates, mainly due to respiratory difficulties, heart disease, and suicide.

Treatment

Despite the absence of a cure for HD, various treatment strategies are available to manage symptoms and enhance the quality of life of affected individuals. Treatment plans are often individualized and based on expert opinions, with pharmacological and non-pharmacological approaches playing crucial roles.

Pharmacological Interventions

- *Motor symptoms*: Common medications for chorea include dopamine receptor-blocking agents such as typical (e.g., haloperidol) and atypical neuroleptics (e.g., risperidone and aripiprazole) and dopamine-depleting agents, mainly Tetrabenazine (Coppen & Roos, 2017). The latter has shown significant efficacy in controlled trials (Duff et al., 2008; Brusa et al., 2009; Mestre et al., 2009; Videnovic, 2013). Benzodiazepines can be temporarily used when chorea transiently worsens due to stressful events. Treatment of hypokinesia and rigidity, especially in juvenile cases, usually includes antiparkinsonian drugs.
- *Psychiatric symptoms*: Selective serotonin reuptake inhibitors and atypical antipsychotics commonly treat depression and behavioral issues, respectively.
- *Emerging treatments*: In 2017, the Food and Drug Administration (FDA) approved Deutetrabenazine, a modified form of Tetrabenazine, for managing chorea. Other experimental treatments, including plant-based therapies and synthetic enzymes, are under investigation but lack conclusive evidence.

Non-Pharmacological Interventions

- *Therapy*: Physical, occupational, and speech therapy are vital in managing motor and cognitive symptoms (Quinn et al., 2020). These interventions improve functional abilities, prevent loss of function, and address risks such as falling and dysphagia. Nutritional management, including thickening agents or percutaneous endoscopic gastrostomy (PEG) tubes, is essential to manage feeding difficulties.
- *Supportive care*: As the disease progresses, multidisciplinary care becomes increasingly important. This includes the involvement of physiotherapists, dietitians, psychologists, and social workers. Palliative care is also considered to minimize symptomatology and improve quality of life. Caregiver support and education are essential, with patient education programs and support groups playing a crucial role in reducing caregiver burnout and distress.
- *Surgical and experimental options*: Surgical interventions, such as deep brain stimulation, have been explored but are not widely practiced in HD. Current research on disease-modifying drugs and embryonic cell implants is also emerging.

In conclusion, managing HD requires a comprehensive approach tailored to the individual's needs, aiming to alleviate symptoms and improve overall well-being and quality of life despite the disease's progressive nature.

Addiction as a Risk Factor for Huntington's Disease

The number of CAG repeats in the HTT gene is a well-established primary determining factor of HD's age of motor onset (AMO). However, genetic factors account for approximately one-third of the variability, leading to the belief that environmental risks mediate the development and progression of the disorder (Schultz et al., 2017).

Substance misuse has been identified as a potential risk factor in AMO. Several scholars have investigated this relationship, and in 2007, the most extensive study at the time found a statistically significant association between AMO and substance misuse, with individuals struggling with substance dependence being more likely to experience AMO (Ehret et al., 2007). However, the authors' failure to control for the primary risk factor—the number of CAG repeats—raised questions about the validity of these findings (Schultz et al., 2017). Consequently, examining studies that have controlled for this variable would provide more reliable insights.

In 2012, Byars and colleagues conducted a study that examined the impact of smoking and drug and alcohol misuse on the age of onset of HD in 136 affected individuals while accounting for the above-mentioned limitation. Results again highlighted an association between lifetime alcohol and drug abuse and an earlier AMO (Byars et al., 2012). However, the authors noted that this association was statistically significant in females only, potentially due to the more pronounced effect of substances on women than men (Byars et al., 2012). Another retrospective study of individuals with HD and tobacco (n = 566), alcohol (n = 374), or drug use (n = 217) observed an association with an earlier AMO in the three groups compared to placebo. The effects were also more amplified in females with HD than in males (Schultz et al., 2017).

This observed association between substance misuse and HD can have several explanations. Individuals who begin noticing the earliest signs of motor symptoms might self-medicate through alcohol and illicit substances (Schultz et al., 2017). Alternatively, common genetic factors and neurobiological pathways might predispose individuals to both substance dependence and HD (Byars et al., 2012). In their recent review, Gardoni and Bellone discuss similar pathophysiological underpinnings in HD and addiction, characterized by aberrant dopaminergic and glutamatergic inputs in the brain, particularly the striatum (Gardoni & Bellone, 2015). Neuronal damage in HD can also result in the deterioration of cortical-striatal circuits, potentially triggering addictive behaviors, inhibitory control, and perseveration (Milton et al., 2020; Waldvogel et al., 2015; Xu et al., 2021). Clinically, and with the lack of disease-modifying pharmacotherapy for HD, preventive measures to delay the onset or severity of symptoms, such as avoidance of tobacco, alcohol, and drug misuse, would be relevant and might have a positive effect on disease progression. Future research is exploring the role of inhibition and perseveration in the development of addictive behaviors in HD (Stevens et al., 2024).

Behavioral Addictions in Huntington's Disease

The hypothesis that individuals with HD are at increased risk of developing behavioral addictions is worth considering. HD, pathological gambling, and other behavioral addictions involve disturbances in the same cortico-striatal circuits and exhibit similar disinhibition-related signs, including altered sensitivity to reward and punishment, impulsivity, and difficulty in considering long-term rewards over short-term ones (Goudriaan et al., 2004; Kalkhoven et al., 2014; Sarafudheen et al., 2022). Furthermore, a recent genome-wide association study of pathological gambling found significant overlap and shared pathology between the condition and HD (Lang et al., 2016).

Unfortunately, the data on behavioral addictions and HD remains scant. In an experimental study, Stout and colleagues discussed how individuals with HD have difficulties learning, remembering, and considering win/loss contingencies when playing card, emphasizing judgment and decision-making problems (Stout et al., 2001). Sarafudheen and colleagues discuss a case of chronic pathological gambling and impulsivity related to HD (Sarafudheen et al., 2022). One review highlighted the possibility of internet gambling being a reason for concern in individuals with HD (Kalkhoven et al., 2014). The authors noted that an increase in gambling problems among individuals afflicted with the condition may not come to attention due to motor symptoms impairing their ability to attend casinos (Kalkhoven et al., 2014). Since internet gambling has eliminated the inconvenience of mobility, there is currently an increased probability of a disproportionate upsurge in gambling-related problems in individuals with HD (Kalkhoven et al., 2014; Sarafudheen et al., 2022). In their extensive review of the topic, Kalkhoven and colleagues conclude that patients with HD may not have an increased propensity to seek risks and engage in gambling. Instead, they have an elevated chance of developing the addictive behavior once they participate in the gambling activity (Kalkhoven et al., 2014). In conclusion, the risk of developing online gambling and other related addictions should be regarded with caution in vulnerable individuals, such as those with HD.

The Potential of Cannabinoids in the Treatment of Huntington's Disease

The use of cannabinoids in the treatment of HD has been subject to investigation for several decades. A clinical trial in 1991 assessed the benefit of cannabidiol (CBD) in reducing chorea and other motor symptoms but found it ineffective, although its use was safe and with minimal side effects (Consroe et al., 1991). One preclinical study using the broad-spectrum cannabinoid Sativex®, an oral spray that delivers equal amounts of delta-9-tetrahydrocannabinol (THC) and CBD concentrations, found it capable of delaying disease progression in HD rat models (Sagredo et al., 2011). A clinical review by the same authors highlights all preclinical evidence supporting the potential use of Sativex® in managing HD (Sagredo et al., 2012). A more recent double-blind, randomized, placebo-controlled clinical trial that included 24 patients found no differences between Sativex® and placebo on motor, cognitive, behavioral, and functional scores. Sativex® was, however, safe and well-tolerated (López-Sendón Moreno et al., 2016).

Alternatively, a systematic review identified 22 studies looking at the effects of medical marijuana or cannabis use on chorea, spasticity, spasms, tremors, and sleep in HD. There was strong evidence for significant improvement in all symptoms following treatment with medical marijuana, with significant improvement particularly noted in rigidity, tremors, and hours of sleep (Akinyemi et al., 2020). Further larger-scale study designs should be conducted while considering the type and dose of cannabinoid combination and more extended treatment periods.

AMYOTROPHIC LATERAL SCLEROSIS

Signs and Symptoms

ALS is a progressive neurological disease that gradually affects most of the muscles of the body, leading to a decrease in quality of life and making affected individuals dependent on others for activities of daily living and survival. Patients with ALS usually present with localized muscle weakness, which can occur in any muscle and is asymmetrical. The most common initial presentation is weakness in the upper or lower limbs, typically starting as hand weakness or foot drop. The gradual weakness of the muscles will lead to their atrophy. With time, more muscles get affected, worsening symptomatology (Goetz, 2000). When bulbar muscles are compromised, patients exhibit swallowing problems, excessive drooling or sialorrhea, facial muscle weakness, and difficulty speaking. Affected respiratory muscles trigger breathing difficulties and respiratory complications, such as pneumonia, which can lead to death (Wijesekera & Leigh, 2009). Patients may also experience cramps and fasciculations with or without muscle weakness (Wijesekera & Leigh, 2009). They may demonstrate inappropriate laughter, crying, or yawning episodes, a phenomenon known as the pseudobulbar effect (Thakore & Pioro, 2017).

Risk Factors

Age, family history, and smoking are the main risk factors for ALS (Gallo et al., 2009). People with occupational exposure have demonstrated a significant risk of ALS as well. For example, exposure to pesticides and metals increases the probability of developing the disease (Malek et al., 2014). Several studies have also noted that military service throughout a person's lifetime increased their risk of developing ALS due to persistent exposure to environmental pollutants (Su et al., 2016; Horner et al., 2003). Another occupational risk factor suspected in athletes is playing football (Daneshvar et al., 2021; Chiò et al., 2005). This association is potentially mediated by the increased likelihood of physical trauma and head injury in this population group (Seals et al., 2016).

Treatment

Unfortunately, there is no cure for ALS. Due to its relentless progression, patients usually survive for 2–5 years after the onset of symptoms. A multidisciplinary approach is typically used to manage symptoms and ensure the comfort and autonomy of patients. Those who receive such care live longer and have a better quality of life than individuals who do not (Van den Berg et al., 2005).

From a pharmacological perspective, patients are typically prescribed Riluzole, a neuroprotective medication that slows functional deterioration and prolongs survival (Miller et al., 2012). However, its mechanism of action is poorly understood (Kennel et al., 2000). Edaravone, a medication targeting free radicals, can also delay the progression of symptoms (Hardiman & van den Berg, 2017). For patients with familial ALS associated with a genetic variant of the gene superoxide dismutase (SOD)-1, the FDA recently approved Tofersen, a gene-silencing therapeutic intervention that downregulates or silences SOD1 (Abati et al., 2020).

Maintaining patient comfort and improving quality of life are cornerstones of treatment. Patients experiencing eating and swallowing difficulties can benefit from a PEG, which also aids in retaining a healthy weight. A mechanical insufflation/exsufflation device can be used to clear secretions. Noninvasive ventilation can also benefit patients with respiratory symptoms (Miller et al., 2009). For refractory sialorrhea, botulinum toxin B or radiation to salivary glands can be considered. Many patients with ALS develop cognitive impairment; therefore, it is important to assess for dementia and manage it accordingly (Miller et al., 2009).

Addiction as a Risk Factor for Amyotrophic Lateral Sclerosis

Recently, a few studies highlighted drug addiction as a potential risk factor for ALS, particularly noteworthy in younger populations (Azhari et al., 2015). For instance, Argyriou et al. (2011) described a case of young-onset ALS in a man who had a history of chronic cocaine use. In a recent literature review, methamphetamine use was found to be a significant risk factor for Parkinson's disease. However, evidence for ALS was weak (Sarlette, 2020). One study noted that a higher intake of alcohol was significantly associated with a decreased risk of ALS, independent of other factors (Huisman et al., 2015). Similarly, a large population-based follow-up study in Sweden (n = 7,965) found heavy alcohol consumption to be inversely associated with the incidence of ALS, suggesting potential protective properties of alcohol through antioxidant effects (Ji et al., 2016).

Smoking is a more studied and established risk factor, as previously mentioned. One study looking at the relationship between smoking and ALS in five large prospective cohorts found cigarette smoking to increase the risk of ALS, with increased risk as the age for smoking initiation decreased (Wang et al., 2011). A meta-analysis of 18 studies on the association between smoking and ALS risk did not support an overall strong association but suggested that the risk is higher in women (Alonso et al., 2010). Although the underlying mechanisms are unclear, this association is potentially mediated by inflammation, oxidative stress, and neurotoxicity from heavy metals and other chemicals present in cigarettes (Alonso et al., 2010).

Opioid Use in Amyotrophic Lateral Sclerosis

Shifting perspectives, pain is a common complaint among individuals with ALS. It is also one of the most impactful symptoms on the quality of life of patients and their caregivers (Pota et al., 2024). Although opioids are not recommended as first-line therapy, they can be used when pain is poorly controlled, refractory, in advanced stages of the disease, in cases of symptoms related to respiratory insufficiency and dyspnea, and to control pain associated with insomnia. However, little evidence exists to support the efficacy and safety of opioids in ALS. Moreover, due to their side effects, including the risk of dependence, opioids tend to be reserved for palliative care, particularly when affected individuals are in hospice units rather than at home (Kwak, 2022; Pota et al., 2024).

The Potential of Cannabinoids in the Treatment of Amyotrophic Lateral Sclerosis

Recently, cannabinoids have gained interest for their potential to manage symptoms of ALS (Raman et al., 2004). A review and meta-analysis of ten studies looking at the effect of cannabinoids in murine ALS models suggested some evidence in prolonging survival time and potentially delaying disease progression, with specific improvements in measures of motor function decline and weight loss (Urbi et al., 2019). Clinically, medical marijuana in ALS has been suggested for analgesia, appetite stimulation, drooling, muscle relaxation, and sleep (Carter and Rosen, 2001). A survey on cannabis use among 131 patients with ALS, 13 of whom reported taking cannabis in the last year, highlighted that it might moderately decrease pain, drooling, appetite loss, spasticity, and depression (Amtmann et al., 2004). One randomized, double-blind, placebo-controlled trial assessed the role of oral THC on cramps in 27 patients with ALS. Although it was well tolerated, there was no evidence of a treatment effect of THC (5 mg twice daily for 2 weeks) on the number of cramps, their intensity, or fasciculations (Weber et al., 2010). Another study highlighted a potential benefit of THC-CBD oromucosal spray in decreasing spasticity in ALS, with improvement depending on the dose used by the patient (Meyer et al., 2019). A more recent survey that included 129 individuals with ALS, of whom 28 reported cannabis use (CBD oil and weed), noted benefits both on the motor (e.g., cramps, fasciculations, rigidity) and non-motor (e.g., emotional state, depression, pain, sleep) symptoms (Lacroix et al., 2023). Lastly, a chart review study found medical marijuana to alleviate pain, poor appetite, and anxiety in the short term, but not insomnia or spasticity, in patients with ALS. Furthermore, usage correlated with faster disease progression (Austin et al., 2024). These findings highlight the need for further research on the potential benefits of cannabinoids in ALS. Future prospective studies and head-to-head comparisons are needed to ascertain the effectiveness of cannabinoids in the management of ALS-related motor and non-motor symptoms.

CONCLUSION

Substance misuse is a significant factor influencing the progression of neurodegenerative diseases, with earlier motor symptom onset in HD linked to substance use, particularly among women. Neurological changes in HD may also increase susceptibility to behavioral addictions, further complicating disease management. In ALS, addiction, particularly to tobacco and potentially to stimulants, is associated with a higher risk of early-onset disease. Emerging evidence supports the potential of cannabinoids in alleviating some symptoms of both HD and ALS, though further research is needed to establish their efficacy and safety. The interplay between addiction and these neurodegenerative disorders reveals complex mechanisms, where some substances accelerate disease progression while others may offer therapeutic benefits under controlled administration.

Future research should prioritize investigating the specific neuropathological pathways linking addiction to HD and ALS to identify actionable targets for intervention. Additionally, large-scale clinical trials are necessary to explore the efficacy, safety, and optimal administration of cannabinoids and other therapeutic agents in managing either condition.

REFERENCES

Abati, E., Bresolin, N., Comi, G., & Corti, S. (2020). Silence superoxide dismutase 1 (SOD1): A promising therapeutic target for amyotrophic lateral sclerosis (ALS). *Expert Opinion on Therapeutic Targets*, *24*(4), 295–310. https://doi.org/10.1080/14728222.2020.1738390

Akinyemi, E., Randhawa, G., Longoria, V., & Zeine, R. (2020). Medical marijuana effects in movement disorders, focus on Huntington disease: A literature review. *Journal of Pharmacy & Pharmaceutical Sciences*, *23*, 389–395. https://doi.org/10.18433/jpps30967

Alonso, A., Logroscino, G., & Hernán, M. A. (2010). Smoking and the risk of amyotrophic lateral sclerosis: A systematic review and meta-analysis. *Journal of Neurology, Neurosurgery & Psychiatry*, *81*(11), 1249–1252.

Amtmann, D., Weydt, P., Johnson, K. L., Jensen, M. P., & Carter, G. T. (2004). Survey of cannabis use in patients with amyotrophic lateral sclerosis. *American Journal of Hospice & Palliative Care*, *21*(2), 95–104. https://doi.org/10.1177/104990910402100206

Argyriou, A. A., Karanasios, P., Makridou, A., Vlachou, M., Giannakopoulou, F., & Makris, N. (2011). Cocaine use and abuse triggering sporadic young-onset amyotrophic lateral sclerosis. *Neuro-degenerative Diseases*, *8*(3), 146–148. https://doi.org/10.1159/000318084

Austin, J. M., Bailey, R., Velazquez, S. G., Sainath, H., & Jackson, C. (2024). Clinical effectiveness of medical marijuana in patients with amyotrophic lateral sclerosis. *Journal of the Neurological Sciences, 466*, 123243.

Azhari, A., Hasanabadi, H., & Rezaei, A. (2015). New cause of death in young adults: Association between illicit drug abuse and amyotrophic lateral sclerosis. *International Journal of Medical Research & Review, 3*(7), 721–725. https://doi.org/10.17511/ijmrr.2015.i7.136

Bates, G. P., Dorsey, R., Gusella, J. F., Hayden, M. R., Kay, C., Leavitt, B. R., Nance, M., Ross, C. A., Scahill, R. I., Wetzel, R., Wild, E. J., & Tabrizi, S. J. (2015). Huntington disease. *Nature Reviews. Disease Primers, 1*, 15005. https://doi.org/10.1038/nrdp.2015.5

Brusa, L., Orlacchio, A., Moschella, V., Iani, C., Bernardi, G., & Mercuri, N. B. (2009). Treatment of the symptoms of Huntington's disease: Preliminary results comparing aripiprazole and tetrabenazine. *Movement Disorders: Official Journal of the Movement Disorder Society, 24*(1), 126–129. https://doi.org/10.1002/mds.22376

Byars, J. A., Beglinger, L. J., Moser, D. J., Gonzalez-Alegre, P., & Nopoulos, P. (2012). Substance abuse may be a risk factor for earlier onset of Huntington disease. *Journal of Neurology, 259*(9), 1824–1831. https://doi.org/10.1007/s00415-012-6415-8

Camacho-Soto, A., Searles Nielsen, S., Faust, I. M., Bucelli, R. C., Miller, T. M., & Racette, B. A. (2022). Incidence of amyotrophic lateral sclerosis in older adults. *Muscle & Nerve, 66*(3), 289–296. https://doi.org/10.1002/mus.27652

Carter, G. T., & Rosen, B. S. (2001). Marijuana in the management of amyotrophic lateral sclerosis. *American Journal of Hospice & Palliative Care, 18*(4), 264–270. https://doi.org/10.1177/104990910101800411

Chiò, A., Benzi, G., Dossena, M., Mutani, R., & Mora, G. (2005). Severely increased risk of amyotrophic lateral sclerosis among Italian professional football players. *Brain: A Journal of Neurology, 128*(Pt 3), 472–476. https://doi.org/10.1093/brain/awh373

Chiò, A., Logroscino, G., Traynor, B. J., Collins, J., Simeone, J. C., Goldstein, L. A., & White, L. A. (2013). Global epidemiology of amyotrophic lateral sclerosis: A systematic review of the published literature. *Neuroepidemiology, 41*(2), 118–130. https://doi.org/10.1159/000351153

Consroe, P., Laguna, J., Allender, J., Snider, S., Stern, L., Sandyk, R., Kennedy, K., & Schram, K. (1991). Controlled clinical trial of cannabidiol in Huntington's disease. *Pharmacology Biochemistry and Behavior, 40*(3), 701–708. https://doi.org/10.1016/0091-3057(91)90386-G

Coppen, E. M., & Roos, R. A. (2017). Current pharmacological approaches to reduce chorea in Huntington's disease. *Drugs, 77*, 29–46.

Daneshvar, D. H., Mez, J., Alosco, M. L., Baucom, Z. H., Mahar, I., Baugh, C. M., Valle, J. P., Weuve, J., Paganoni, S., Cantu, R. C., Zafonte, R. D., Stern, R. A., Stein, T. D., Tripodis, Y., Nowinski, C. J., & McKee, A. C. (2021). Incidence of and mortality from amyotrophic lateral sclerosis in national football league athletes. *JAMA Network Open, 4*(12), e2138801. https://doi.org/10.1001/jamanetworkopen.2021.38801

Dayalu, P., & Albin, R. L. (2015). Huntington disease: Pathogenesis and treatment. *Neurologic Clinics, 33*(1), 101–114. https://doi.org/10.1016/j.ncl.2014.09.003

Di Maio, L., Squitieri, F., Napolitano, G., Campanella, G., Trofatter, J. A., & Conneally, P. M. (1993). Suicide risk in Huntington's disease. *Journal of Medical Genetics, 30*(4), 293–295. https://doi.org/10.1136/jmg.30.4.293

Duff, K., Beglinger, L. J., O'Rourke, M. E., Nopoulos, P., Paulson, H. L., & Paulsen, J. S. (2008). Risperidone and the treatment of psychiatric, motor, and cognitive symptoms in Huntington's disease. *Annals of Clinical Psychiatry: Official Journal of the American Academy of Clinical Psychiatrists, 20*(1), 1–3. https://doi.org/10.1080/10401230701844802

D'Ovidio, F., d'Errico, A., Farina, E., Calvo, A., Costa, G., & Chiò, A. (2016). Amyotrophic lateral sclerosis incidence and previous prescriptions of drugs for the nervous system. *Neuroepidemiology, 47*(1), 59–66. https://doi.org/10.1159/000448618

Ehret, J. C., Day, P. S., Wiegand, R., Wojcieszek, J., & Chambers, R. A. (2007). Huntington disease as a dual diagnosis disorder: Data from the National Research Roster for Huntington Disease Patients and Families. *Drug and Alcohol Dependence, 86*(2–3), 283–286. https://doi.org/10.1016/j.drugalcdep.2006.06.009

Gallo, V., Bueno-De-Mesquita, H. B., Vermeulen, R., Andersen, P. M., Kyrozis, A., Linseisen, J., Kaaks, R., Allen, N. E., Roddam, A. W., Boshuizen, H. C., Peeters, P. H., Palli, D., Mattiello, A., Sieri, S., Tumino, R., Jiménez-Martín, J. M., Díaz, M. J., Suarez, L. R., Trichopoulou, A., Agudo, A., … & Riboli, E. (2009). Smoking and risk for amyotrophic lateral sclerosis: Analysis of the EPIC cohort. *Annals of Neurology, 65*(4), 378–385. https://doi.org/10.1002/ana.21653

Gardoni, F., & Bellone, C. (2015). Modulation of the glutamatergic transmission by dopamine: A focus on Parkinson, Huntington, and addiction diseases. *Frontiers in Cellular Neuroscience, 9*, 25. https://doi.org/10.3389/fncel.2015.00025

Goetz, C. G. (2000). Amyotrophic lateral sclerosis: Early contributions of Jean-Martin Charcot. *Muscle & Nerve, 23*(3), 336–343. https://doi.org/10.1002/(sici)1097-4598(200003)23:3<336::aid-mus4>3.0.co;2-l

Goudriaan, A. E., Oosterlaan, J., de Beurs, E., & Van den Brink, W. (2004). Patho- logical gambling: A comprehensive review of biobehavioral findings. *Neuroscience & Biobehavioral Reviews, 28*, 123–141. https://doi.org/10.1016/j.neubiorev.2004.03.001

Guo, C., Chen, L., & Wang, Y. (2023). Substance abuse and neurodegenerative diseases: Focus on ferroptosis. *Archives of Toxicology, 97*(6), 1519–1528. https://doi.org/10.1007/s00204-023-03505-4

Hardiman, O., & van den Berg, L. H. (2017). Edaravone: A new treatment for ALS on the horizon? *Lancet Neurology, 16*(7), 490–491. https://doi.org/10.1016/S1474-4422(17)30163-1

Horner, R. D., Kamins, K. G., Feussner, J. R., Grambow, S. C., Hoff-Lindquist, J., Harati, Y., Mitsumoto, H., Pascuzzi, R., Spencer, P. S., Tim, R., Howard, D., Smith, T. C., Ryan, M. A., Coffman, C. J., & Kasarskis, E. J. (2003). Occurrence of amyotrophic lateral sclerosis among Gulf War veterans. *Neurology, 61*(6), 742–749. https://doi.org/10.1212/01.wnl.0000069922.32557.ca

Huisman, M. H., Seelen, M., van Doormaal, P. T., de Jong, S. W., de Vries, J. H., van der Kooi, A. J., ... & Veldink, J. H. (2015). Effect of presymptomatic body mass index and consumption of fat and alcohol on amyotrophic lateral sclerosis. *JAMA Neurology, 72*(10), 1155–1162.

Jensen, R. N., Bolwig, T., & Sørensen, S. A. (2018). Psychiatric symptoms in patients with Huntington's disease. *Ugeskrift for laeger, 180*(13), V07160499.

Ji, J., Sundquist, J., & Sundquist, K. (2016). Association of alcohol use disorders with amyotrophic lateral sclerosis: A Swedish national cohort study. *European Journal of Neurology, 23*(2), 270–275. https://doi.org/10.1111/ene.12667

Kalkhoven, C., Sennef, C., Peeters, A., & van den Bos, R. (2014). Risk-taking and pathological gambling behavior in Huntington's disease. *Frontiers in Behavioral Neuroscience, 8*, 103. https://doi.org/10.3389/fnbeh.2014.00103

Kennel, P., Revah, F., Bohme, G. A., Bejuit, R., Gallix, P., Stutzmann, J. M., Imperato, A., & Pratt, J. (2000). Riluzole prolongs survival and delays muscle strength deterioration in mice with progressive motor neuronopathy (PMN). *Journal of the Neurological Sciences, 180*(1–2), 55–61. https://doi.org/10.1016/s0022-510x(00)00423-8

Kwak, S. (2022). Pain in amyotrophic lateral sclerosis: a narrative review. *Journal of Yeungnam Medical Science, 39*(3), 181–189.

Lacroix, C., Guilhaumou, R., Micallef, J., Bruneteau, G., Desnuelle, C., & Blin, O. (2023). Cannabis for the treatment of amyotrophic lateral sclerosis: What is the patients' view? *Revue Neurologique, 179*(9), 967–974.

Lang, M., Lemènager, T., Streit, F., Fauth-Bühler, M., Frank, J., Juraeva, D., ... & Mann, K. F. (2016). Genome-wide association study of pathological gambling. *European Psychiatry, 36*, 38–46.

López-Sendón Moreno, J. L., García Caldentey, J., Trigo Cubillo, P., Ruiz Romero, C., García Ribas, G., Alonso Arias, M. A., ... & García de Yébenes Prous, J. (2016). A double-blind, randomized, cross-over, placebo-controlled, pilot trial with Sativex in Huntington's disease. *Journal of Neurology, 263*, 1390–1400.

Malek, A. M., Barchowsky, A., Bowser, R., Heiman-Patterson, T., Lacomis, D., Rana, S., Youk, A., Stickler, D., Lackland, D. T., & Talbott, E. O. (2014). Environmental and occupational risk factors for amyotrophic lateral sclerosis: A case-control study. *Neuro-Degenerative Diseases, 14*(1), 31–38. https://doi.org/10.1159/000355344

McAllister, B., Gusella, J. F., Landwehrmeyer, G. B., Lee, J. M., MacDonald, M. E., Orth, M., Rosser, A. E., Williams, N. M., Holmans, P., Jones, L., Massey, T. H., & REGISTRY Investigators of the European Huntington's Disease Network (2021). Timing and impact of psychiatric, cognitive, and motor abnormalities in Huntington disease. *Neurology, 96*(19), e2395–e2406. https://doi.org/10.1212/WNL.0000000000011893

McCusker, E. A., Gunn, D. G., Epping, E. A., Loy, C. T., Radford, K., Griffith, J., Mills, J. A., Long, J. D., Paulsen, J. S., & PREDICT-HD Investigators of the Huntington Study Group. (2013). Unawareness of motor phenoconversion in Huntington disease. *Neurology, 81*(13), 1141–1147. https://doi.org/10.1212/WNL.0b013e3182a55f05

Medina, A., Mahjoub, Y., Shaver, L., & Pringsheim, T. (2022). Prevalence and incidence of Huntington's disease: An updated systematic review and meta-analysis. *Movement Disorders: Official Journal of the Movement Disorder Society, 37*(12), 2327–2335. https://doi.org/10.1002/mds.29228

Mestre, T., Ferreira, J., Coelho, M. M., Rosa, M., & Sampaio, C. (2009). Therapeutic interventions for symptomatic treatment in Huntington's disease. *Cochrane Database of Systematic Reviews*, (3), CD006456. https://doi.org/10.1002/14651858.CD006456.pub2

Meyer, T., Funke, A., Münch, C., Kettemann, D., Maier, A., Walter, B., ... & Spittel, S. (2019). Real world experience of patients with amyotrophic lateral sclerosis (ALS) in the treatment of spasticity using tetrahydrocannabinol: Cannabidiol (THC: CBD). *BMC Neurology*, *19*, 1–13.

Miller, R. G., Jackson, C. E., Kasarskis, E. J., England, J. D., Forshew, D., Johnston, W., Kalra, S., Katz, J. S., Mitsumoto, H., Rosenfeld, J., Shoesmith, C., Strong, M. J., Woolley, S. C., & Quality Standards Subcommittee of the American Academy of Neurology. (2009). Practice parameter update: The care of the patient with amyotrophic lateral sclerosis: drug, nutritional, and respiratory therapies (an evidence-based review): Report of the quality standards subcommittee of the American Academy of Neurology. *Neurology*, *73*(15), 1218–1226. https://doi.org/10.1212/WNL.0b013e3181bc0141

Miller, R. G., Jackson, C. E., Kasarskis, E. J., England, J. D., Forshew, D., Johnston, W., Kalra, S., Katz, J. S., Mitsumoto, H., Rosenfeld, J., Shoesmith, C., Strong, M. J., Woolley, S. C., & Quality Standards Subcommittee of the American Academy of Neurology. (2009). Practice parameter update: The care of the patient with amyotrophic lateral sclerosis: Multidisciplinary care, symptom management, and cognitive/behavioral impairment (an evidence-based review): Report of the quality standards Subcommittee of the American Academy of Neurology. *Neurology*, *73*(15), 1227–1233. https://doi.org/10.1212/WNL.0b013e3181bc01a4

Miller, R. G., Mitchell, J. D., & Moore, D. H. (2012). Riluzole for amyotrophic lateral sclerosis (ALS)/motor neuron disease (MND). *Cochrane Database of Systematic Reviews*, *2012*(3), CD001447. https://doi.org/10.1002/14651858.CD001447.pub3

Milton, L. K., Mirabella, P. N., Greaves, E., Spanswick, D.C., Buuse M van den, Oldfield, B. J., & Foldi, C. J. (2020). Suppression of cortico-striatal circuit activity improves cognitive flexibility and prevents body weight loss in activity-based anorexia in rats. *bioRxiv*. https://doi.org/10.1101/2020.03.29.015115.

Pota, V., Sansone, P., De Sarno, S., Aurilio, C., Coppolino, F., Barbarisi, M., ... & Pace, M. C. (2024). Amyotrophic lateral sclerosis and pain: A narrative review from pain assessment to therapy. *Behavioural Neurology*, *2024*(1), 1228194.

Quinn, L., Kegelmeyer, D., Kloos, A., Rao, A. K., Busse, M., & Fritz, N. E. (2020). Clinical recommendations to guide physical therapy practice for Huntington disease. *Neurology*, *94*(5), 217–228. https://doi.org/10.1212/WNL.0000000000008887

Raman, C., McAllister, S. D., Rizvi, G., Patel, S. G., Moore, D. H., & Abood, M. E. (2004). Amyotrophic lateral sclerosis: Delayed disease progression in mice by treatment with a cannabinoid. *Amyotrophic Lateral Sclerosis and Other Motor Neuron Disorders: Official Publication of the World Federation of Neurology, Research Group on Motor Neuron Diseases*, *5*(1), 33–39. https://doi.org/10.1080/14660820310016813

Rowland, L. P. (1994). Amyotrophic lateral sclerosis. *Current Opinion in Neurology*, *7*, 310–315.

Sagredo, O., Pazos, M. R., Satta, V., Ramos, J. A., Pertwee, R. G., & Fernández-Ruiz, J. (2011). Neuroprotective effects of phytocannabinoid-based medicines in experimental models of Huntington's disease. *Journal of Neuroscience Research*, *89*(9), 1509–1518.

Sagredo, O., Pazos, M. R., Valdeolivas, S., & Fernandez-Ruiz, J. (2012). Cannabinoids: Novel medicines for the treatment of Huntington's disease. *Recent Patents on CNS Drug Discovery*, *7*(1), 41–48. https://doi.org/10.2174/157488912798842278

Sarafudheen, S., Shoka, A., & Kathirgamachelvam, J. (2022). Pathological gambling and impulsivity related to Huntington's disease. *Progress in Neurology and Psychiatry*, *26*(3), 12–15.

Sarlette, K. (2020). Methamphetamine: A potential risk factor for neurodegenerative conditions. Physician Assistant Scholarly Project Papers. 85. https://commons.und.edu/pas-grad-papers/85

Schultz, J. L., Kamholz, J. A., Moser, D. J., Feely, S. M. E., Paulsen, J. S., & Nopoulos, P. C. (2017). Substance abuse may hasten motor onset of Huntington disease: Evaluating the Enroll-HD database. *Neurology*, *88*(9), 909–915. https://doi.org/10.1212/WNL.0000000000003661

Seals, R. M., Hansen, J., Gredal, O., & Weisskopf, M. G. (2016). Physical trauma and amyotrophic lateral sclerosis: A population-based study using Danish national registries. *American Journal of Epidemiology*, *183*(4), 294–301. https://doi.org/10.1093/aje/kwv169

Stevens, M., Corley, E., Burke, T., & HD-West Consortium. (2024). F071 Substance use behaviour in Huntington's disease: Investigating the role of inhibitory control and perseveration. *Journal of Neurology, Neurosurgery & Psychiatry*, *95*, A97-A98.

Stout, J. C., Rodawalt, W. C., & Siemers, E. R. (2001). Risky decision making in Huntington's disease. *Journal of the International Neuropsychological Society*, *7*(1), 92–101.

Su, F. C., Goutman, S. A., Chernyak, S., Mukherjee, B., Callaghan, B. C., Batterman, S., & Feldman, E. L. (2016). Association of environmental toxins with amyotrophic lateral sclerosis. *JAMA Neurology*, *73*(7), 803–811. https://doi.org/10.1001/jamaneurol.2016.0594

Thakore, N. J., & Pioro, E. P. (2017). Laughter, crying and sadness in ALS. *Journal of Neurology, Neurosurgery, and Psychiatry*, *88*(10), 825–831. https://doi.org/10.1136/jnnp-2017-315622

Urbi, B., Owusu, M. A., Hughes, I., Katz, M., Broadley, S., & Sabet, A. (2019). Effects of cannabinoids in amyotrophic lateral sclerosis (ALS) murine models: A systematic review and meta-analysis. *Journal of Neurochemistry*, *149*(2), 284–297.

Van den Berg, J. P., Kalmijn, S., Lindeman, E., Veldink, J. H., de Visser, M., Van der Graaff, M. M., Wokke, J. H., & Van den Berg, L. H. (2005). Multidisciplinary ALS care improves quality of life in patients with ALS. *Neurology*, *65*(8), 1264–1267. https://doi.org/10.1212/01.wnl.0000180717.29273.12

van der Burg, J. M., Björkqvist, M., & Brundin, P. (2009). Beyond the brain: Widespread pathology in Huntington's disease. *Lancet Neurology*, *8*(8), 765–774. https://doi.org/10.1016/S1474-4422(09)70178-4

van Duijn, E., Kingma, E. M., & van der Mast, R. C. (2007). Psychopathology in verified Huntington's disease gene carriers. *J Neuropsychiatry Clin Neurosci*, *19*, 441–448.

Videnovic, A. (2013). Treatment of Huntington disease. *Current Treatment Options in Neurology*, *15*(4), 424–438. https://doi.org/10.1007/s11940-013-0219-8

Waldvogel, H. J., Kim, E. H., Tippett, L. J., Vonsattel, J.-PG., & Faull, R. L. (2015). The neuropathology of Huntington's Disease. In H. H. P. Nguyen & M. A. Cenci (Eds.), *Behavioral neurobiology of Huntington's disease and Parkinson's disease* (pp. 33–80). Springer. https://doi.org/10.1007/7854_2014_354

Walker, F. O. (2007). Huntington's disease. *Lancet* (London, England), *369*(9557), 218–228. https://doi.org/10.1016/S0140-6736(07)60111-1

Wang, H., O'Reilly, É. J., Weisskopf, M. G., Logroscino, G., McCullough, M. L., Thun, M. J., ... & Ascherio, A. (2011). Smoking and risk of amyotrophic lateral sclerosis: A pooled analysis of 5 prospective cohorts. *Archives of Neurology*, *68*(2), 207–213.

Weber, M., Goldman, B., & Truniger, S. (2010). Tetrahydrocannabinol (THC) for cramps in amyotrophic lateral sclerosis: A randomised, double-blind crossover trial. *Journal of Neurology, Neurosurgery & Psychiatry*, *81*(10), 1135–1140.

Wijesekera, L. C., & Leigh, P. N. (2009). Amyotrophic lateral sclerosis. *Orphanet Journal of Rare Diseases*, *4*, 3. https://doi.org/10.1186/1750-1172-4-3

Wolfson, C., Gauvin, D. E., Ishola, F., & Oskoui, M. (2023). Global prevalence and incidence of amyotrophic lateral sclerosis: A systematic review. *Neurology*, *101*(6), e613–e623. https://doi.org/10.1212/WNL.0000000000207474

Xu, T., Gu, Q., Zhao, Q., Wang, P., Liu, Q., Fan, Q., Chen, J., & Wang, Z. (2021). Impaired cortico-striatal functional connectivity is related to trait impulsivity in unmedicated patients with obsessive-compulsive disorder. *Journal of Affective Disorders*, *281*, 899–907. https://doi.org/10.1016/j.jad.2020.11.037

5 Parkinson's Disease and Addiction

Wassim Hassan, Noreen Hassan, Ahmad Hassan, and Luke Petry

INTRODUCTION

Parkinson's disease (PD) is a debilitating neurodegenerative disease characterized by the loss of dopaminergic neurons in the substantia nigra (SN) pars compacta and the presence of Lewy bodies—cytoplasmic inclusions consisting of misfolded α-synuclein protein aggregates (Zafar & Yaddanapudi, 2023). Clinically, PD manifests as bradykinesia—the generalized slowing of movements—as well as a resting tremor and limb rigidity. PD affects more than 6 million individuals worldwide and is expected to double in prevalence by 2040 (Day & Mullin, 2021). Unfortunately, no disease-modifying therapies currently exist for the disorder. Rather, treatment focuses primarily on dopamine replacement to alleviate symptoms.

Although PD is classically considered a motor disorder, non-motor symptoms such as mood dysregulation, dementia, sleep disturbances, and behavioral issues are common among patients and have a more significant negative impact on quality of life than motor symptomatology (Müller et al., 2013). Neuropsychiatric symptoms are common in PD, with up to one-third of patients experiencing depression, up to one-quarter experiencing anxiety, and up to one-quarter experiencing impulse control disorders (ICDs) (Costello et al., 2022). Table 5.1 provides an overview of common neuropsychiatric symptoms seen in PD.

In this chapter, we discuss the relationship between PD and addiction, examining the role of dopamine in PD, the intricate association between ICDs, various drugs of misuse, and PD, and how dopaminergic medications can lead to addiction in this patient population.

THE ROLE OF DOPAMINE IN PARKINSON'S DISEASE AND ADDICTION

Dopamine is a neurotransmitter produced in the SN, ventral tegmental area, and hypothalamus (Juárez Olguín et al., 2016). It plays a vital role in motor function and the reward pathway, influencing impulsivity, motivation, and reinforcement learning (Grall-Bronnec et al., 2018). Dysfunction of the dopaminergic pathways in the central nervous system (CNS) is implicated in various pathologies such as PD, drug dependence, and endocrine disorders (Rang et al., 2012). There are three main dopaminergic pathways in the CNS:

- *Nigrostriatal pathway*: Implicated in motor control, this pathway consists of cell bodies in the SN with axons terminating in the corpus striatum.
- *Mesocorticolimbic pathway*: Also known as the reward system, this pathway consists of cell bodies located in the ventral tegmental area with axons terminating in various parts of the limbic system.
- *Tuberoinfundibular pathway*: Implicated in the secretion of hormones, such as prolactin, this pathway consists of cell bodies in the ventral hypothalamus with axons that project to the median eminence and pituitary gland.

DOI: 10.1201/9781003478485-5

TABLE 5.1
Common Neuropsychiatric Symptoms in Parkinson's Disease

Symptom	Description	Lifetime Prevalence	Management Strategies
Depression	Persistent sadness, loss of interest, feelings of hopelessness.	38% (Cong et al., 2022).	Antidepressants (e.g., fluoxetine, duloxetine), psychotherapy, physical activity.
Anxiety	Excessive worry, panic attacks, general unease.	30–40% (Aarsland et al., 2009).	Anti-anxiety medications (fluoxetine, buspirone), cognitive-behavioral therapy.
Cognitive impairment	Memory problems, difficulty concentrating, executive dysfunction.	40% (Baiano et al., 2020).	Cognitive training, cholinesterase inhibitors (e.g., donepezil), structured routines.
Psychosis	Hallucinations (visual more common than auditory), delusions.	38.4% (Stang et al., 2022).	Antipsychotic medications (e.g., quetiapine, clozapine), reducing dopaminergic medications.
Apathy	Lack of motivation, decreased initiative, indifference.	17–70% (Aarsland et al., 2009).	Behavioral interventions, stimulant medications (e.g., methylphenidate), engaging activities.
Impulse control disorders	Compulsive behaviors (e.g., gambling, shopping, eating, sexual activity).	46% (Corvol et al., 2018).	Reducing or discontinuing dopamine agonists, behavioral therapy.
Sleep disturbances	Insomnia, REM sleep behavior disorder, excessive daytime sleepiness.	71% (Dodet et al., 2024).	Sleep hygiene, medications (e.g., melatonin for REM sleep disorder), CPAP for sleep apnea.
Fatigue	Persistent tiredness, lack of energy.	33–70% (Nassif and Pereira, 2018).	Exercise, sleep optimization, medications (e.g., modafinil).
Dementia	Severe cognitive decline affecting daily functioning.	24–50% (Anang et al., 2014).	Cholinesterase inhibitors, NMDA receptor antagonists (e.g., memantine), supportive care.
Mood swings	Rapid changes in mood, irritability.	Up to two-thirds (Richard et al., 2005).	Medication adjustments, mood stabilizers (e.g., lamotrigine, Depakote), psychotherapy.

Abbreviations: CPAP = continuous positive airway pressure; REM = rapid eye movement.

A hallmark of PD is the loss of dopaminergic neurons in the SN pars compacta due to the presence of Lewy bodies (Surmeier, 2018). This results in altered dopaminergic signaling in the pathways listed above. The nigrostriatal and mesocorticolimbic pathways, which regulate reward-driven behavior, are the focus of many studies investigating the pathophysiological basis of addictive behaviors (Haber & Knutson, 2010). Their disturbance would explain addictive symptomatology in PD.

IMPULSE CONTROL DISORDERS AND PARKINSON'S DISEASE

Definition and Prevalence of Impulse Control Disorders

ICDs are defined as an inability to resist a drive, impulse, or temptation to perform an act that is harmful to the person or others (Carbone & Djamshidian, 2024). ICDs are often seen as comorbidities in various neurodegenerative disorders such as PD, frontotemporal dementia, multiple system atrophy, and progressive supranuclear palsy (Weintraub et al., 2010; Pompanin et al., 2014; Cilia et al., 2014; Kok et al., 2021). Patients with PD have three times the odds of developing ICDs and related compulsive behaviors compared to those without the condition (Erga et al., 2018). ICDs are not directly caused by PD pathology. However, they are thought to be associated with dopamine agonist medications (Weintraub et al., 2013; Grosset & Cardoso, 2011). In most patients diagnosed with PD, these addictive behaviors arise after the initiation of dopaminergic therapy, primarily involving dopamine agonists. ICDs have been documented in patients with PD as far back as the 1960s and 1970s, shortly after the introduction of levodopa (Barbeau, 1969; Bowers et al., 1971; Shapiro et al., 1973).

According to the latest edition of the *Diagnostic and Statistical Manual of Mental Disorders* (DSM-5-TR), ICDs progress through five stages (American Psychiatric Association, 2022). Initially, individuals experience an increased sense of tension. This is followed by an inability to resist the urge to act. During the act, arousal reaches its peak, and upon completion, a sense of relief or release is felt. Finally, individuals may experience feelings of remorse or guilt about their behavior. These patients often persist in their addictive behaviors despite experiencing negative consequences. Efforts to stop these behaviors frequently result in dysphoria, anxiety, and depression, akin to withdrawal symptoms seen in drug misuse (Garcia & Thibaut, 2010). The most commonly described ICDs in PD include compulsive sexual disorder, gambling disorder, compulsive shopping, and compulsive eating (Weintraub et al., 2010). Drug addiction, excessive hoarding, and dopamine dysregulation syndrome (DDS) are also seen (Barbosa et al., 2022; Friedman & Chang, 2013).

DDS is similar to ICD in that it is also a dopaminergic complication of PD treatment (Friedman and Chang, 2013). DDS, however, is characterized by the compulsive overuse of dopaminergic medications—particularly levodopa—exceeding the dosage necessary for controlling PD symptoms (Katzenschlager, 2011). Patients with DDS often experience a "high" from each dose of levodopa, which initially provides euphoric sensations but later leads to an uncontrollable craving for higher doses without the same enjoyment or therapeutic benefit (Barbosa et al., 2018). This behavior closely mirrors the patterns seen in chronic cocaine and amphetamine users (Volkow et al., 2004).

Impulse Control Disorder Progression and Risk Factors

Several studies have shed light on the behavioral changes that accompany dopaminergic therapy. It has been repeatedly shown that medication-naïve patients with PD show improved learning from negative feedback, while medicated patients learn better from positive feedback (Frank et al., 2004; Voon et al., 2010; Cools et al., 2010; Bodi et al., 2009). This enhanced positive feedback and diminished negative feedback learning are thought to contribute to the development of addictive behaviors and ICDs in medicated patients with PD.

Patients with PD who develop ICDs experience a more significant burden of non-motor symptoms than those without addictive behaviors (Carbone & Djamshidian, 2024). These patients frequently suffer from increased rates of depression, poorer quality of life (Antonini et al., 2017), reduced social well-being (Phu et al., 2014), increased rates of apathy (Scott et al., 2020), worse sleep, increased anxiety, higher mania scores (O'Sullivan et al., 2011), psychosis (Hinkle et al., 2018), and a higher frequency of rapid eye movement-sleep behavior disturbances (Fantini et al., 2015). Additionally, they exhibit higher levels of aggressiveness, irritability, disinhibition, denial, and poorer insight (Latella et al., 2019). Physical symptoms such as urinary dysfunction, fatigue, cardiovascular problems, and poorer working memory also negatively impact their quality of life (Jesus et al., 2020; Djamshidian et al., 2010). According to data from the National Danish Patient Registry, patients with PD who develop ICDs have a longer disease duration than those without (9.3 years versus 7.5 years) (Callesen et al., 2014).

It remains unclear why some patients with PD develop ICDs while others do not. Therefore, it is unlikely that a single mechanism is responsible for developing ICDs. It is well-established that the use of dopaminergic medications, particularly dopamine receptor agonists, in susceptible patients with PD leads to the development of ICDs (Augustine et al., 2021). Several factors have been discovered that increase the risk of developing ICDs, such as younger age, younger onset of PD, depression, and anxiety (Carbone & Djamshidan, 2024; Marín-Lahoz et al., 2019). Other risk factors include a history of alcoholism or smoking, a higher novelty-seeking personality, increased caffeine consumption, and a personal or family history of addictive behavior (Averbeck et al., 2014; Bastiaens et al., 2013). Alexithymia—a difficulty to express, define, or identify emotions—is also a risk factor for developing ICDs and is associated with increased impulsivity in drug-naïve patients with PD (Poletti et al., 2012; Goerlich-Dobre et al., 2014). Genetic factors may also elevate the risk of developing ICDs, as a recent genome-wide association study of a large cohort of 5,770 patients with PD identified four loci (DAB1, PRKAG2, MEFV, and PRKCE) associated with ICDs (Weintraub et al., 2022). Studies have also shown an association between ICD development in patients with PD and genetic polymorphisms in several genes, such as dopamine receptor (*DRD1–3*), glutamate receptor (*GRIN2B*), serotonin receptor (*HTR2A*), dopamine transporter (*DAT1*), and tryptophan hydroxylase 2 (*TPH2*) (Erga et al., 2018).

THE ROLE OF DOPAMINERGIC MEDICATIONS IN ADDICTION AMONG PATIENTS WITH PARKINSON'S DISEASE

Dopamine Agonists and Impulse Control Disorders

Dopaminergic medications prescribed for the motor dysfunction associated with PD can induce or intensify various neuropsychiatric issues such as delirium, hallucinations, delusions, paranoia, psychosis, compulsive behaviors, and ICDs (Weiss et al., 2012). As previously discussed, ICDs are not uncommon in patients with PD and can occur as a complication of dopamine replacement therapy (Drew et al., 2020). The most common examples of ICDs include compulsive buying, gambling, eating, disinhibited sexual behavior, and drug addiction. Up to a quarter of patients with PD experience ICDs as a neuropsychiatric complication, often leading to severe personal, financial, and social consequences (Baig et al., 2019). ICDs are not considered to be a direct consequence of PD; instead, they are thought to develop in response to dopaminergic therapy (Drew et al., 2020). Studies have shown that the prevalence of ICDs among healthy controls is similar to drug-naïve patients with PD, and ICDs only increase in prevalence among patients with PD who have received dopaminergic therapy (Antonini et al., 2011). ICDs primarily arise in patients treated with dopamine agonists such as pramipexole, ropinirole, rotigotine, or pergolide (Seeman, 2015). Moreover, the risk of developing an ICD is even higher in patients receiving a combination of a dopamine agonist and levodopa (Weiss et al., 2012).

Levodopa and Motor Complications

Levodopa is the gold standard therapy for PD, but its chronic use often leads to motor complications known as L-DOPA–induced dyskinesia (Bargiotas & Konitsiosis, 2013). Currently, dopamine receptor agonists are the preferred initial treatment for patients with PD, deferring the use of levodopa therapy until later stages. Pramipexole and ropinirole, which are dopamine D2/D3 agonists with a preference for D3 receptors, improve the motor symptoms of PD (Stark et al., 2018). However, these medications can also induce impulsive and compulsive behaviors characterized by an abnormal focus on reward-driven activities. A recent meta-analysis confirmed that the transdermal dopamine agonist rotigotine was three times less likely to induce addictive behaviors than pramipexole and ropinirole (Soileau et al., 2024). It is thought that transdermal drug application can lead to more continuous drug delivery, thus avoiding plasma level peaks and troughs associated with more pulsatile oral drug delivery. While plasma concentrations of oral dopamine agonists typically drop after 6–12 hours, plasma levels during rotigotine therapy remain stable for up to 24 hours. Additionally, the transdermal application of rotigotine allows direct access to the bloodstream, bypassing the hepatic first-pass effect associated with oral dopamine agonists (Elshoff et al., 2015).

Clinical findings highlight the causal relationship between dopamine receptor agonists and the development of impulsive behaviors, emphasizing that reducing or discontinuing dopamine receptor agonists in individuals with ICDs often results in symptom improvement (Claassen et al., 2013). Dopamine agonists have a higher propensity to induce ICDs compared to levodopa due to their pharmacological properties (Weiss et al., 2012). D1 and D2 receptors are greatly expressed in striatal motor regions, with significant motor effects attributed to D2 receptor stimulation. D1 receptor stimulation synergizes with D2 activity to promote motor responses and may contribute to dyskinesias. D3 receptors, involved in motor responses and located in limbic nuclei, are targeted by psychiatric drugs and implicated in the addiction process.

Dopamine agonists exhibit a specific affinity for D3 receptors—approximately 20–100 times greater than D2 receptors—and little to no affinity for D1 receptors (Weiss et al., 2012). Consequently, dopamine agonists tend to stimulate pathways associated with reward behavior more strongly than dopamine derived from levodopa therapy. This difference in receptor affinity may explain why dopamine agonists are more likely to cause behavioral changes such as hallucinations, psychosis, and ICDs compared to levodopa. Additionally, their lower affinity for D2 and D1 receptors accounts for their reduced efficacy in alleviating motor symptoms in PD and their lower risk of triggering dyskinesias. A recent pharmacovigilance-pharmacodynamic study supported this role of D3 receptor agonism in inducing impulsivity (Fusaroli et al., 2023).

MANAGEMENT OF IMPULSE CONTROL DISORDERS IN PARKINSON'S DISEASE

Prevention and Education

The best approach to managing ICDs in patients with PD is through preemption and prevention (Weiss et al., 2012). Patients and their caregivers should be counseled about the risks of developing ICDs before initiating dopaminergic therapies. Dopamine agonists should be avoided or used with extreme caution in patients with a history of obsessive-compulsive disorder, alcoholism, drug misuse, smoking, or other addictive behaviors (Stamy & Jankovic, 2008). The vast majority of patients with PD who develop ICDs will experience a significant reduction or elimination of these behaviors through the gradual tapering and discontinuation of dopamine agonist therapy (Weiss et al., 2012). Typically, tapering dopamine agonists requires increasing levodopa doses to manage motor symptoms, yet ICD symptoms still improve despite this adjustment (Mamikonyan et al., 2008).

Pharmacological and Non-Pharmacological Approaches

Pharmacological options for treating ICDs in patients with PD showed varied results (Carbone and Djamshidian, 2024). Naltrexone, an opioid receptor antagonist effective in treating alcohol and opioid use disorders, did not improve ICDs in patients with PD (Papay et al., 2014). Clonidine, an α2-adrenergic agonist, was found to result in a non-significant reduction of impulsivity compared to placebo in a recent randomized, controlled, double-blind, phase IIb trial involving patients with PD with ICDs treated over 8 weeks (Laurencin et al., 2023). Atomoxetine was shown to decrease impulsivity and risk-taking among patients with PD, although none of the patients included in the study had a diagnosis of ICD (Kehagia et al., 2014). Currently, pimavanserin, a selective serotonin 5-HT_{2A} inverse agonist, is being studied in a phase II trial (NCT03947216), with results expected in 2025.

Deep brain stimulation of the subthalamic nucleus can improve ICDs secondary to the consequent reduction in dopaminergic therapy (Lhommee et al., 2012). As previously discussed, discontinuation or reduction in dopamine agonist dose is linked with better outcomes in ICD (Barbosa et al., 2019). In addition, switching a patient from a dopamine agonist to levodopa can improve impulsive behavior within a few months (Lee et al., 2019). However, patients should be advised that dopamine agonist withdrawal syndrome may occur, which is typically characterized by feelings of anxiety, depression, fatigue, pain, and the feeling of being undertreated (Rabinak & Nirenberg, 2010). Remission of ICD can be achieved in about 40–80% of patients (Carbone & Djamshidian, 2024).

DRUGS OF ADDICTION AND PARKINSON'S DISEASE

The discussion of addiction in PD would be incomplete without addressing the role of drug addiction. As discussed previously, pharmacologic treatments for Parkinson's, particularly dopamine agonists, increase the propensity for ICDs, which have a high rate of concurrent substance use disorders (Schreiber et al., 2011). The literature on drug addiction in PD largely falls into two categories. The first focuses on substances potentially linked to the onset of PD. The second category explores illicit drugs that may exhibit therapeutic benefits. Understanding both the neurotoxic and neuroprotective aspects of various substances is essential to advancing therapeutic approaches and managing the risk of addiction in patients with PD.

Neurotoxic Drugs Linked to Parkinson's Disease

Several drugs of abuse have been linked to neurotoxicity, particularly within the dopamine pathways affected in PD. One of the most well-known examples is 1-methyl-4-phenyl-1,2,3,6-tetrahydr opyridine (MPTP), a by-product of synthetic heroin production (Langston, 2017). In the early 1980s, individuals in northern California developed Parkinson-like symptoms after using heroin contaminated with MPTP (AlShimemri et al., 2021). Langston's insightful clinical observation and subsequent work led to the identification of MPTP as a dopaminergic neurotoxin, which facilitated the development of MPTP rodent models and advanced our understanding of PD pathophysiology and treatments.

Methamphetamine is a stimulant that has been strongly associated with the development of PD (Mursaleen & Stamford, 2016). Neuronal damage caused by methamphetamine is primarily mediated via its role as a dopamine reuptake inhibitor, resulting in excessive increases in synaptic dopamine and the subsequent generation of neurotoxic reactive oxygen species (Jang et al., 2017). A retrospective population-based cohort study found that individuals with a history of methamphetamine abuse are 1.76 times more likely to develop PD later in life (Callaghan et al., 2012). A study comparing SN morphology in stimulant users to non-drug users found that the former group exhibited significantly higher SN echogenicity (Todd et al., 2013). Hyperechogenicity of the SN is an established risk factor for developing PD later in life (Berg et al., 2011). Interestingly, cocaine use does not appear to significantly increase the risk of a subsequent PD diagnosis (Callaghan et al., 2012).

Neuroprotective Drugs Linked to Parkinson's Disease

Regarding drugs that exhibit therapeutic effects in PD, nicotine stands out as having protective properties against the development of Parkinson's. The association between smoking and a reduced risk of PD is one of the most robust environmental links in neuroepidemiology (Rose et al., 2024). Several studies have demonstrated that smokers have a substantially lower relative risk for PD than nonsmokers (Hernán et al., 2002; Scott et al., 2005; Ritz et al., 2007; Gallo et al., 2019). Brain autopsy studies have also demonstrated that heavy smokers have a significantly lower risk of developing Lewy-related pathology, particularly in the SN (Tsuang et al., 2010). Interestingly, it has been hypothesized that the increasing rates of PD are at least partially linked to reduced rates of smoking (Dorsey et al., 2018; Willis et al., 2022). Several small studies involving patients with PD demonstrated that nicotine provides transient improvement in motor symptoms such as tremor, rigidity, and bradykinesia, suggesting a role for nicotine as a potential therapeutic drug in PD (Thiriez et al., 2011). It is believed that the therapeutic benefits of smoking in PD are mediated by nicotine or nicotinic receptor agonists, which stimulate dopamine release and enhance dopamine function (Quik et al., 2008).

Similarly to nicotine, caffeine has been associated with a reduced risk of developing PD. A recent meta-analysis of 13 studies found that individuals with regular caffeine consumption had a significantly lower risk of PD during follow-up evaluation (Hong et al., 2020). It was also concluded that individuals with a PD diagnosis who consumed caffeine demonstrated a significantly lower rate of disease progression. However, clinical studies have produced mixed results when evaluating the therapeutic effect of caffeine on motor symptoms in individuals with PD (Mursaleen & Stamford, 2016).

The association between PD and alcohol consumption is less clear. For example, it has been shown that the risk of developing PD is similar between those who abstain from alcohol and those with moderate consumption (Hernán et al., 2003). However, a more recent study identifying over 600,000 patients through the Swedish National Inpatient Register found that a history of alcohol dependence was associated with an increased risk of a PD diagnosis with a hazard ratio of 1.38 (Eriksson et al., 2013). Interestingly, a study using MPTP-lesioned mice as a PD model found that daily administration of resveratrol, a polyphenol with antioxidant properties found in red wine, prevented striatal dopamine depletion (Blanchet et al., 2008). Overall, more research is needed to determine the benefits or harms that alcohol consumption confers in PD.

The legalization of cannabis in many countries across the world has resulted in patients with PD turning to cannabis as a potential therapeutic option (Urbi et al., 2022). Cannabis has been implicated in the pharmacotherapy of several diseases, and various studies have found that cannabinoid agonists result in the improvement of motor symptoms in mouse models of PD (Giuffrida & McMahon, 2010). Animal studies provide evidence that tetrahydrocannabinol (THC) and cannabidiol (CBD) exhibit antioxidant properties, which may offer neuroprotection against the progressive degeneration of dopaminergic neurons of the SN in patients with PD (García-Arencibia et al., 2007). In the central nervous system, CBD binds to CB1 receptors, and CB1 agonists in animal models were found to reduce tremors, while CB1 antagonists were shown to reduce bradykinesia (Mohanty & Lippman, 2019). It is unclear if human subjects derive the same benefits. Clinical studies on cannabis use in PD are inconclusive due to various limitations, such as small sample sizes, expectancy bias, and the lack of standardized outcome measures (Stampanoni et al., 2017). For example, two survey studies showed that patients with PD-derived motor and non-motor benefit from smoking cannabis (Venderová et al., 2004; Finseth et al., 2015). In contrast, a 4-week randomized, placebo-controlled crossover study showed that cannabis provided no significant changes in motor symptoms, quality of life, or sleep (Carroll et al., 2004). A more recent retrospective chart review including patients with PD treated with medical cannabis found that 87% of patients noted an improvement in PD symptoms such as tremors, spasticity, dyskinesia, and lack of appetite (Aladeen et al., 2023). Additionally, 56% of opioid users in the study were able to decrease or discontinue opioid use. Overall, more research is needed to elucidate the benefits that cannabis can have in PD.

CONCLUSION

Dopaminergic medications, while essential for managing motor symptoms in PD, pose significant risks for developing ICDs. The likelihood of developing ICDs depends on several risk factors, such as age, personality traits, and a history of addictive behaviors. Early intervention is vital in the management of ICDs due to the potentially catastrophic effects they can have on a patient's life. Management of ICDs can involve a reduction in medication, switching from dopamine receptor agonists to levodopa, and cognitive-behavioral therapy. Educating patients and families on the potential development of ICDs prior to initiating dopaminergic therapy is paramount to allow for early intervention. Future research on ICDs in PD primarily focuses on identifying further genetic risk factors and potential pharmacologic treatment options for ICDs. Additionally, understanding the neurotoxic and neuroprotective roles of drugs of abuse, such as stimulants, is essential for advancing therapeutic approaches that address both motor and non-motor symptoms of PD.

REFERENCES

Aarsland D, Marsh L, Schrag A. Neuropsychiatric symptoms in Parkinson's disease. *Mov Disord.* 2009 Nov 15;24(15):2175–2186. https://doi.org/10.1002/mds.22589.

Aladeen TS, Mattle AG, Zelen K, Mesha M, Rainka MM, Geist T, Myers B, Mechtler L. Medical cannabis in the treatment of Parkinson's disease. *Clin Neuropharmacol.* 2023 May-Jun 01;46(3):98–104. https://doi.org/10.1097/WNF.0000000000000550.

AlShimemeri S, Di Luca DG, Fox SH. MPTP Parkinsonism and implications for understanding Parkinson's disease. *Mov Disord Clin Pract.* 2021 Sep 25;9(1):42–47. https://doi.org/10.1002/mdc3.13344.

American Psychiatric Association. *Diagnostic and statistical manual of mental disorders* (5th ed., text rev.). 2022. https://doi.org/10.1176/appi.books.9780890425787.

Anang JB, Gagnon JF, Bertrand JA, Romenets SR, Latreille V, Panisset M, Montplaisir J, Postuma RB. Predictors of dementia in Parkinson disease: A prospective cohort study. *Neurology.* 2014 Sep 30;83(14):1253–1260. https://doi.org/10.1212/WNL.0000000000000842.

Antonini A, Siri C, Santangelo G, Cilia R, Poletti M, Canesi M, Caporali A, Mancini F, Pezzoli G, Ceravolo R, Bonuccelli U, Barone P. Impulsivity and compulsivity in drug-naïve patients with Parkinson's disease. *Movement Disorders: Official Journal of the Movement Disorder Society.* 2011 Mar;26(3):464–468. https://doi.org/10.1002/mds.23501.

Antonini A, Barone P, Bonuccelli U, Annoni K, Asgharnejad M, Stanzione P. ICARUS study: Prevalence and clinical features of impulse control disorders in Parkinson's disease. *J Neurol Neurosurg Psychiatry.* 2017;88(4):317–324. https://doi.org/10.1136/jnnp-2016-315277.

Augustine A, Winstanley CA, Krishnan V. Impulse control disorders in Parkinson's disease: From bench to bedside. *Front Neurosci.* 2021;15:654238. https://doi.org/10.3389/fnins.2021.654238.

Averbeck BB, O'Sullivan SS, Djamshidian A. Impulsive and compulsive behaviors in Parkinson's disease. *Annu Rev Clin Psychol.* 2014;10:553–580. https://doi.org/10.1146/annurev-clinpsy-032813-153705.

Baiano C, Barone P, Trojano L, Santangelo G. Prevalence and clinical aspects of mild cognitive impairment in Parkinson's disease: A meta-analysis. *Mov Disord.* 2020 Jan;35(1):45–54. https://doi.org/10.1002/mds.27902.

Baig F, Kelly MJ, Lawton MA, Ruffmann C, Rolinski M, Klein JC, Barber T, Lo C, Ben-Shlomo Y, Okai D, Hu MT. Impulse control disorders in Parkinson disease and RBD: A longitudinal study of severity. *Neurology.* 2019 Aug 13;93(7):e675–e687. https://doi.org/10.1212/WNL.0000000000007942.

Barbeau A. L-DOPA therapy in Parkinson's disease: A critical review of nine years' experience. *Canadian Medical Association Journal*, 1969; 101(13), 59–68.

Barbosa P, O'Sullivan SS, Joyce E, Lees AJ, Warner TT, Djamshidian A. Neuropsychiatric features of punding and hobbyism in Parkinson's disease. *Mov Disord Clin Pract.* 2022;9(1):82–86. https://doi.org/10.1002/mdc3.13363.

Barbosa P, Djamshidian A, Lees AJ, Warner TT. The outcome of dopamine dysregulation syndrome in Parkinson's disease: a retrospective postmortem study. *Movement Disorders Clinical Practice* 2018;5(5):519–522. https://doi.org/10.1002/mdc3.12671

Barbosa PM, Djamshidian A, O'Sullivan SS, de Pablo-Fernandez E, Korlipara P, Morris HR, et al. The long-term outcome of impulsive compulsive behaviours in Parkinson's disease. *J Neurol Neurosurg Psychiatry.* 2019;90(11):1288–1289.

Bargiotas P, Konitsiotis S. Levodopa-induced dyskinesias in Parkinson's disease: Emerging treatments. *Neuropsychiatr Dis Treat.* 2013;9:1605–1617.

Bastiaens J, Dorfman BJ, Christos PJ, Nirenberg MJ. Prospective cohort study of impulse control disorders in Parkinson's disease. *Mov Disord.* 2013;28(3):327–333. https://doi.org/10.1002/mds.25291.

Berg D, Seppi K, Behnke S, Liepelt I, Schweitzer K, Stockner H, Wollenweber F, Gaenslen A, Mahlknecht P, Spiegel J, Godau J, Huber H, et al. Enlarged substantia nigra hyperechogenicity and risk for Parkinson disease: A 37-month 3-center study of 1847 older persons. *Arch Neurol.* 2011 Jul;68(7):932–937. https://doi.org/10.1001/archneurol.2011.141.

Blanchet J, Longpré F, Bureau G, Morissette M, DiPaolo T, Bronchti G, Martinoli MG. Resveratrol, a red wine polyphenol, protects dopaminergic neurons in MPTP-treated mice. *Prog Neuropsychopharmacol Biol Psychiatry.* 2008 Jul 1;32(5):1243–1250. https://doi.org/10.1016/j.pnpbp.2008.03.024.

Bodi N, Keri S, Nagy H, Moustafa A, Myers CE, Daw N, et al. Reward-learning and the novelty-seeking personality: A between- and within-subjects study of the effects of dopamine agonists on young Parkinson's patients. *Brain.* 2009;132(Pt 9):2385–2395. https://doi.org/10.1093/brain/awp094.

Bowers MB Jr, Van Woert M, Davis L. Sexual behavior during L-DOPA treatment for parkinsonism. *Am J Psychiatry.* 1971 Dec;127(12): 1691–1693. https://doi.org/10.1176/ajp.127.12.1691.

Callaghan RC, Cunningham JK, Sykes J, Kish SJ. Increased risk of Parkinson's disease in individuals hospitalized with conditions related to the use of methamphetamine or other amphetamine-type drugs. *Drug Alcohol Depend.* 2012 Jan 1;120(1–3):35–40. https://doi.org/10.1016/j.drugalcdep.2011.06.013.

Callesen MB, Weintraub D, Damholdt MF, Moller A. Impulsive and compulsive behaviors among Danish patients with Parkinson's disease: Prevalence, depression, and personality. *Parkinsonism Relat Disord.* 2014;20(1):22–26. https://doi.org/10.1016/j.parkreldis.2013.09.006.

Carbone F, Djamshidian A. Impulse control disorders in Parkinson's disease: An overview of risk factors, pathogenesis and pharmacological management. *CNS Drugs.* 2024;38(6):443–457. https://doi.org/10.1007/s40263-024-01087-y

Carroll CB, Bain PG, Teare L, Liu X, Joint C, Wroath C, Parkin SG, Fox P, Wright D, Hobart J, Zajicek JP. Cannabis for dyskinesia in Parkinson disease: A randomized double-blind crossover study. *Neurology.* 2004 Oct 12;63(7):1245–1250. https://doi.org/10.1212/01.wnl.0000140288.48796.8e.

Cilia R, Siri C, Colombo A, Pezzoli G. Multiple compulsive behaviors in multiple system atrophy: The importance of predisposition to addiction. *Parkinsonism Relat Disord.* 2014;20(3):355–357. https://doi.org/10.1016/j.parkreldis.2013.12.010.

Claassen DO, Kanoff K, Wylie SA. Dopamine agonists and impulse control disorders in Parkinson's disease. *US Neurol.* 2013;9:13–16. https://doi.org/10.17925/USN.2013.09.01.13.

Cong S, Xiang C, Zhang S, Zhang T, Wang H, Cong S. Prevalence and clinical aspects of depression in Parkinson's disease: A systematic review and meta-analysis of 129 studies. *Neurosci Biobehav Rev.* 2022 Oct;141:104749. https://doi.org/10.1016/j.neubiorev.2022.104749.

Cools R, Rogers R, Barker RA, Robbins TW. Top-down attentional control in Parkinson's disease: Salient considerations. *J Cogn Neurosci.* 2010;22(5):848–859. https://doi.org/10.1162/jocn.2009.21227.

Corvol JC, Artaud F, Cormier-Dequaire F, Rascol O, Durif F, Derkinderen P, Marques AR, Bourdain F, Brandel JP, Pico F, Lacomblez L, et al. DIGPD Study Group. Longitudinal analysis of impulse control disorders in Parkinson disease. *Neurology.* 2018 Jul 17;91(3):e189–e201. https://doi.org/10.1212/WNL.0000000000005816.

Costello H, Berry AJ, Reeves S, et al. Disrupted reward processing in Parkinson's disease and its relationship with dopamine state and neuropsychiatric syndromes: A systematic review and meta-analysis. *J Neurol Neurosurg Psychiatry.* 2022;93:555–562.

Day JO, Mullin S. The genetics of Parkinson's disease and implications for clinical practice. *Genes* (Basel). 2021 Jun 30;12(7):1006. https://doi.org/10.3390/genes12071006.

Djamshidian A, Jha A, O'Sullivan SS, Silveira-Moriyama L, Jacobson C, Brown P, et al. Risk and learning in impulsive and nonimpulsive patients with Parkinson's disease. *Mov Disord.* 2010;25(13):2203–2210. https://doi.org/10.1002/mds.23247.

Dodet P, Houot M, Leu-Semenescu S, et al. Sleep disorders in Parkinson's disease, an early and multiple problem. *NPJ Parkinsons Dis.* 2024;10(46). https://doi.org/10.1038/s41531-024-00642-0.

Dorsey ER, Sherer T, Okun MS, Bloem BR. The emerging evidence of the Parkinson pandemic. *J Parkinson's Dis.* 2018;8:S3–S8.

Drew DS, et al. Dopamine and reward hypersensitivity in Parkinson's disease with impulse control disorder. *Brain: A Journal of Neurology.* 2020;143(8):2502–2518. https://doi.org/10.1093/brain/awaa198

Elshoff JP, Cawello W, Andreas JO, Mathy FX, Braun M. An update on pharmacological, pharmacokinetic properties and drug-drug interactions of rotigotine transdermal system in Parkinson's disease and restless legs syndrome. *Drugs.* 2015 Apr;75(5):487–501. https://doi.org/10.1007/s40265-015-0377-y.

Erga AH, Dalen I, Ushakova A, Chung J, Tzoulis C, Tysnes OB, Alves G, Pedersen KF, Maple-Grødem J. Dopaminergic and opioid pathways associated with impulse control disorders in Parkinson's disease. *Front Neurol.* 2018 Feb 28;9:109. https://doi.org/10.3389/fneur.2018.00109.

Eriksson AK, Löfving S, Callaghan RC, Allebeck P. Alcohol use disorders and risk of Parkinson's disease: Findings from a Swedish national cohort study 1972-2008. *BMC Neurol.* 2013 Dec 5;13:190. https://doi.org/10.1186/1471-2377-13-190.

Fantini ML, Macedo L, Zibetti M, Sarchioto M, Vidal T, Pereira B, et al. Increased risk of impulse control symptoms in Parkinson's disease with REM sleep behaviour disorder. *J Neurol Neurosurg Psychiatry.* 2015;86(2):174–179. https://doi.org/10.1136/jnnp-2014-307904.

Finseth TA, Hedeman JL, Brown RP 2nd, et al. Self-reported efficacy of cannabis and other complementary medicine modalities by Parkinson's disease patients in Colorado. *Evid Based Complement Alternat Med.* 2015;2015:87484–87489.

Frank MJ, Seeberger LC, O'reilly RC. By carrot or by stick: Cognitive reinforcement learning in parkinsonism. *Science.* 2004;306(5703):1940–1943. https://doi.org/10.1126/science.1102941.

Friedman JH, Chang V. Crack cocaine use due to dopamine agonist therapy in Parkinson disease. *Neurology*. 2013;80(24):2269–2270. https://doi.org/10.1212/WNL.0b013e318296e9d5.

Fusaroli M, Giunchi V, Battini V, Gringeri M, Rimondini R, Menchetti M, et al. Exploring the underlying mechanisms of drug-induced impulse control disorders: A pharmacovigilance-pharmacodynamic study. *Psychiatry Clin Neurosci*. 2023;77(3):160–167. https://doi.org/10.1111/pcn.13511.

Gallo V, et al. Exploring causality of the association between smoking and Parkinson's disease. *Int J Epidemiol*. 2019;48:912–925.

Garcia FD, Thibaut F. Sexual addictions. *Am J Drug Alcohol Abuse*. 2010;36(5):254–260. https://doi.org/10.3109/00952990.2010.503823.

García-Arencibia M, González S, de Lago E, et al. Evaluation of the neuroprotective effect of cannabinoids in a rat model of Parkinson's disease: Importance of antioxidant and cannabinoid receptor-independent properties. *Brain Res*. 2007;1134:162–170. https://doi.org/10.1016/j.brainres.2006.11.063.

Giuffrida A, McMahon LR. In vivo pharmacology of endocannabinoids and their metabolic inhibitors: Therapeutic implications in Parkinson's disease and abuse liability. *Prostaglandins & Other Lipid Mediators*. 2010;91(3–4):90–103. https://doi.org/10.1016/j.prostaglandins.2009.05.004.

Goerlich-Dobre KS, Probst C, Winter L, Witt K, Deuschl G, Moller B, et al. Alexithymia: An independent risk factor for impulsive-compulsive disorders in Parkinson's disease. *Mov Disord*. 2014;29(2):214–220. https://doi.org/10.1002/mds.25679.

Grall-Bronnec, M, et al. Dopamine agonists and impulse control disorders: A complex association. *Drug Safety*. 2018;41(1):19–75. https://doi.org/10.1007/s40264-017-0590-6

Grosset DG, Cardoso F, Lees A. Dopamine agonists vs levodopa in impulse control disorders. *Arch Neurol*. 2011;68(4):544–546. https://doi.org/10.1001/archneurol.2011.41.

Haber SN, Knutson B. The reward circuit: Linking primate anatomy and human imaging. *Neuropsychopharmacology*. 2010;35:4–26. https://doi.org/10.1038/npp.2009.129.

Hernán MA, Chen H, Schwarzschild MA, Ascherio A. Alcohol consumption and the incidence of Parkinson's disease. *Ann Neurol*. 2003 Aug;54(2):170–175. https://doi.org/10.1002/ana.10611.

Hernán MA, Takkouche B, Caamaño-Isorna F, Gestal-Otero JJ. A meta-analysis of coffee drinking, cigarette smoking, and the risk of Parkinson's disease. *Ann Neurol*. 2002;52:276–284.

Hinkle JT, Perepezko K, Rosenthal LS, Mills KA, Pantelyat A, Mari Z, et al. Markers of impaired motor and cognitive volition in Parkinson's disease: Correlates of dopamine dysregulation syndrome, impulse control disorder, and dyskinesias. *Parkinsonism Relat Disord*. 2018;47:50–56. https://doi.org/10.1016/j.parkreldis.2017.11.338.

Hong, C. T., Chan, L., & Bai, C. H. (2020). The effect of caffeine on the risk and progression of Parkinson's disease: A meta-analysis. *Nutrients, 12*(6), 1860. https://doi-org.ezproxy.aub.edu.lb/10.3390/nu12061860

Jang EY, Yang CH, Hedges DM, Kim SP, Lee JY, Ekins TG, Garcia BT, Kim HY, Nelson AC, Kim NJ, Steffensen SC. The role of reactive oxygen species in methamphetamine self-administration and dopamine release in the nucleus accumbens. *Addict Biol*. 2017 Sep;22(5):1304–1315. https://doi.org/10.1111/adb.12419.

Jesus S, Labrador-Espinosa MA, Adarmes AD, Mendel-Del Barrio C, Martinez-Castrillo JC, Alonso-Canovas A, et al. Non-motor symptom burden in patients with Parkinson's disease with impulse control disorders and compulsive behaviours: Results from the COPPADIS Cohort. *Sci Rep*. 2020;10(1):16893. https://doi.org/10.1038/s41598-020-73756-z.

Juárez Olguín H et al. The role of dopamine and its dysfunction as a consequence of oxidative stress. *Oxidative Medicine and Cellular Longevity*. 2016(2016):9730467. https://doi.org/10.1155/2016/9730467.

Katzenschlager R. Dopaminergic dysregulation syndrome in Parkinson's disease. *J Neurol Sci*. 2011;310:271–275.

Kehagia AA, Housden CR, Regenthal R, Barker RA, Muller U, Rowe J et al. Targeting impulsivity in Parkinson's disease using atomoxetine. *Brain*. 2014;137:1986–1997. https://doi.org/10.1093/brain/awu117.

Kok ZQ, Murley AG, Rittman T, Rowe J, Passamonti L. Co-occurrence of apathy and impulsivity in progressive supranuclear palsy. *Mov Disord Clin Pract*. 2021;8(8):1225–1233. https://doi.org/10.1002/mdc3.13339.

Langston JW. The MPTP story. *J Parkinsons Dis*. 2017;7(s1):S11–S19. https://doi.org/10.3233/JPD-179006.

Latella D, Maggio MG, Maresca G, Saporoso AF, Le Cause M, Manuli A, et al. Impulse control disorders in Parkinson's disease: A systematic review on risk factors and pathophysiology. *J Neurol Sci*. 2019;398:101–106. https://doi.org/10.1016/j.jns.2019.01.034.

Laurencin C, Timestit N, Marques A, Duchez DD, Giordana C, Meoni S, et al. Efficacy and safety of clonidine for the treatment of impulse control disorder in Parkinson's disease: A multicenter, parallel, randomised, double-blind, phase 2b clinical trial. *J Neurol.* 2023;270(10):4851–4859. https://doi.org/10.1007/s00415-023-11814-y.

Lee JY, Jeon B, Koh SB, Yoon WT, Lee HW, Kwon OD, et al. Behavioural and trait changes in parkinsonian patients with impulse control disorder after switching from dopamine agonist to levodopa therapy: Results of REIN-PD trial. *J Neurol Neurosurg Psychiatry.* 2019;90(1):30–37. https://doi.org/10.1136/jnnp-2018-318942.

Lhommee E, Klinger H, Thobois S, Schmitt E, Ardouin C, Bichon A, et al. Subthalamic stimulation in Parkinson's disease: Restoring the balance of motivated behaviours. *Brain.* 2012;135(Pt 5):1463–1477. https://doi.org/10.1093/brain/aws078.

Mamikonyan E, Siderowf AD, Duda JE, et al. Long-term follow-up of impulse control disorders in Parkinson's disease. *Mov Disord.* 2008;23:75–80.

Marín-Lahoz, J., Sampedro, F., Martinez-Horta, S., Pagonabarraga, J., & Kulisevsky, J. Depression as a risk factor for impulse control disorders in Parkinson disease. *Annals of Neurology.* 2019;86(5):762–769. https://doi.org/10.1002/ana.25581

Mohanty D, Lippmann S. Marijuana for Parkinson's disease? *Innovations in Clinical Neuroscience.* 2019;16(1–2):33–34.

Müller B, Assmus J, Herlofson K, Larsen JP, Tysnes OB. Importance of motor vs. non-motor symptoms for health-related quality of life in early Parkinson's disease. *Parkinsonism Relat Disord.* 2013 Nov;19(11):1027–1032. https://doi.org/10.1016/j.parkreldis.2013.07.010.

Mursaleen LR, Stamford JA. Drugs of abuse and Parkinson's disease. *Prog Neuropsychopharmacol Biol Psychiatry.* 2016 Jan 4;64:209–217. https://doi.org/10.1016/j.pnpbp.2015.03.013.

Nassif DV, Pereira JS. Fatigue in Parkinson's disease: Concepts and clinical approach. *Psychogeriatrics.* 2018 Mar;18(2):143–150. https://doi.org/10.1111/psyg.12302.

O'Sullivan SS, Loane CM, Lawrence AD, Evans AH, Piccini P, Lees AJ. Sleep disturbance and impulsive-compulsive behaviours in Parkinson's disease. *J Neurol Neurosurg Psychiatry.* 2011;82(6):620–622. https://doi.org/10.1136/jnnp.2009.186874.

Papay K, Xie SX, Stern M, Hurtig H, Siderowf A, Duda JE, et al. Naltrexone for impulse control disorders in Parkinson disease: A placebo-controlled study. *Neurology.* 2014;83(9):826–833. https://doi.org/10.1212/WNL.0000000000000729.

Phu AL, Xu Z, Brakoulias V, Mahant N, Fung VS, Moore GD, et al. Effect of impulse control disorders on disability and quality of life in Parkinson's disease patients. *J Clin Neurosci.* 2014;21(1):63–66. https://doi.org/10.1016/j.jocn.2013.02.032.

Poletti M, Frosini D, Pagni C, Claudio L, del Paolo D, Roberto C, et al. Alexithymia is associated with impulsivity in newly-diagnosed, drug-naive patients with Parkinson's disease: An affective risk factor for the development of impulse-control disorders? *J Neuropsychiatry Clin Neurosci.* 2012;24(4):E36–E37 https://doi.org/10.1176/appi.neuropsych.11110326.

Pompanin S, Jelcic N, Cecchin D, Cagnin A. Impulse control disorders in frontotemporal dementia: Spectrum of symptoms and response to treatment. *Gen Hosp Psychiatry.* 2014;36(6):760.e5–7. https://doi.org/10.1016/j.genhosppsych.2014.06.005.

Quik M, O'Leary K, Tanner CM. Nicotine and Parkinson's disease: Implications for therapy. *Mov Disord.* 2008 Sep 15;23(12):1641–1652. https://doi.org/10.1002/mds.21900.

Rabinak CA, Nirenberg MJ. Dopamine agonist withdrawal syndrome in Parkinson disease. *Arch Neurol.* 2010;67(1):58–63. https://doi.org/10.1001/archneurol.2009.294.

Rang HP, Dale MM, Ritter JM, Flower RJ, Henderson G. *Rang & Dale's pharmacology.* 7. Edinburgh: Elsevier Churchill Livingstone; 2012.

Richard IH, Frank S, LaDonna KA, Wang H, McDermott MP, Kurlan R. Mood fluctuations in Parkinson's disease: A pilot study comparing the effects of intravenous and oral levodopa administration. *Neuropsychiatr Dis Treat.* 2005 Sep;1(3):261–268.

Ritz B, et al. Pooled analysis of tobacco use and risk of Parkinson disease. *Arch Neurol.* 2007;64:990.

Rose KN, Schwarzschild MA, Gomperts SN. Clearing the smoke: What protects smokers from Parkinson's disease? *Mov Disord.* 2024 Feb;39(2):267–272. https://doi.org/10.1002/mds.29707.

Schreiber L, Odlaug BL, Grant JE. Impulse control disorders: Updated review of clinical characteristics and pharmacological management. *Front Psychiatry.* 2011 Feb 21;2:1. https://doi.org/10.3389/fpsyt.2011.00001.

Scott BM, Eisinger RS, Burns MR, Lopes J, Okun MS, Gunduz A, et al. Co-occurrence of apathy and impulse control disorders in Parkinson disease. *Neurology.* 2020;95(20):e2769–e2780. https://doi.org/10.1212/WNL.0000000000010965.

Scott WK et al. Family-based case-control study of cigarette smoking and Parkinson disease. *Neurology.* 2005;64:442–447.

Seeman P. Parkinson's disease treatment may cause impulse-control disorder via dopamine D3 receptors. *Synapse.* 2015 Apr;69(4):183–189. https://doi.org/10.1002/syn.21805.

Shapiro DY, Sadowsky DA, Henderson WG, Van Buren JM. An assessment of cognitive function in post-thalamotomy Parkinson patients. *Confinia Neurologica.* 1973 Mar;35(3):144–166. https://doi.org/10.1159/000102839.

Soileau LG, Talbot NC, Storey NR, Spillers NJ, D'Antoni JV, Carr PC, et al. Impulse control disorders in Parkinson's disease patients treated with pramipexole and ropinirole: A systematic review and meta-analysis. *Neurol Sci.* 2024;45(4):1399–1408. https://doi.org/10.1007/s10072-023-07254-1.

Stampanoni Bassi M, Sancesario A, Morace R, Centonze D, Iezzi E. Cannabinoids in Parkinson's disease. *Cannabis and Cannabinoid Research.* 2017;2(1):21–29. https://doi.org/10.1089/can.2017.0002.

Stamy W, Jankovic J. Impulse control disorders and pathological gambling in patients with Parkinson disease. *Neurologist.* 2008;14:89–99.

Stang CD, Mullan AF, Camerucci E, Hajeb M, Turcano P, Martin P, Mielke MM, Josephs KA, Splett M, Abler V, Boeve BF, Bower JH, Savica R. Incidence, prevalence, and mortality of psychosis associated with Parkinson's disease (1991–2010). *J Parkinsons Dis.* 2022;12(4):1319–1327. https://doi.org/10.3233/JPD-213035.

Stark, AJ, et al. Nigrostriatal and mesolimbic $D_{2/3}$ receptor expression in Parkinson's disease patients with compulsive reward-driven behaviors. *Journal of Neuroscience: The Official Journal of the Society for Neuroscience.* 2018;38(13):3230–3239. https://doi.org/10.1523/JNEUROSCI.3082-17.2018.

Surmeier DJ. Determinants of dopaminergic neuron loss in Parkinson's disease. *FEBS Journal.* 2018;285(19):3657–3668. https://doi.org/10.1111/febs.14607.

Thiriez C, Villafane G, Grapin F, Fenelon G, Remy P, Cesaro P. Can nicotine be used medicinally in Parkinson's disease? *Expert Rev Clin Pharmacol.* 2011 Jul;4(4):429–436. https://doi.org/10.1586/ecp.11.27.

Tsuang D, Larson EB, Li G, Shofer JB, Montine KS, Thompson ML, Sonnen JA, Crane PK, Leverenz JB, Montine TJ. Association between lifetime cigarette smoking and Lewy body accumulation. *Brain pathol.* 2010 Mar;20(2):412–418. https://doi.org/10.1111/j.1750-3639.2009.00296.x.

Todd G, Noyes C, Flavel SC, Della Vedova CB, Spyropoulos P, Chatterton B, Berg D, White JM. Illicit stimulant use is associated with abnormal substantia nigra morphology in humans. *PLoS One.* 2013;8(2):e56438. https://doi.org/10.1371/journal.pone.0056438.

Urbi B, Corbett J, Hughes I, Owusu MA, Thorning S, Broadley SA, Sabet A, Heshmat S. Effects of cannabis in Parkinson's disease: A systematic review and meta-analysis. *J Parkinsons Dis.* 2022;12(2):495–508. https://doi.org/10.3233/JPD-212923.

Venderová K, Ruzicka E, Vorísek V, et al. Survey on cannabis use in Parkinson's disease: Subjective improvement of motor symptoms. *Mov Disord.* 2004;19:1102–1106.

Volkow ND, Fowler JS, Wang GJ, Swanson JM. Dopamine in drug abuse and addiction: Results from imaging studies and treatment implications. *Mol Psychiatry.* 2004;9:557–569.

Voon V, Reynolds B, Brezing C, Gallea C, Skaljic M, Ekanayake V, et al. Impulsive choice and response in dopamine agonist-related impulse control behaviors. *Psychopharmacology.* 2010;207(4):645–659. https://doi.org/10.1007/s00213-009-1697-y.

Weintraub D, Koester J, Potenza MN, Siderowf AD, Stacy M, Voon V, et al. Impulse control disorders in Parkinson disease: A cross-sectional study of 3090 patients. *Arch Neurol.* 2010;67(5):589–595. https://doi.org/10.1001/archneurol.2010.65.

Weintraub D, Papay K, Siderowf AD. Screening for impulse control symptoms in patients with *de novo* Parkinson disease—A case–control study. *Neurology.* 2013;80:176–180. https://doi.org/10.1212/WNL.0b013e31827b915c.

Weintraub D, Posavi M, Fontanillas P, Tropea TF, Mamikonyan E, Suh E, et al. Genetic prediction of impulse control disorders in Parkinson's disease. *Ann Clin Transl Neurol.* 2022;9(7):936–949. https://doi.org/10.1002/acn3.51569.

Weiss HD, Marsh L. Impulse control disorders and compulsive behaviors associated with dopaminergic therapies in Parkinson disease. *Neurol. Clin Pract.* 2012 Aug;2(4):267–274.

Willis AW et al. Incidence of Parkinson disease in North America. *NPJ Parkinson's Dis.* 2022;8:170.

Zafar S, Yaddanapudi SS. *Parkinson disease.* [Updated 2023 Aug 7]. In: StatPearls [Internet]. Treasure Island (FL): StatPearls Publishing; 2024 Jan-. Available from: https://www.ncbi.nlm.nih.gov/books/NBK470193/.

6 Alcohol and Neurogenerative Disorders

Victor Olufolahan Lasebikan, Victor Makanjuola, Tiwatayo Olufolahan Lasebikan, Oluremi Oladipo Oladele, Morufat Abosede Alabi, and Oyeyemi Afolabi

INTRODUCTION

With the global increase in the aging population, greater attention must be given to age-related neurodegenerative disorders (NDDs) due to their profound negative impact on patients, families, caregivers, and society (Wyss-Coray, 2016). NDDs are a group of diseases in which essential proteins necessary for cellular homeostasis accumulate in neurons, leading to neuroinflammation, loss of neuronal cells, and eventual degeneration (Domínguez-Fernández et al., 2023). NDDs include Alzheimer's disease (AD), Parkinson's disease (PD), amyotrophic lateral sclerosis (ALS), Huntington's disease (HD), motor neuron disease, spinal muscular atrophy, and spinocerebellar ataxia (Grinberg et al., 2011). NDDs have subtle differences yet share the commonality of widespread intracellular or extracellular aggregates formed by distinct deposits. The common neuropathological deposits or changes are amyloid plaques, tau neurofibrillary tangles, alpha-synuclein inclusions, TDP-43 inclusions, neuronal loss, angiopathy, and gliosis (Cornblath et al., 2020). Neurodegeneration is also associated with synaptic and neural network dysfunction (Hoover et al., 2010; Milnerwood & Raymond, 2010). Alcohol-related dementia (ARD) is a diagnosis clinically different from NDDs, with distinct clinicopathological features (Cheng et al., 2017).

Alcohol consumption is widespread and considered an integral part of socialization worldwide. The consequences of drinking are shaped by complex genetic and environmental factors (Guerri & Pascual, 2010). While mild-to-moderate regular drinking could confer health benefits (Kohno et al., 2019), excessive and chronic alcohol consumption has social, economic, psychological, and physical health consequences (Hillemacher et al., 2018; Simpson et al., 2019),

This chapter explores the role of alcohol in the pathogenesis of NDDs, examining neuroanatomical changes and neuropathological mechanisms and highlighting the latest research findings and their implications for clinical practice. By understanding these interactions, clinicians can better address prevention strategies and therapeutic approaches for at-risk individuals.

ALCOHOL AND NEUROPATHOLOGY

The mechanisms underlying the pronounced impact of alcohol on the central nervous system are complex. Alcohol-induced neurodegeneration stems from several pathways: alcohol-induced oxidative stress, neuroimmune response, glutamate excitotoxicity and cellular injury, an interplay between oxidation, neuroinflammation, and excitotoxicity, and direct induction of NDDs (Kamal et al., 2020). Alcohol use disorder (AUD) is associated with significant brain volume loss, including in the hippocampus and white matter (Chen et al., 2012; Ozsoy et al., 2013), as well as necrotic and necroptotic cell death, as evidenced by the widespread presence of "Fluoro-Jade B" and "pyknotic nuclei" (biomarkers of cell death) in the corticolimbic circuit (M. E. Maynard et al., 2018). Thus, excessive alcohol consumption leads to permanent neuronal damage, which is a characteristic feature of dementia.

 DOI: 10.1201/9781003478485-6

Alcohol-Related Dementia

Research and clinical evidence indicate that excessive alcohol consumption may induce a clinical form of dementia distinct from other types. This form of dementia, known as ARD, is directly caused by chronic alcohol consumption or its associated risks (Ritchie & Villebrun, 2008). ARD is a common form of dementia, particularly in early-onset cases. ARD is typically characterized as frontal or subcortical dementia, presenting with impairments in executive function, cognitive decline, and behavioral changes. These manifestations differ from those seen in AD, vascular dementia, or Wernicke–Korsakoff syndrome (WKS) (Cheng et al., 2017). However, since thiamine deficiency is the central pathogenesis of WKS and a potential pathway for ARD development, some researchers have suggested that ARD may be a form of WKS, a syndrome marked by ophthalmoplegia, cerebellar dysfunction, impaired mental state, and severe memory impairment (Cheng et al., 2017). In this context, the 2020 Lancet Commission report identified excessive alcohol use or AUD in midlife as a significant "modifiable risk factor" for dementia (Livingston et al., 2020), alongside other known risk factors such as hypertension and diabetes mellitus (Schwarzinger et al., 2018). The report corroborates previous evidence highlighting the toxic effects of alcohol on the brain (Rao & Topiwala, 2020; Scarmeas et al., 2018; Topiwala & Ebmeier, 2018).

EPIDEMIOLOGY OF NEURODEGENERATIVE DISORDERS AND ALCOHOL USE IN OLDER ADULTS

NDDs are characterized by progressive neuropathological changes or deposits in specific brain areas, including tau neurofibrillary tangles, amyloid plaques, alpha-synuclein inclusions, TDP-43 inclusions, angiopathy, gliosis, and neuronal loss (Cornblath et al., 2020). These proteins may be deposited simultaneously and are not specific to a particular NDD. They can also be present in individuals without neurodegeneration (Jellinger, 2012). Different biomolecules interact through various mechanisms to drive these deposits or changes. The same overlapping dynamics also apply to the epidemiology of NDDs, particularly regarding their age of onset, supporting the view that neurodegeneration is a unified process rather than a fragmented phenomenon (de Pedro-Cuesta et al., 2015). Unfortunately, with advancing age, these processes become less specific and more complex (Crary et al., 2014). For this and other reasons, such as inadequate study sample sizes, meaningful epidemiological studies on NDDs are difficult to conduct (Araujo et al., 2021).

AD is the most prevalent NDD, accounting for 60% to 80% of all dementia cases. Despite this, the global prevalence of NDDs is relatively low, at 1.03% (Alonso-Sardón et al., 2023), with PD affecting approximately 0.3% of the population (Erkkinen et al., 2018). Despite the low prevalence, NDDs pose significant survival risks and contribute to a high medical and socioeconomic burden. For example, only 40% of those diagnosed with AD live more than 5 years after diagnosis (Akushevich et al., 2021). The average survival for PD patients is about 9 years, while 75% of individuals diagnosed with ALS survive only 5 years (Newell et al., 2024).

Alternatively, epidemiological data in the United States show that approximately 40% of older adults drink alcohol (Mental Health Services Administration, 2013), and 50% of individuals in geriatric inpatient services have alcohol misuse or dependence (Caputo et al., 2012). AUD is becoming more prevalent among older adults, mainly due to the rapidly aging population. In addition, in this vulnerable group, the adverse effects of alcohol are often experienced at lower levels of consumption (Caputo et al., 2012). While men typically have higher rates of alcohol use during adulthood, this trend tends to diminish with age. In contrast, alcohol consumption among older women is more commonly associated with isolation or late-life depression (Helvik et al., 2020; Roche et al., 2020). Nonetheless, older men are generally more likely to have AUD than women, as the latter group tends to experience more negative outcomes at the same level of alcohol consumption, prompting earlier discontinuation (McHugh et al., 2018). There are also sex differences in the likelihood of developing alcohol-induced pathology later in life (Savage et al., 2000), including neurodegenerative changes (Mark E. Maynard et al., 2018).

RISK FACTORS FOR NEURODEGENERATIVE DISORDERS

Although a small proportion of NDDs are caused by genetic mutations or disturbances in the metabolic or signaling pathways (Bayassi-Jakowicka et al., 2022), research evidence shows that a wide range of environmental factors is also linked to their development (Al-Chalabi & Hardiman, 2013; Ball et al., 2019). Risk factors include non-modifiable elements such as family history, aging, and brain injury. Certain genetic factors, such as the apolipoprotein E (APOE4) gene, are particularly implicated (Peng et al., 2020). Modifiable risk factors include hypertensive heart disease, vascular diseases, stroke, uncontrolled diabetes mellitus, hypercholesterolemia, smoking, a sedentary lifestyle, increased body mass index, poor cognitive activity, and low socioeconomic status—all of which are potential targets for dementia prevention (Serrano-Pozo & Growdon, 2019).

In the brain, two proteins, β-amyloid (Aβ) and tau, aggregate and form deposits in critical areas, leading to cognitive decline and behavioral changes (Medeiros et al., 2011). There are also reports of an increased likelihood of developing AD with chronic heavy alcohol use (Venkataraman et al., 2017). Specifically, long-term daily consumption of three or more units of alcohol is associated with a higher risk of AD (Xu et al., 2017) but not PD (Kim et al., 2020) or ALS (D'Ovidio et al., 2019). Given that all NDDs share many pathological features, the role of alcohol in their broader etiology cannot be ignored (Araujo et al., 2021).

Neuroanatomical Considerations

Convincing evidence suggests that the limbic system plays a central role in the pathomorphological basis of NDDs and addiction (Salgado & Kaplitt, 2015). The nucleus accumbens, a central component of the limbic system, is a critical area affected by negative stimuli and is a crucial component of the brain's reward system. This region coordinates the vegetative, motor cortex, sensory, and limbic systems and is also involved in all major neurotransmitter systems and functional synaptic plasticity mechanisms (Salgado & Kaplitt, 2015). Table 6.1 and Figure 6.1 illustrate the brain areas affected by alcohol, which are also implicated in various NDDs.

ALCOHOL, ALZHEIMER'S DISEASE, AND ALZHEIMER'S DEMENTIA

Preclinical Evidence

It has been documented that quercetin, an alcohol-derived polyphenol, increases the levels of protective stress-induced oxidative biomarkers, such as glutathione, peroxidase, and catalase, in the hypothalamus of adult male Sprague-Dawley rats. This suggests that quercetin may potentially prevent and treat stress-induced oxidative damage in the brain (Haleagrahara et al., 2009). Similarly, Dong et al. (2014) found that quercetin significantly altered the progression of ischemic brain injuries and NDDs among adult Sprague-Dawley rats by protecting against oxidative stress. Specifically,

TABLE 6.1
Sites of Action of Alcohol in the Brain

Site	Function
Prefrontal cortex	Working memory Executive functions
Basal ganglia	Site of initiation of motor activities and habitual behavior
Mammillary bodies	Integrating emotion and memory
Hippocampus	Memory consolidation
Cerebellum	Motor coordination and balance

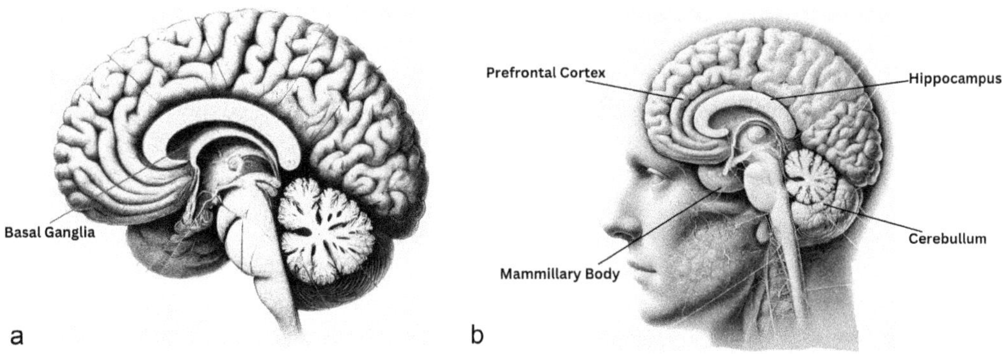

FIGURE 6.1 Brain areas where alcohol exerts its effects: (a) basal ganglia, (b) other involved brain structures.

administering quercetin to these rats reduced Caspase-3 activity and brain edema, improving neurobehavioral deficits.

In another in vitro experiment, Wang and colleagues administered red wine (Cabernet Sauvignon, containing 6% alcohol) to Tg2576 mice, which model Alzheimer's-type amyloid beta-protein neuropathology (Wang et al., 2006). The researchers found that the red wine significantly halted cognitive decline, as tested by impairment in spatial memory, in the mice that received it compared to the control group. Another animal study also showed that ethanol concentrations ranging from 0.02% to 0.08% (low concentration) offered partial protection against Aβ-induced synapse toxicity in cultured cortical and hippocampal neurons (Bate & Williams, 2011).

Alternatively, Huang et al. (2018) found that the Aβ precursor protein increased when neuronal cells were exposed to high ethanol concentrations, indicating that excessive ethanol promotes AD pathogenesis. Another in vivo study demonstrated that the exposure of the 3xTg-AD mouse brain to alcohol for one month led to hyperphosphorylation at the glycogen synthase kinase-3 site on tau protein in the CA1 area of the hippocampus, highlighting the negative effect of chronic alcohol consumption on AD (Hoffman et al., 2019).

Clinical Evidence

Some clinical studies have corroborated preclinical findings regarding the neuroprotective properties of alcohol. For example, the PAQUID (Personnes Agées QUID) study, an epidemiological investigation that followed a cohort of 4,000 older adults (65 years and older) living at home to study normal and pathological brain aging, found that moderate drinkers were less likely to develop dementia compared to abstainers (Dartigues et al., 1992). This study accounted for modifiable factors such as age, gender, education, occupation, and baseline cognition (Orgogozo et al., 1997). While 90% of alcohol users in the PAQUID study consumed wine, the Rotterdam study, another prospective study, enrolled over 5,000 participants aged 50 and above who had been drinking beer, wine, or spirits for over 6 years. The major finding of this study was that the risk of developing dementia was significantly lower among male drinkers compared to male non-drinkers (Ruitenberg et al., 2002). Similarly, a 4-year prospective study of around 1,000 participants in New York found that consuming three units of alcohol daily was associated with a lower risk of Alzheimer's dementia, although liquor and beer were not protective (Luchsinger et al., 2004). These findings suggest that polyphenols in alcohol may have neuroprotective actions, possibly through interactions with specific plasma membrane binding sites (Matias et al., 2016).

Several meta-analyses have also shown that AD is less likely among light or moderate drinkers than abstainers (Anstey et al., 2019; Xu et al., 2015). One prospective study revealed that the association between alcohol consumption in middle age and cognitive decline follows a U-shaped curve, suggesting that excessive consumption is linked to a higher likelihood of dementia. However, this effect may only be significant in individuals with the APOE4 gene (Anttila et al., 2004).

The link between alcohol consumption and AD appears to depend on several factors, including the quantity, pattern, and frequency of drinking, the type of beverage consumed, nutritional status, and the individual's genotype. For example, a 23-year cohort study by Sabia et al. (2018) found that abstainers were more likely to develop dementia than those who consumed more than 14 units of alcohol per week during midlife. Along the same lines, a magnetic resonance imaging study indicated that wine consumption was associated with better cognitive function, as evidenced by significantly lower white matter pathology and improved cerebral blood flow (Haller et al., 2018). Three high-quality systematic reviews, including two meta-analyses, also indicated that light-to-moderate alcohol consumption, compared to non-drinking, was associated with a 25% reduction in the risk of developing AD, vascular dementia, or any dementia (Anstey et al., 2019).

However, the "sick quitter" hypothesis challenges the notion that light or moderate alcohol consumption is neuroprotective. This is due to several limitations in the published studies on the relationship between alcohol consumption and dementia. One limitation is that the "abstainer" group often includes both lifetime abstainers and former drinkers who quit due to health issues. Additionally, there is no universal consensus on the definitions of "light," "moderate," and "heavy" alcohol use across studies (Stockwell et al., 2016). In this regard, a more recent study by Joshi et al. (2023) failed to find an association between drinking levels and progression from mild cognitive impairment to AD, suggesting the need for hematological biomarkers to address this issue better. Besides, most data on alcohol and dementia come from high-income countries, with limited data from low- and middle-income countries, where alcohol consumption is on the rise and the burden of dementia is growing (Livingston et al., 2020).

ALCOHOL AND PARKINSON'S DISEASE

PD is one of the most common age-related NDDs, characterized by bradykinesia, resting tremors, rigidity, and postural problems (Marras et al., 2019). Its cause remains unclear in most cases. Several environmental risk factors have been identified for developing PD (Ascherio & Schwarzschild, 2016; Kalia & Lang, 2015), including chronic pesticide exposure (Narayan et al., 2017), high iron intake (Powers et al., 2009), and brain injury (Eric Nyam et al., 2018). Lifestyle factors, such as alcohol consumption, may also increase the likelihood of developing PD (Bettiol et al., 2015). However, a systematic review and meta-analysis described the association between alcohol and PD as weak (Mitchell et al., 2022).

PRECLINICAL EVIDENCE

Animal models show that ethanol is neurotoxic to distinct brain areas in mice, including the cerebellum, hippocampus, and cortex. In PD, there is a loss of dopaminergic neurons in the pars compacta of the substantia nigra, the ventral tegmental area, the caudate nucleus, and the caudate putamen, leading to motor disturbances (Peng et al., 2020). As a result, the risk of developing PD may be increased by alcohol consumption. However, Blanchet et al. (2008) found that the administration of phytoestrogens to mice inhibited the cytotoxic effects of the dopaminergic neurotoxin 1-methyl-4-phenyl-1,2,3,6-tetrahydropyridine (MPTP), preventing the depletion of nigrostriatal dopaminergic neurons. These findings suggest a potential therapeutic effect of alcohol-derived polyphenols in preventing PD. The disparity between these findings is difficult to explain, but it could be related to factors such as dosage, frequency of consumption, the type of alcoholic beverage used, or inherent characteristics of the experimental animals (Sabia et al., 2018).

CLINICAL EVIDENCE

Research on alcohol consumption and PD shows mixed findings. One study found that abstainers and heavy alcohol users had an elevated risk of PD compared to moderate drinkers (Palacios et al., 2012). Among light and moderate drinkers, beer—of all alcoholic beverage types—offered the highest protection against PD, especially for males (Zhang et al., 2014). A study by Paganini-Hill (2001) also reported that alcohol use of two units per day reduced the risk of PD, particularly among those who preferred wine. Palacios et al. (2012) could not link any specific type of alcoholic beverage with the development of PD. In contrast, a large Swedish study found that chronic alcohol consumption increased the likelihood of PD (Eriksson et al., 2013).

Epidemiologic evidence suggests an inverse relationship between alcohol and PD, among other lifestyle behaviors (Bellou et al., 2016). Given that PD has a long prodromal phase (Elbaz, 2016; Savica et al., 2018), the possibility of reverse causation, in which the outcome precedes and influences the exposure, should be considered. This is exemplified by the loss of appetite for smoking during the prodrome of PD (Ritz & Rhodes, 2010), as well as the increased success of smoking cessation attempts among individuals who later develop PD, which may be due to reduced dopamine reward (Ritz et al., 2014).

On the other hand, parkinsonism has been reported among individuals with problem drinking due to reductions in dopamine, 3,4-dihydroxyphenylacetic acid, and tyrosine hydroxylase in the ventral striatum (Chen & Wang, 2015). This activity is mediated by stimulating cytochrome P450 2E1 (CYP2E1), a major enzyme responsible for oxidizing alcohol (Heit et al., 2013). Additionally, exposure to alcohol or MPTP has been shown to increase the production of α-synuclein by CYP2E1, a protein involved in the pathogenesis of PD and often used to experimentally induce the disease (Janakiraman et al., 2017).

ALCOHOL AND AMYOTROPHIC LATERAL SCLEROSIS

ALS is a motor neuron disease that presents with muscle weakness, progressing to paralysis and respiratory failure. While about 10% of ALS cases have a genetic origin, the sporadic variant is particularly devastating, with most patients dying between the third and fourth year after disease onset. Critical factors in ALS pathogenesis include lipids, proteins, and DNA damage caused by oxidation, leading to the degeneration of motor neurons and abnormal astrocyte function (D'Amico et al., 2013).

Given that familial factors account for only 1 in 10 cases of ALS, ecological factors have garnered attention in understanding the etiology and progression of the disease. These factors include smoking (Peters et al., 2020), pesticides (Vinceti et al., 2017), and high-density metals (Cavaleri, 2015), all of which increase the likelihood of ALS. The relationship between alcohol and ALS remains controversial. For example, D'Ovidio et al. (2019) found no link between alcohol and ALS. The very low incidence of ALS and limited follow-up studies are the main reasons why it has been challenging to establish a clear association (Logroscino et al., 2010).

Despite the limited data on the alcohol–ALS relationship, a meta-analysis including various methodological studies, such as follow-up and case–control studies, indicated that the likelihood of developing ALS was significantly lower among alcohol users. However, a dose–response relationship could not be established, as only one study investigated this concept (Meng et al., 2016). Conversely, in the Euro-Motor study, alcohol was not associated with the risk of developing ALS (D'Ovidio et al., 2019).

Alternatively, research shows that in ALS, there is dysfunction in glutamate excitation, which is also observed with chronic alcohol intake (Rao & Sari, 2012). For this reason, drugs that stimulate glutamate transporter expression, such as β-lactam antibiotics, are currently being tested for their efficacy in ALS treatment (Rao & Sari, 2012). Ceftriaxone, a β-lactam antibiotic, stimulates the expression of glutamate transporter 1 (GLT1) in both the hippocampus and spinal cord, providing neuroprotective effects (Ramos et al., 2010). In a mouse model of HD, a 5-day course of intraperitoneal ceftriaxone at a dose of 200 mg/kg led to significantly elevated levels of plasma GLT1 and glutamate uptake in the corpus striatum, thereby attenuating HD symptoms (Miller et al., 2008). Ceftriaxone was also found to decrease alcohol consumption by increasing GLT1 expression in the prefrontal cortex and nucleus accumbens (Rao & Sari, 2012), as demonstrated in experimental animal models (Bell et al., 2006).

CONCLUSION

In summary, the pathways through which alcohol consumption predisposes individuals to NDDs are highly varied, and this holds for AD, PD, and ALS. For example, in AD, while preclinical and clinical studies suggest that low or moderate alcohol consumption may be protective against Aβ toxicity and tau phosphorylation in hippocampal neurons (Anstey et al., 2019; Xu et al., 2015), chronic alcohol misuse increases the accumulation of these proteins in neurons, leading to their degeneration (Hoffman et al., 2019). The role of alcohol in PD remains controversial (Mitchell et al., 2022). While some studies have found alcohol to be protective against PD, others have found no such association. However, experimental animal models demonstrate that chronic and excessive alcohol use leads to the induction of CYP2E1 and αSYN, causing toxic damage to neurons similar to what is observed in PD (Heit et al., 2013). Regarding ALS, its association with alcohol use is inconsistent, with the consensus being that alcohol does not play a role in the development of ALS. There is a need for more prospective studies on this subject due to the lack of consistency in some of the findings in the literature. This is particularly important for some of the less common NDDs, such as ALS. Clinically, given the inconclusive evidence surrounding alcohol's role in NDDs, clinicians should prioritize a personalized risk assessment approach in older adults that considers both genetic and environmental factors. This will help guide alcohol consumption recommendations and reduce the potential for neurodegeneration, particularly in individuals at risk for NDDs.

Disclaimer: This chapter is exclusively for information and education. It was not supported by any alcohol or wine-producing company. The authors are not in any way promoting the use of alcohol of any beverage type, given documented evidence of its negative social and health implications.

REFERENCES

Akushevich, I., Yashkin, A. P., Kravchenko, J., & Yashin, A. (2021). Analysis of time trends in Alzheimer's disease and related dementias using partitioning approach. *J Alzheimers Dis, 82*(3), 1277–1289. https://doi.org/10.3233/JAD-210273

Al-Chalabi, A., & Hardiman, O. (2013). The epidemiology of ALS: A conspiracy of genes, environment and time. *Nat Rev Neurol*, *9*(11), 617–628. https://doi.org/10.1038/nrneurol.2013.203

Alonso-Sardón, M., Paulino, M., Fernandes, A., Martín-Delgado, M. A., Pérez-Martín, R. N., Magalhães, C. P., Pimentel, H., Mata, A., Silvano, J., Corte, A., Paiva, T., & Marques, E., … & Marques, E. (2023). Epidemiological impact of neurodegenerative diseases in the rural Spanish-Portuguese cross-border region. *Neurodegener Dis*, *23*(3–4), 25–34. https://doi.org/10.1159/000535683

Anstey, K. J., Ee, N., Eramudugolla, R., Jagger, C., & Peters, R. (2019). A systematic review of meta-analyses that evaluate risk factors for dementia to evaluate the quantity, quality, and global representativeness of evidence. *J Alzheimers Dis*, *70*(s1), S165–s186. https://doi.org/10.3233/jad-190181

Anttila, T., Helkala, E. L., Viitanen, M., Kåreholt, I., Fratiglioni, L., Winblad, B., Soininen, H., Tuomilehto, J., Nissinen, A., & Kivipelto, M. (2004). Alcohol drinking in middle age and subsequent risk of mild cognitive impairment and dementia in old age: A prospective population based study. *BMJ*, *329*(7465), 539. https://doi.org/10.1136/bmj.38181.418958.BE

Araujo, I., Henriksen, A., Gamsby, J., & Gulick, D. (2021). Impact of alcohol abuse on susceptibility to rare neurodegenerative diseases [Review]. *Front Mol Biosci, 8*. https://doi.org/10.3389/fmolb.2021.643273

Ascherio, A., & Schwarzschild, M. A. (2016). The epidemiology of Parkinson's disease: Risk factors and prevention. *Lancet Neurol*, *15*(12), 1257–1272. https://doi.org/10.1016/s1474-4422(16)30230-7

Ball, N., Teo, W. P., Chandra, S., & Chapman, J. (2019). Parkinson's disease and the environment. *Front Neurol*, *10*, 218. https://doi.org/10.3389/fneur.2019.00218

Bate, C., & Williams, A. (2011). Ethanol protects cultured neurons against amyloid-β and α-synuclein-induced synapse damage. *Neuropharmacology*, *61*(8), 1406–1412. https://doi.org/10.1016/j.neuropharm.2011.08.030

Bayassi-Jakowicka, M., Lietzau, G., Czuba, E., Patrone, C., & Kowiański, P. (2022). More than addiction-the nucleus accumbens contribution to development of mental disorders and neurodegenerative diseases. *International Journal of Molecular Sciences*, *23*(5). https://doi.org/10.3390/ijms23052618

Bell, R. L., Rodd, Z. A., Lumeng, L., Murphy, J. M., & McBride, W. J. (2006). The alcohol-preferring P rat and animal models of excessive alcohol drinking. *Addict Biol*, *11*(3–4), 270–288. https://doi.org/10.1111/j.1369-1600.2005.00029.x

Bellou, V., Belbasis, L., Tzoulaki, I., Evangelou, E., & Ioannidis, J. P. (2016). Environmental risk factors and Parkinson's disease: An umbrella review of meta-analyses. *Parkinsonism Relat Disord*, *23*, 1–9. https://doi.org/10.1016/j.parkreldis.2015.12.008

Bettiol, S. S., Rose, T. C., Hughes, C. J., & Smith, L. A. (2015). Alcohol consumption and Parkinson's disease risk: A review of recent findings. *J Parkinsons Dis*, *5*(3), 425–442. https://doi.org/10.3233/jpd-150533

Blanchet, J., Longpré, F., Bureau, G., Morissette, M., DiPaolo, T., Bronchti, G., & Martinoli, M. G. (2008). Resveratrol, a red wine polyphenol, protects dopaminergic neurons in MPTP-treated mice. *Prog Neuropsychopharmacol Biol Psychiatry*, *32*(5), 1243–1250. https://doi.org/10.1016/j.pnpbp.2008.03.024

Caputo, F., Vignoli, T., Leggio, L., Addolorato, G., Zoli, G., & Bernardi, M. (2012). Alcohol use disorders in the elderly: A brief overview from epidemiology to treatment options. *Exp Gerontol*, *47*(6), 411–416. https://doi.org/10.1016/j.exger.2012.03.019

Cavaleri, F. (2015). Review of amyotrophic lateral sclerosis, Parkinson's and Alzheimer's diseases helps further define pathology of the novel paradigm for Alzheimer's with heavy metals as primary disease cause. *Med Hypotheses*, *85*(6), 779–790. https://doi.org/10.1016/j.mehy.2015.10.009

Chen, C. H., Walker, J., Momenan, R., Rawlings, R., Heilig, M., & Hommer, D. W. (2012). Relationship between liver function and brain shrinkage in patients with alcohol dependence. *Alcohol Clin Exp Res*, *36*(4), 625–632. https://doi.org/10.1111/j.1530-0277.2011.01662.x

Chen, G. C., & Wang, H. C. (2015). Acute parkinsonism in alcohol withdrawal: A case report. *Parkinsonism Relat Disord*, *21*(11), 1385–1386. https://doi.org/10.1016/j.parkreldis.2015.09.046

Cheng, C., Huang, C. L., Tsai, C. J., Chou, P.-H., Lin, C. C., & Chang, C. K. (2017). Alcohol-related dementia: A systemic review of epidemiological studies. *Psychosomatics*, *58*(4), 331–342. https://doi.org/https://doi.org/10.1016/j.psym.2017.02.012

Cornblath, E. J., Robinson, J. L., Irwin, D. J., Lee, E. B., Lee, V. M., Trojanowski, J. Q., & Bassett, D. S. (2020). Defining and predicting transdiagnostic categories of neurodegenerative disease. *Nat Biomed Eng*, *4*(8), 787–800. https://doi.org/10.1038/s41551-020-0593-y

Crary, J. F., Trojanowski, J. Q., Schneider, J. A., Abisambra, J. F., Abner, E. L., Alafuzoff, I., Arnold, S. E., Attems, J., Beach, T. G., Bigio, E. H., Cairns, N. J., Dickson, D. W., Gearing, M., Grinberg, L.T., Hof, P. R., Hyman, B. T., Jellinger, K., Jicha, G. A ., kovasc, G.G., … & Nelson, P. T. (2014). Primary age-related tauopathy (PART): A common pathology associated with human aging. *Acta Neuropathol*, *128*(6), 755–766. https://doi.org/10.1007/s00401-014-1349-0

D'Amico, E., Factor-Litvak, P., Santella, R. M., & Mitsumoto, H. (2013). Clinical perspective on oxidative stress in sporadic amyotrophic lateral sclerosis. *Free Radic Biol Med*, *65*, 509–527. https://doi.org/10.1016/j.freeradbiomed.2013.06.029

D'Ovidio, F., Rooney, J. P. K., Visser, A. E., Manera, U., Beghi, E., Logroscino, G.,… & Chiò, A. (2019). Association between alcohol exposure and the risk of amyotrophic lateral sclerosis in the Euro-MOTOR study. *J Neurol Neurosurg Psychiatry*, *90*(1), 11–19. https://doi.org/10.1136/jnnp-2018-318559

Dartigues, J. F., Gagnon, M., Barberger-Gateau, P., Letenneur, L., Commenges, D., Sauvel, C., Michel, P., & Salamon, R. (1992). The Paquid epidemiological program on brain ageing. *Neuroepidemiology*, *11 Suppl 1*, 14–18. https://doi.org/10.1159/000110955

de Pedro-Cuesta, J., Rábano, A., Martínez-Martín, P., Ruiz-Tovar, M., Alcalde-Cabero, E., Almazán-Isla, J., Avellanal, F., & Calero, M. (2015). Comparative incidence of conformational, neurodegenerative disorders. *PLoS One*, *10*(9), e0137342. https://doi.org/10.1371/journal.pone.0137342

Domínguez-Fernández, C., Egiguren-Ortiz, J., Razquin, J., Gómez-Galán, M., De Las Heras-García, L., Paredes-Rodríguez, E., Astigarraga, E., Miguélez, C., & Barreda-Gómez, G. (2023). Review of technological challenges in personalised medicine and early diagnosis of neurodegenerative disorders. *International Journal of Molecular Sciences*, *24*(4). https://doi.org/10.3390/ijms24043321

Dong, Y., Wang, J., Feng, D., Qin, H., Wen, H., Yin, Z., Gao, J., & Li, C. (2014). Protective effect of Quercetin against oxidative stress and brain edema in an experimental rat model of subarachnoid hemorrhage [Research Paper]. *International Journal of Medical Sciences*, *11*(3), 282–290. https://doi.org/10.7150/ijms.7634

Elbaz, A. (2016). Prodromal symptoms of Parkinson's disease: Implications for epidemiological studies of disease etiology. *Rev Neurol (Paris)*, *172*(8–9), 503–511. https://doi.org/10.1016/j.neurol.2016.07.001

Eric Nyam, T. T., Ho, C. H., Wang, Y. L., Lim, S. W., Wang, J. J., Chio, C. C., Kuo, J. R., & Wang, C. C. (2018). The risk of traumatic brain injury occurring among patients with Parkinson disease: A 14-year population-based study. *World Neurosurg*, *113*, e328–e335. https://doi.org/10.1016/j.wneu.2018.02.027

Eriksson, A. K., Löfving, S., Callaghan, R. C., & Allebeck, P. (2013). Alcohol use disorders and risk of Parkinson's disease: Findings from a Swedish national cohort study 1972–2008. *BMC Neurol*, *13*, 190. https://doi.org/10.1186/1471-2377-13-190

Erkkinen, M. G., Kim, M. O., & Geschwind, M. D. (2018). Clinical neurology and epidemiology of the major neurodegenerative diseases. *Cold Spring Harb Perspect Biol*, *10*(4). https://doi.org/10.1101/cshperspect.a033118

Grinberg, L. T., Rueb, U., & Heinsen, H. (2011). Brainstem: Neglected locus in neurodegenerative diseases. *Front Neurol*, *2*, 42. https://doi.org/10.3389/fneur.2011.00042

Guerri, C., & Pascual, M. (2010). Mechanisms involved in the neurotoxic, cognitive, and neurobehavioral effects of alcohol consumption during adolescence. *Alcohol*, *44*(1), 15–26. https://doi.org/10.1016/j.alcohol.2009.10.003

Haleagrahara, N., Radhakrishnan, A., Lee, N., & Kumar, P. (2009). Flavonoid quercetin protects against swimming stress-induced changes in oxidative biomarkers in the hypothalamus of rats. *Eur J Pharmacol*, *621*(1), 46–52. https://doi.org/https://doi.org/10.1016/j.ejphar.2009.08.030

Haller, S., Montandon, M. L., Rodriguez, C., Herrmann, F. R., & Giannakopoulos, P. (2018). Impact of coffee, wine, and chocolate consumption on cognitive outcome and MRI parameters in old age. *Nutrients*, *10*(10). https://doi.org/10.3390/nu10101391

Heit, C., Dong, H., Chen, Y., Thompson, D. C., Deitrich, R. A., & Vasiliou, V. K. (2013). The role of CYP2E1 in alcohol metabolism and sensitivity in the central nervous system. *Subcell Biochem*, *67*, 235–247. https://doi.org/10.1007/978-94-007-5881-0_8

Helvik, A. S., Engedal, K., & Johannessen, A. (2020). Elevated alcohol consumption among geriatric psychiatric in-patients. *Nordisk Alkohol Nark*, *37*(4), 400–410. https://doi.org/10.1177/1455072520936813

Hillemacher, T., Bachmann, O., Kahl, K. G., & Frieling, H. (2018). Alcohol, microbiome, and their effect on psychiatric disorders. *Prog Neuropsychopharmacol Biol Psychiatry*, *85*, 105–115. https://doi.org/10.1016/j.pnpbp.2018.04.015

Hoffman, J. L., Faccidomo, S., Kim, M., Taylor, S. M., Agoglia, A. E., May, A. M., Smith, E. N., Wong, L. C. & Hodge, C. W. (2019). Alcohol drinking exacerbates neural and behavioral pathology in the 3xTg-AD mouse model of Alzheimer's disease. *Int Rev Neurobiol*, *148*, 169–230. https://doi.org/10.1016/bs.irn.2019.10.017

Hoover, B. R., Reed, M. N., Su, J., Penrod, R. D., Kotilinek, L. A., Grant, M. K., R. Pitstick, R., Carlson, G. A., Lanier, L. M., Yuan, L. L., Ashe, K. H., & Liao, D. (2010). Tau mislocalization to dendritic spines mediates synaptic dysfunction independently of neurodegeneration. *Neuron*, *68*(6), 1067–1081. https://doi.org/10.1016/j.neuron.2010.11.030

Huang, D., Yu, M., Yang, S., Lou, D., Zhou, W., Zheng, L., Wang, Z., Cai, F., Zhou, W., Li, T., & Song, W. (2018). Ethanol alters APP processing and aggravates Alzheimer-associated phenotypes. *Mol Neurobiol*, *55*(6), 5006–5018. https://doi.org/10.1007/s12035-017-0703-3

Janakiraman, U., Manivasagam, T., Justin Thenmozhi, A., Dhanalakshmi, C., Essa, M. M., Song, B. J., & Guillemin, G. J. (2017). Chronic mild stress augments MPTP induced neurotoxicity in a murine model of Parkinson's disease. *Physiol Behav*, *173*, 132–143. https://doi.org/10.1016/j.physbeh.2017.01.046

Jellinger, K. A. (2012). Interaction between pathogenic proteins in neurodegenerative disorders. *J Cell Mol Med*, *16*(6), 1166–1183. https://doi.org/10.1111/j.1582-4934.2011.01507.x

Joshi, P., Chen, Y., Su, Y., & Reiman, E. M. (2023). Effect of alcohol use on progression to mild cognitive impairment and dementia. *19*(S18), e074355. https://doi.org/https://doi.org/10.1002/alz.074355

Kalia, L. V., & Lang, A. E. (2015). Parkinson's disease. *Lancet*, *386*(9996), 896–912. https://doi.org/10.1016/s0140-6736(14)61393-3

Kamal, H., Tan, G. C., Ibrahim, S. F., Shaikh, M. F., Mohamed, I. N., Mohamed, R. M. P., Hamid, H.A., Ugusman, A., & Kumar, J. (2020). Alcohol use disorder, Neurodegeneration, Alzheimer's and Parkinson's disease: Interplay between oxidative stress, neuroimmune response and excitotoxicity [Review]. *Front Cell Neurosci, 14*. https://doi.org/10.3389/fncel.2020.00282

Kim, I. Y., Yang, T. O., Heath, A. K., Simpson, R. F., Reeves, G. K., Green, J., Floud, S., Brown, A., Hunter, D.J., & Beral, V. J. M. D. (2020). Alcohol intake and Parkinson's disease risk in the million women study. *Mov Disord, 35*(3), 443–449. https://doi.org/10.1002/mds.27933

Kohno, K., Niihara, H., Hamano, T., Takeda, M., Nakagawa, Y., Shiwaku, K., Nabika, T., Zöller, B., Li, X., Sundquist, K., Sundquist, J., & Morita, E. (2019). J-curve association between alcohol intake and varicose veins in Japan: The Shimane CoHRE study. *J Dermatol, 46*(10), 902–906. https://doi.org/10.1111/1346-8138.15022

Livingston, G., Huntley, J., Sommerlad, A., Ames, D., Ballard, C., Banerjee, S., Brayne, C., Burns, A., Cohen-Mansfield, J., Cooper, C., Costafreda, S.G., Dias, A., Fox, N., Gitlin, L.N., Howard, R., Kales, H.C., Kivimäki, M., Larson, E.B., Ogunniyi, A., … & Mukadam, N. (2020). Dementia prevention, intervention, and care: 2020 report of the Lancet Commission. *Lancet, 396*(10248), 413–446. https://doi.org/10.1016/s0140-6736(20)30367-6

Logroscino, G., Traynor, B. J., Hardiman, O., Chiò, A., Mitchell, D., Swingler, R.J., Millul, A., Benn, E., & Beghi, E. (2010). Incidence of amyotrophic lateral sclerosis in Europe. *J Neurol Neurosurg Psychiatry, 81*(4), 385–390. https://doi.org/10.1136/jnnp.2009.183525

Luchsinger, J. A., Tang, M. X., Siddiqui, M., Shea, S., & Mayeux, R. (2004). Alcohol intake and risk of dementia. *J Am Geriatr Soc, 52*(4), 540–546. https://doi.org/10.1111/j.1532-5415.2004.52159.x

Marras, C., Canning, C. G., & Goldman, S. M. (2019). Environment, lifestyle, and Parkinson's disease: Implications for prevention in the next decade. *Mov Disord, 34*(6), 801–811. https://doi.org/10.1002/mds.27720

Matias, I., Buosi, A. S., & Gomes, F. C. A. (2016). Functions of flavonoids in the central nervous system: Astrocytes as targets for natural compounds. *Neurochem Int, 95*, 85–91. https://doi.org/https://doi.org/10.1016/j.neuint.2016.01.009

Maynard, M. E., Barton, E. A., Robinson, C. R., Wooden, J. I., & Leasure, J. L. (2018). Sex differences in hippocampal damage, cognitive impairment, and trophic factor expression in an animal model of an alcohol use disorder. *Brain Struct Funct, 223*(1), 195–210. https://doi.org/10.1007/s00429-017-1482-3

McHugh, R. K., Votaw, V. R., Sugarman, D. E., & Greenfield, S. F. (2018). Sex and gender differences in substance use disorders. *Clin Psychol Rev, 66*, 12–23. https://doi-org.ezproxy.aub.edu.lb/10.1016/j.cpr.2017.10.012.

Medeiros, R., Baglietto-Vargas, D., & LaFerla, F. M. (2011). The role of tau in Alzheimer's disease and related disorders. *CNS Neurosci Ther, 17*(5), 514–524. https://doi.org/10.1111/j.1755-5949.2010.00177.x

Meng, E., Sug=fang, Y., Jianrui, D., Wu, J., Xiang, C., Yiyang, M., Diojian, Z., & Rumei, R. (2016). Association between alcohol consumption and amyotrophic lateral sclerosis: A meta-analysis of five observational studies. *Neurol Sci, 37*(8), 1203–1208. https://doi.org/10.1007/s10072-016-2575-0

Mental Health Services Administration. (2013). Substance Abuse—Results from the 2013 Greenlake Extension. *2*(013), 55–68. https://scholar.google.com/scholar_lookup?title=Rockville,+MD:+Substance+Abuse+and+Mental+Health+Services+Administration&publication_year=2015&

Miller, B. R., Dorner, J. L., Shou, M., Sari, Y., Barton, S. J., Sengelaub, D. R., Kennedy, R. T., & Rebec, G. V. (2008). Up-regulation of GLT1 expression increases glutamate uptake and attenuates the Huntington's disease phenotype in the R6/2 mouse. *Neuroscience, 153*(1), 329–337. https://doi.org/10.1016/j.neuroscience.2008.02.004

Milnerwood, A. J., & Raymond, L. A. (2010). Early synaptic pathophysiology in neurodegeneration: Insights from Huntington's disease. *Trends Neurosci, 33*(11), 513–523. https://doi.org/10.1016/j.tins.2010.08.002

Mitchell, E., Chohan, H., Bestwick, J. P., & Noyce, A. J. (2022). Alcohol and Parkinson's disease: A systematic review and meta-analysis. *J Parkinsons Dis, 12*(8), 2369–2381. https://doi.org/10.3233/jpd-223522

Narayan, S., Liew, Z., Bronstein, J. M., & Ritz, B. (2017). Occupational pesticide use and Parkinson's disease in the Parkinson environment gene (PEG) study. *Environ Int, 107*, 266–273. https://doi.org/10.1016/j.envint.2017.04.010

Newell, M. E., Babbrah, A., Aravindan, A., Rathnam, R., Kiernan, R., Driver, E. M., Bowes, D. A., & Halden, R. U. (2024). Prevalence rates of neurodegenerative diseases versus human exposures to heavy metals across the United States. *Sci Total Environ, 928*, 172260. https://doi.org/https://doi.org/10.1016/j.scitotenv.2024.172260

Orgogozo, J.M., Dartigues, J.F., Lafont, S., Letenneur, L., Commenges, D., Salamon, R., Renaud, S., & Breteler, M.B. (1997). Wine consumption and dementia in the elderly: A prospective community study in the Bordeaux area. *Rev Neurol (Paris), 153*(3), 185–192.

Ozsoy, S., Durak, A. C., & Esel, E. (2013). Hippocampal volumes and cognitive functions in adult alcoholic patients with adolescent-onset. *Alcohol, 47*(1), 9–14. https://doi.org/10.1016/j.alcohol.2012.09.002

Paganini-Hill, A. (2001). Risk factors for Parkinson's disease: The Leisure World Cohort Study. *Neuroepidemiology, 20*(2), 118–124. https://doi.org/10.1159/000054770

Palacios, N., Gao, X., O'Reilly, E., Schwarzschild, M., McCullough, M. L., Mayo, T., Gapstur, S. M., & Ascherio, A. A. (2012). Alcohol and risk of Parkinson's disease in a large, prospective cohort of men and women. *Mov Disord*, *27*(8), 980–987. https://doi.org/10.1002/mds.25050

Peng, B., Yang, Q., R, B. J., Liu, Y., Akbar, M., Song, B. J., Zhou, S., & Wang, X. (2020). Role of alcohol drinking in Alzheimer's disease, Parkinson's disease, and amyotrophic lateral sclerosis. *International Journal of Molecular Sciences*, *21*(7). https://doi.org/10.3390/ijms21072316

Peters, S., Visser, A. E., D'Ovidio, F., Vlaanderen, J., Portengen, L., Beghi, E., Chio, A., Logroscino, G., Hardiman, O., Pupillo, E., Veldink, J. H., Vermeulen, R., & van den Berg, L. H. (2020). Effect modification of the association between total cigarette smoking and ALS risk by intensity, duration and time-since-quitting: Euro-MOTOR. *J Neurol Neurosurg Psychiatry*, *91*(1), 33–39. https://doi.org/10.1136/jnnp-2019-320986

Powers, K. M., Smith-Weller, T., Franklin, G. M., Longstreth, W. T., Jr., Swanson, P. D., & Checkoway, H. (2009). Dietary fats, cholesterol and iron as risk factors for Parkinson's disease. *Parkinsonism Relat Disord*, *15*(1), 47–52. https://doi.org/10.1016/j.parkreldis.2008.03.002

Ramos, K. M., Lewis, M. T., Morgan, K. N., Crysdale, N. Y., Kroll, J. L., Taylor, F. R., Harrison, J. A., Sloane, E. M., Maier, S. F., & Watkins, L. R. (2010). Spinal upregulation of glutamate transporter GLT-1 by ceftriaxone: Therapeutic efficacy in a range of experimental nervous system disorders. *Neuroscience*, *169*(4), 1888–1900. https://doi.org/10.1016/j.neuroscience.2010.06.014

Rao, P. S., & Sari, Y. (2012). Glutamate transporter 1: Target for the treatment of alcohol dependence. *Curr Med Chem*, *19*(30), 5148–5156. https://doi.org/10.2174/092986712803530511

Rao, R., & Topiwala, A. (2020). Alcohol use disorders and the brain. *Addiction*, *115*(8), 1580–1589. https://doi.org/10.1111/add.15023

Ritchie, K., & Villebrun, D. (2008). Epidemiology of alcohol-related dementia. In *Handbook of clinical neurology* (Vol. 89, pp. 845–850). Elsevier. https://doi.org/10.1016/S0072-9752(07)01273-0

Ritz, B., Lee, P. C., Lassen, C. F., & Arah, O. A. (2014). Parkinson disease and smoking revisited: Ease of quitting is an early sign of the disease. *Neurology*, *83*(16), 1396–1402. https://doi.org/10.1212/wnl.0000000000000879

Ritz, B., & Rhodes, S. L. (2010). After half a century of research on smoking and PD, where do we go now? *Neurology*, *74*(11), 870–871. https://doi.org/10.1212/WNL.0b013e3181d63aa8

Roche, A. M., Harrison, N. J., Chapman, J., Kostadinov, V., Woodman, R. J. (2020). Ageing and alcohol: Drinking typologies among older adults. *J Aging Health*, *32*(10), 1486–1497.

Ruitenberg, A., van Swieten, J. C., Witteman, J. C., Mehta, K. M., van Duijn, C. M., Hofman, A., & Breteler, M. M. (2002). Alcohol consumption and risk of dementia: The Rotterdam Study. *Lancet*, *359*(9303), 281–286. https://doi.org/10.1016/s0140-6736(02)07493-7

Sabia, S., Fayosse, A., Dumurgier, J., Dugravot, A., Akbaraly, T., Britton, A., Kivimäki, M., & Singh-Manoux, A. (2018). Alcohol consumption and risk of dementia: 23-year follow-up of Whitehall II Cohort Study. *BMJ*, *362*, k2927. https://doi.org/10.1136/bmj.k2927

Salgado, S., & Kaplitt, M. G. (2015). The nucleus accumbens: A comprehensive review. *Stereotact Funct Neurosurg*, *93*(2), 75–93. https://doi.org/10.1159/000368279

Savage, L. M., Candon, P. M., Hohmann, H. L. (2000). Alcohol-induced brain pathology and behavioral dysfunction: Using an animal model to examine sex differences. *Alcohol Clin Exp Res*, *24*(4), 465–475. https://doi.org/10.1111/j.1530-0277.2000.tb02013.x

Savica, R., Boeve, B. F., & Mielke, M. M. (2018). When do α-Synucleinopathies start? An epidemiological timeline: A review. *JAMA Neurol*, *75*(4), 503–509. https://doi.org/10.1001/jamaneurol.2017.4243

Scarmeas, N., Anastasiou, C. A., & Yannakoulia, M. (2018). Nutrition and prevention of cognitive impairment. *Lancet Neurol*, *17*(11), 1006–1015. https://doi.org/10.1016/s1474-4422(18)30338-7

Schwarzinger, M., Pollock, B. G., Hasan, O. S. M., Dufouil, C., & Rehm, J. (2018). Contribution of alcohol use disorders to the burden of dementia in France 2008–13: A nationwide retrospective cohort study. *Lancet Public Health*, *3*(3), e124–e132. https://doi.org/10.1016/s2468-2667(18)30022-7

Serrano-Pozo, A., & Growdon, J. H. (2019). Is Alzheimer's disease risk modifiable? *J Alzheimers Dis*, *67*(3), 795–819. https://doi.org/10.3233/jad181028

Simpson, R. F., Hermon, C., Liu, B., Green, J., Reeves, G. K., Beral, V., & Floud, S. (2019). Alcohol drinking patterns and liver cirrhosis risk: analysis of the prospective UK Million Women Study. *Lancet Public Health*, *4*(1), e41–e48. https://doi.org/10.1016/s2468-2667(18)30230-5

Stockwell, T., Zhao, J., Panwar, S., Roemer, A., Naimi, T., & Chikritzhs, T. (2016). Do "moderate" drinkers have reduced mortality risk? A systematic review and meta-analysis of alcohol consumption and all-cause mortality. *J Stud Alcohol Drugs*, *77*(2), 185–198. https://doi.org/10.15288/jsad.2016.77.185

Topiwala, A., & Ebmeier, K. P. (2018). Effects of drinking on late-life brain and cognition. *Evid Based Ment Health*, *21*(1), 12–15. https://doi.org/10.1136/eb-2017-102820

Venkataraman, A., Kalk, N., Sewell, G., Ritchie, C. W., & Lingford-Hughes, A. J. A. (2017). Alcohol and Alzheimer's disease—Does alcohol dependence contribute to beta-amyloid deposition, neuroinflammation and neurodegeneration in Alzheimer's disease? *Alcohol Alcohol, 52*(2), 151–158. https://doi.org/10.1093/alcalc/agw092

Vinceti, M., Violi, F., Tzatzarakis, M., Mandrioli, J., Malagoli, C., Hatch, E. E., Fini, N., Fasano, A., Rakitskii, V.N., Kalantzi, O.I., & Tsatsakis, A. (2017). Pesticides, polychlorinated biphenyls and polycyclic aromatic hydrocarbons in cerebrospinal fluid of amyotrophic lateral sclerosis patients: A case-control study. *Environ Res*, *155*, 261–267. https://doi.org/10.1016/j.envres.2017.02.025

Wang, J., Ho, L., Zhao, Z., Seror, I., Humala, N., Dickstein, D.L., Thiyagarajan, M., Percival, S.S., Talcott, S.T., & Pasinetti, G.M. (2006). Moderate consumption of Cabernet Sauvignon attenuates Abeta neuropathology in a mouse model of Alzheimer's disease. *FASEB J*, *20*(13), 2313–2320. https://doi.org/10.1096/fj.06-6281com

Wyss-Coray, T. (2016). Ageing, neurodegeneration and brain rejuvenation. *Nature*, *539*(7628), 180–186. https://doi.org/10.1038/nature20411

Xu, W., Tan, L., Wang, H. F., Jiang, T., Tan, M. S., Tan, L., Zhao, Q. F., Li, J. Q., Wang, J., Yu, J. T. (2015). Meta-analysis of modifiable risk factors for Alzheimer's disease. *J Neurol Neurosurg Psychiatry*, *86*(12), 1299–1306. https://doi.org/10.1136/jnnp-2015-310548

Xu, W., Wang, H., Wan, Y., Tan, C., Li, J., Tan, L., & Yu, J. (2017). Alcohol consumption and dementia risk: A dose–response meta-analysis of prospective studies. *Eur J Epidemiol*, *32*, 31–42. https://doi.org/10.1007/s10654-017-0225-3

Zhang, D., Jiang, H., & Xie, J. (2014). Alcohol intake and risk of Parkinson's disease: A meta-analysis of observational studies. *Mov Disord*, *29*(6), 819–822. https://doi.org/10.1002/mds.25863

7 Cannabis and Neurodegenerative Disorders

Elie Atallah, Sahar Obeid, Myriam El Khoury-Malhame, and Michel Sfeir

INTRODUCTION

Cannabis and Its Derivatives

Cannabis, a psychoactive plant commonly known as marijuana, raises significant societal concerns due to its addictive potential and its deleterious effects on cognitive functions. Research has indicated that cannabis use, mainly when initiated in adolescence or consumed daily, can lead to addiction and mental health disorders (Connor et al., 2021; Volkow et al., 2014). In this regard, it is important to note that the term "cannabis" is an umbrella concept that includes a variety of substances, each with unique origins and effects, as detailed below.

Cannabis can be classified into different categories based on various factors, such as its chemical composition, effects, and usage. The two main distinctions commonly made are between indica and sativa types of cannabis. Indica cannabis strains are known for their relaxing effects, while sativa types are characterized by energizing and uplifting properties (Hazekamp & Fischedick, 2012). Modern cannabis varieties often do not fit into these classifications and are more likely hybrids of the two, making it challenging to differentiate cannabis products accurately based on phenotypic, aromatic, or chemical characteristics (Small, 2015).

A second type of cannabis classification is based on its chemical composition, particularly in terms of cannabinoids. Marijuana comprises approximately 60 pharmacologically active compounds referred to as "cannabinoids," which interact with the human body's endocannabinoid system (ECS) (Barbara S. Koppel et al., 2014). These compounds, such as tetrahydrocannabinol (THC) and cannabidiol (CBD), have different effects on the body and are often used for medicinal purposes (Hazekamp & Fischedick, 2012; Lafaye et al., 2017). There are also synthetic cannabinoids, artificially created compounds that mimic the effects of natural cannabinoids. These products are often developed for research purposes to probe the ECS or as potential therapeutic agents (Castaneto et al., 2014).

In this chapter, the term "cannabis" is used to encompass all substances derived from the plant, including marijuana, synthetic cannabinoids, and other derivatives. However, when discussing specific nuances and properties, we use precise terminology to refer to the substance. Understanding the differences between these categories is essential due to variations in their origins, compositions, and effects on the body, which allows for a more precise understanding of the potential health impacts and therapeutic implications associated with each compound.

For instance, THC is renowned for its psychoactive effects, including hypolocomotion, hypothermia, catalepsy, and analgesia. Conversely, CBD exhibits neuroprotective, antispasmodic, and anti-inflammatory properties, mediated by activating receptors such as cannabinoid (CB)-2 and PPARγ (Atakan, 2012; Pagano et al., 2022). While the medical utility of marijuana was initially centered on THC, recent interest has shifted toward the potential therapeutic applications of CBD. Unlike THC, the latter does not induce psychoactive effects and appears to offer several therapeutic benefits (Arachchige, 2023). Indeed, CBD is recognized for its therapeutic potential in addressing neurodegenerative and hyperkinetic disorders as a result of its neuroprotective properties (Singh

DOI: 10.1201/9781003478485-7

et al., 2023). In this context, preclinical studies indicate that cannabinoids hold promise for symptomatic relief and neuroprotection across a range of neurological conditions, including movement disorders (Iuvone et al., 2009; Kluger et al., 2015). Moreover, research indicates that CBD is well tolerated and safe for humans, even at high doses (Savage et al., 2016).

The Endocannabinoid System

The ECS is a network of receptors and signaling molecules in the brain that regulate various functions, including memory, learning, and mood (Coleman et al., 2022). Cannabis interacts with the ECS via its main psychoactive component, THC, and other cannabinoids.

Cannabinoid receptors CB1 modulate neurotransmitter release across nearly all brain regions (Matsuda et al., 1990; Munro et al., 1993). Cannabinoid ligands and receptors are particularly prevalent in basal ganglia structures, especially nuclei that receive striatal efferents like the internal and external globi pallidi and the substantia nigra (Bisogno et al., 1999; Herkenham et al., 1991). The cannabinoid system significantly influences basal ganglia function by regulating he release of neurotransmitters within basal ganglia circuits (Kano et al., 2009; Lovinger, 2008). Furthermore, cannabinoids that activate CB1 receptors are crucial in motor responses (Fernández-Ruiz & Gonzáles, 2005). CB1 receptor activation decreases locomotion while antagonizing CB1 receptors induces hyperlocomotion (Compton et al., 1996; Ledent et al., 1999). CB1 receptors also bind to TRPV1 receptors located on dopamine terminals, potentially reducing hyperactivity associated with hyperdopaminergia (Tzavara et al., 2006). CB1 receptors may offer potential utility in mitigating neurodegenerative effects (van der Stelt et al., 2005), as they have been shown to protect neurons by inhibiting glutamate exocytosis via CB1 receptor activation, closing voltage-sensitive calcium channels, and reducing oxidative stress (Abood et al., 2001; Shen & Thayer, 1998). Consequently, cannabinoid-based therapies could hold promise for treating basal ganglia disorders (Fernández-Ruiz, 2009).

Cannabis Consumption

Chronic cannabis consumption can lead to addiction, characterized by a loss of control over cannabis use despite negative consequences, including impaired social and occupational functioning, as well as physical health issues (Terry-McElrath et al., 2022). This dependence is associated with long-term brain changes, potentially increasing vulnerability to the development of other substance-related mental disorders. Notably, studies suggest that cannabis consumption and schizophrenia may share underlying vulnerabilities that affect subcortical brain structures, leading to observable differences in neural activation patterns (Løberg et al., 2012). In addition, functional magnetic resonance imaging (fMRI) studies reveal brain function patterns associated with cannabis cue reactivity, particularly in the orbitofrontal cortex, implying a potential link with addictive behaviors (Sehl et al., 2021). Chronic cannabis consumption poses significant health risks (Volkow et al., 2014); such exposure, particularly to high doses and synthetic forms, disrupts brain metabolism and function, impacting memory, attention, and executive control (Bouter et al., 2023; Sagar & Gruber, 2018). Chronic exposure can also lead to neurotoxic actions in brain areas enriched with cannabinoid receptors (Rocchetti et al., 2013). Furthermore, recent research highlighted possible cognitive deficits and neural changes secondary to frequent recreational cannabis use, as well as possible links between cannabis consumption and neurodegenerative disorders (NDDs). Alternatively, novel preclinical studies suggest that CBD may provide neuroprotective benefits due to its antioxidant and anti-inflammatory properties (Cassano et al., 2020).

This chapter aims to highlight some of the existing literature on NDDs and cannabis. We will first introduce NDDs and the limitations of current treatments. We will then explore a few animal models' contributions to understanding cannabinoid effects on neurodegeneration pathogenesis. We will additionally present how NDDs affect cognition and then delve into the characteristics of the use of cannabis in the geriatric population. Finally, we will discuss the potential role, benefits, and harms of cannabinoid-based treatments and interventions in managing NDDs and associated conditions.

NEURODEGENERATIVE DISORDERS: AN OVERVIEW

NDDs involve progressive neuronal dysfunction and structural loss of the brain and spinal cord. They involve the gradual deterioration of neurons in the central nervous system (CNS), resulting in severe symptoms and diminishing cognitive and motor abilities (Arachchige, 2023). Common examples include Alzheimer's disease (AD), Parkinson's disease (PD), Huntington's disease (HD), and amyotrophic lateral sclerosis (ALS). Although the exact causes are not fully understood, research indicates that neuroinflammation and oxidative stress may be important in developing NDDs. Table 7.1 reviews some of the characteristics and treatment limitations of common NDDs.

Despite considerable investigation, viable therapies capable of halting or reversing these diseases are still lacking. Recently, there has been growing curiosity regarding the potential therapeutic application of marijuana in addressing NDDs (Bhunia et al., 2022). Studies have highlighted the importance of understanding the ECS in the context of NDDs, emphasizing the role of glial cells in chronic neurodegenerative and inflammatory disorders (Bernal-Chico et al., 2023). As the endocannabinoid signaling pathway is involved in the modulation of neuroinflammation and neuronal survival, this suggests a potential avenue for developing new therapies for NDDs.

The Contribution of Animal Models to Understanding the Link between Cannabinoids and Neurodegeneration

Many studies have investigated the neuroprotective properties of cannabinoids in animal models. We highlight some of them in Table 7.2. These findings might pave the way for cannabinoid-based therapeutic strategies in treating CNS conditions such as epilepsy or pain and, more specifically, HD and PD, which are known to involve striatal damage and progressive neurodegeneration.

Exploring Cognitive Effects and Links to Neurodegenerative Disorders

The relationship between cannabis consumption and cognitive function is complex and multifaceted. While some studies suggest that chronic cannabis consumption may lead to short-term adverse effects on memory, attention, and decision-making (Ożarowski et al., 2021), others indicate that medical cannabis use in older individuals could improve cognitive performance by indirectly addressing factors like pain, sleep, and mood (Contino et al., 2017). These seemingly contradictory pieces of evidence highlight the intricate interaction between cannabis, the brain, and various cognitive processes. To further complicate matters, research suggests that cannabis may have a biphasic effect on myelination, the process by which nerve fibers are insulated with a fatty substance. While some studies hint at potential benefits for myelination (e.g., increased nerve impulse velocity), others associate excessive cannabis consumption, especially during adolescence, with demyelination and cognitive decline (Menendez-Miranda et al., 2019).

In this regard, studies have linked cannabis consumption to structural changes in brain regions crucial for cognitive function, such as the prefrontal cortex and hippocampus (Daniju et al., 2022). These changes could potentially contribute to the development and progression of NDDs. One potential mechanism for this link involves the ECS. While the exact mechanisms remain under investigation, excessive and prolonged cannabis consumption has been implicated in disrupting the delicate balance of the ECS, potentially resulting in long-term cognitive deficits (Wanke et al., 2022). Chronic consumption or high doses of cannabis can impair learning

TABLE 7.1
Review of the Most Common Neurodegenerative Disorders

Neurodegenerative Disorder	Characteristics

Huntington's disease	• Autosomal dominant NDD arising from a mutation in the gene responsible for producing a protein called huntingtin (htt).
	• Characterized by two main symptoms: Motor abnormalities and cognitive decline (Ross & Tabrizi, 2011).
	• Gradual progression of neurodegeneration with motor symptoms displaying a biphasic pattern:
	  • Initially, the disease manifests with involuntary movements known as "chorea," which result from the impairment of striatal GABA neurons projecting to the globus pallidus (Cowan & Raymond, 2006; Rosas et al., 2008).
	  • Alongside chorea, patients may experience other involuntary movements, including tics, myoclonus, and dystonia (Ashizawa & Jankovic, 1996; Watt, 1990).
	• No cure; patients typically succumb to the illness within 10–20 years after diagnosis:
	  • Relief therapy is available to manage certain symptomatic features, such as the use of antidopaminergic drugs to alleviate hyperkinesia observed in the early stages (Factor & Friedman, 1997).
	  • Specific cannabinoid-related compounds have been proposed for mitigating hyperkinetic involuntary movements in HD (Sagredo et al., 2012).

Alzheimer's disease	• AD poses a 1% risk in the general population at age 60, doubling every 5 years after that (Alzheimer Society of Canada, 2010).
	• Based on its pathophysiology, AD is categorized into sporadic or late-onset and familial or early-onset:
	  • Sporadic AD: Accounts for about 95% of cases, a complex disease influenced by genetic and epigenetic factors, as well as environmental and lifestyle variables (Hval et al., 2012).
	  • Familial AD: Less than 5% of cases, linked to mutations in genes such as amyloid precursor protein (APP), presenilin-1 (PSEN1), and presenilin-2 (PSEN2).
	• BPSD are prevalent in various forms of dementia, particularly in AD, occurring in more than 95% of cases (Cerejeira et al., 2012; Hval et al., 2012):
	  • BPSD can worsen cognitive decline and physical impairment in patients (Mintzer et al., 1998), with anxiety being a common neuropsychiatric symptom (NPS) associated with BPSD in AD (Benoit et al., 1999).
	  • Other symptoms include agitation, aggression, depression, apathy, delusions, hallucinations, as well as changes in sleep and appetite (Cerejeira et al., 2012).
	• Currently, there is no cure for AD, and available treatments only alleviate symptoms in the early stages (Girgin et al., 2023).

Parkinson's disease	• The second most common NDD consists of motor symptoms such as rest tremor, rigidity, bradykinesia, and akinesia (Tolosa et al., 2021). • Also features non-motor symptoms, most frequently NPS, such as depression, anxiety, apathy, psychosis, cognitive decline, and disrupted impulse control (Weintraub et al., 2022). • Pathological findings include progressive loss of dopaminergic neurons in the substantia nigra pars compacta and abnormal intra-cytoplasmic accumulation of α-synuclein protein (Poewe et al., 2017). • No single cause has been identified; associated with a complex interplay of genetic and environmental risk factors (Hayes, 2019; Kalia & Lang, 2015). • Available dopaminergic treatments cannot stop the progression and may lead to several adverse effects (Vallée et al., 2021).

TABLE 7.2
Preclinical Data Highlighting the Neuroprotective Effects of Cannabinoids

Studies	Findings
Costa et al. (2022); Sieradzan et al. (2001)	CBD promoted a reduction in proinflammatory markers, such as COX-2 expression, in motor regions of the striatum and decreased the dyskinesia associated with dopaminergic drug treatment.
Dearborn et al. (2022)	Chronic administration of CBD resulted in immunomodulatory effects in mouse models of neuroinflammation and neurodegeneration.
Duncan et al. (2024)—Review	The neuroprotective effects of CBD, demonstrated in both *in vitro* and *in vivo* animal models, suggest promising therapeutic applications. This is mediated by molecular and cellular mechanisms through which CBD mitigates excitotoxicity, oxidative stress, hypoxia, and inflammation.
Eubanks et al. (2006)	THC blocked β-amyloid aggregation by competitively inhibiting the enzyme acetylcholinesterase, a therapeutic target for AD drugs.
Harvey et al. (2012); Janefjord et al. (2014)	While cannabinoids do not alter preformed β-amyloid peptides and variably disrupt the formation of new ones, these molecules exert neuroprotective effects by inhibiting activated microglia neurotoxicity.
Heumann et al. (2014); Junior et al. (2020)—Review	Cannabinoid antidyskinetic properties may be accounted for by the modulation of CB1 receptors and vanilloid TRPV1 receptors, which are co-expressed in several brain areas and seem to exert opposing effects on excitatory and inhibitory neurotransmissions. Indeed, it appears that CB1 receptor activation and TRPV1 blockade play a role in alleviating L-DOPA-induced dyskinesia.
Kuharic et al. (2021)—Systematic Review	A Cochrane review reported that cannabinoids could have specific disease-modifying actions in AD by reducing excitotoxicity, mitochondrial dysfunction, oxidative stress, neuroinflammation, and the formation of amyloid plaques and neurofibrillary tangles.
Lastres-Becker et al. (2001); van Vliet et al. (2008)	THC treatment improved locomotor activities in marmoset monkeys with drug-induced PD. One explanation for this positive effect is that PD marmoset models exhibit higher CB1 concentrations in the basal ganglia.
Patel et al. (2023)	CBD was found to protect striatal neurons, independent of adenosine secretion, mainly through its antioxidant and anti-inflammatory properties. The authors further documented the protective effect of CBD pre-exposure on cell death signaling.
Perez et al. (2013)	CBD action was potentiated by reversing inhibited GABAergic secretions and increasing plummeting mRNA levels of striatal markers, such as substance P, initially blunted by substances inhibiting mitochondrial complex.
Quinn et al. (2008)	Cognitive impairments associated with cannabis were only sustained when young rodents' developing brains were exposed.
Ramírez et al. (2005)	Synthetic cannabinoids exerted neuroprotective effects in rats by counteracting the activation of microglial cells induced by β-amyloid peptides, a key pathological marker of AD, and by reducing tumor necrosis factor-α (TNF-α) release in cultured microglia. They also prevented cognitive deficits and loss of neuronal markers in treated rats.
Sagredo et al. (2007)	CBD exerted neuroprotective effects preferentially on the striatal neurons projecting to the substantia nigra in rats.
Urbi, Lee et al. (2022)—Systematic Review and Meta-Analysis	CBD seems promising in alleviating neuropathology and improving motor function in rodents. However, the authors noted the heterogeneity of studies and possible bias in study methods, models used, and cannabinoid types.
Vallée et al. (2021)—Review	CBD downregulated GSK-3β, the primary inhibitor of the WNT/β-catenin pathway, which is deregulated in PD pathogenesis and may be associated with controlling oxidative stress and neuroinflammation.
Walther and Halpern (2010)—Review	CB1 receptor agonists may reduce glutamate release, thus preventing excitotoxicity, while CB2 receptor agonists could have anti-inflammatory and β-amyloid plaque-removal effects. In particular, CBD has been found to have antioxidant and antiapoptotic properties and to target tau phosphorylation.

(*Continued*)

TABLE 7.2 (CONTINUED)

Studies	Findings
Wang et al. (2022)	CBD has a therapeutic effect on the performance of PD mice models in cognitive tasks and spontaneous locomotion. This phenomenon could be accounted for by inhibiting dopaminergic neuron apoptosis and decreasing inflammation. On a molecular level, CBD increased the levels of serotonin, dopamine, and interleukin (IL)-10 in treated mice, as well as the expression of tyrosine hydroxylase in the substantia nigra. It also reduced the levels of TNF-α, IL-1β, and IL-6. Additionally, CBD upregulated the expression of the Bcl-2 gene, thus preventing the expression of the NLRP3/caspase-1/IL-1β inflammasome pathway.
Xu et al. (2016)	Activation of cannabinoid receptors alleviated neuropathy by reducing local neuroinflammation in the spines of adult rats.

speed and information processing due to effects on motor and cognitive abilities, as well as cerebral networks (Daniju et al., 2022). Furthermore, research indicates that chronic cannabis use may be associated with declines in attention and memory (Grant et al., 2012), underscoring the potential risks for individuals vulnerable to neurodegeneration, where cognitive decline is a defining characteristic.

However, the relationship between cannabis consumption and cognition is not solely negative. For example, as previously discussed, CBD exhibits neuroprotective properties mediated via its anti-inflammatory and antioxidant actions (Cooper & Abrams, 2019). These properties suggest potential benefits for NDDs characterized by chronic inflammation and oxidative stress. The neuroprotective potential of CBD encompasses other mechanisms of action. These include regulating intracellular calcium levels (Kim et al., 2021) and improving glucose metabolism (Bhunia et al., 2022). Additionally, CBD has antipsychotic, anxiolytic, and anti-inflammatory effects, reinforcing its prospective therapeutic utility in NDDs (Peres et al., 2018) by modulating neurotrophic signaling pathways, which are crucial for neuronal survival and growth (Karimi-Haghighi et al., 2022). CBD was also found to modulate proteinopathies, protect neural cells from mitochondrial dysfunction, and enhance autophagy, all of which may help combat NDDs (Dash et al., 2021). Its effects on neuropathic and inflammatory pain further suggest utility in managing symptoms associated with these disorders (Karimi-Haghighi et al., 2022). In conclusion, CBD's neuroprotective, anti-inflammatory, and antioxidant properties provide a strong foundation for therapeutic benefits in NDDs (Fernández-Ruiz et al., 2013; Landucci et al., 2021; Peres et al., 2018).

Lastly, it is noteworthy to mention that the relationship between cannabis use and cognitive function is influenced by a variety of factors, including an individual's genetic predisposition, shared environmental influences, and the presence of preexisting psychiatric disorders (Curran et al., 2016).

Characteristics of the Use of Cannabis and Its Derivatives among Older Adults

Recent studies document a significant rise in cannabis use among older adults, increasing from 0.6% in the United States (US) in 2006 to approximately 2.9% in 2015 and reaching 4–6% by 2018, reflecting a striking 75% relative increase (Han & Palamar, 2020). This trend is particularly notable among subgroups such as women, individuals with prior mental health conditions, and those who consume alcohol. COVID-19 appears to have further driven this trend, with an estimated 12% of adults aged 50–80 in the US now using cannabis for medical or recreational purposes. The increase is more pronounced among ethnic minorities, unmarried individuals, and those living alone (Fernandez et al., 2024). The authors highlight this exponential rise to be driven by post-pandemic stress and the legalization of cannabis in additional states in the country.

Research has documented both potential benefits and risks of cannabis use in this demographic. On the one hand, cannabis appears to alleviate pain, particularly from medical and surgical complications common in older adults, and may help manage other conditions and behavioral symptoms. On the other hand, it can potentially exacerbate cognitive decline and impair overall functioning (Solomon et al., 2021). The impact of cannabis and cannabinoids on the brain health of older adults remains to be fully explored. Preliminary findings suggest protective effects of cannabis in older adults (Weinstein & Sznitman, 2020) in contrast to its lasting harmful effects when used during early adolescence (Scheyer et al., 2023). This age-dependent impact is mirrored in animal studies, where cognitive impairments associated with cannabis exposure are observed primarily in young rodents with developing brains (Quinn et al., 2008). Differences in neurobiological pathways, brain structures, and epigenetic changes between these age groups have been documented, underscoring the need for further research.

Cannabis use in older adults is associated with significant risks, particularly regarding mental health (Hudson & Hudson, 2021). These risks are influenced by factors such as the reasons for use (e.g., reliance on cannabis as a sleep aid), preexisting psychological vulnerabilities or conditions, demographics (e.g., younger males within the older population), and patterns of use. Even in medical applications, studies have observed that treatments often begin with higher CBD concentrations, with THC doses increasing over time, which complicates its safety profile and therapeutic use in this age group (Kaufmann et al., 2022).

Medical cannabis use, approximately 35% of which involves individuals aged 65 years and older, also necessitates close monitoring for side effects, including blood pressure changes, confusion, hallucinations, and short-term cognitive impairments (Whiting et al., 2015). While some findings suggest benefits of cannabis for older adults in specific conditions, such as end-stage cancer, AD, PD, and dementia, these results should be interpreted cautiously. Outside of these contexts, evidence remains inconsistent, with some studies reporting no significant benefits or even adverse effects, including episodes of depression, anxiety, substance use, and injury (Wolfe et al., 2023).

The overall impact of cannabis on older adults is further complicated by age-related physiological factors, such as slowed metabolism and declining cognition, as well as environmental factors, including an increased risk of falls, injuries, and polypharmacy, which heighten the potential for drug interactions (Solomon et al., 2021).

Significant gaps remain in the research on cannabis use in older adults. Methodological limitations, including the lack of randomized controlled trials (RCTs), hinder a comprehensive understanding of cannabis's effects on older adults. Future studies should address key areas such as modes of administration, types of cannabinoids used, baseline cognitive performance, duration of abstinence, and residual confounding variables like socioeconomic status, cultural background, and personality traits (Weinstein & Sznitman, 2020). Robust evidence is essential to guide policymaking, minimize the risks of medical and psychiatric disorders, and optimize individualized benefits while addressing specific clinical considerations for older adults.

CANNABIS USE IN THE TREATMENT OF NEURODEGENERATIVE DISORDERS

Cannabinoids: Novel Medicines for the Treatment of Huntington's Disease

Over the past decade, there has been increasing interest in exploring the therapeutic potential of cannabinoids in movement disorders (Kluger et al., 2015; Barbara S. Koppel et al., 2014). However, data on their effects in HD remain scarce and inconsistent. Several case reports using nabilone, a synthetic cannabinoid capsule acting as a selective agonist for CB1 and CB2 receptors, suggested benefits in managing chorea and irritability, though other cases reported a worsening of chorea (Armstrong et al., 2012; Curtis & Rickards, 2006). Only three available RCTs in this population group deemed cannabinoids safe (Kluger et al., 2015; López-Sendón Moreno et al., 2016; Pidgeon & Rickards, 2013). Notably, in a study involving the administration of nabilone at 1 and 2 mg daily compared to a placebo (n = 44) (Curtis et al., 2009), a statistically significant reduction in chorea according to the Unified Huntington's Disease Rating Scale (UHDRS) chorea subscale was observed, along with a trend toward improvement in the UHDRS behavior score as a secondary endpoint, although no significant change was noted in the primary endpoint, the "UHDRS motor score." Another randomized, placebo-controlled, crossover pilot trial (n = 26) found that up to 12 sprays/day of Sativex spray, a combination of delta-9-THC and CBD, did not demonstrate any effects on motor, cognitive, behavioral, or functional scores, although it was deemed safe and well tolerated (López-Sendón Moreno et al., 2016). A more extended treatment period may be necessary to allow the benefits derived to be visible in the progression of the disease. In an older double-blind, randomized crossover design (n = 15), oral CBD (10 mg/kg/day for 6 weeks) was found to be safe, yet no significant effects on chorea severity or other outcome variables were observed (Consroe et al., 1991).

More recently, a systematic review aimed at assessing the evidence regarding the efficacy of medical marijuana in mitigating symptoms prevalent in movement disorders, particularly HD, revealed therapeutic benefits (Akinyemi et al., 2020). Medical marijuana, including the use of Sativex and nabilone, demonstrated efficacy in managing symptomatology associated with movement disorders (Kluger et al., 2015; Wissel et al., 2006). It is presumed that the effects observed in studies utilizing standardized pharmaceutical formulations are more clinically reliable than those derived from survey studies based on self-reported cannabis use. Additionally, given the prevalence of pain in patients with HD, medical marijuana could provide relief, as evidence suggests that cannabinoid-based pharmacotherapies may effectively manage chronic pain (Yanes et al., 2019).

Cannabinoids in Alzheimer's Disease

Effect of Cannabinoids in Patients with Alzheimer's Disease

Preclinical research suggests that cannabinoids, such as THC and CBD, may have pharmacological effects on the cholinergic system and amyloid-beta aggregation, both of which contribute to the irreversible and progressive neuronal dysfunction and cell death in AD ("2021 Alzheimer's Disease Facts and Figures," 2021).

Results from clinical research also indicate that cannabis use may lead to improvements in behavioral and psychological symptoms of dementia (BPSD), including agitation and sleep disturbances, as well as aid in weight and pain management among patients with AD (Sherman et al., 2018). Disturbed and agitated behavior in dementia was evaluated across four RCTs, revealing an overall statistically significant effect favoring cannabis use (Bilbao & Spanagel, 2022). However, the evidence regarding specific cannabinoids is either limited or absent. While the combined results of three studies involving dronabinol (Van den Elsen, Ahmed, Verkes, Feuth, et al., 2015; Van Den Elsen, Ahmed, Verkes, Kramers, et al., 2015; Volicer et al., 1997) did not collectively demonstrate statistical significance, a singular study utilizing nabilone (Herrmann et al., 2019) reported a significant reduction in agitation and aggression. Furthermore, a 2020 systematic review and meta-analysis found that cannabinoids are efficacious and exhibit good tolerance in the treatment of neuropsychiatric symptoms (NPS) of dementia. Nonetheless, the authors mention that their findings are not robust due to the small sample sizes, and cannabinoid use in patients with dementia should be considered experimental rather than routinely recommended (Bahji et al., 2020).

Overall, previous clinical trials demonstrated that the adverse effects of cannabis in AD are generally well tolerated at the studied doses (Sherman et al., 2018), further increasing interest in investigating the role of CBD in the field of Alzheimer's research.

Limits of Cannabinoids in Alzheimer's Disease Therapy

As of now, the US Food and Drug Administration (FDA) has not sanctioned the use of cannabis or cannabis-derived substances for treating AD or managing its associated symptoms. Only a handful of relevant clinical trials have been conducted or are presently underway (Abate et al., 2021), and future studies are recommended to replicate current positive findings (Herrmann et al., 2019).

Considering cannabinoids as a therapeutic option for AD presents challenges, particularly in determining the appropriate dosage and treatment duration (Abate et al., 2021). It is widely acknowledged that molecular changes associated with AD begin several years before symptoms become apparent (Abate et al., 2021). Hence, hypothetically, it would be crucial to administer neuroprotective cannabinoids before AD progresses and becomes chronic. Screening older adults with mild symptoms to ensure early diagnosis and management might be considered (Liss et al., 2021). However, the particular use of cannabinoids for treating mild symptoms of AD remains an area of dire research, and robust clinical trials are needed to establish clear guidelines while ensuring safety (Benninghoff & Perneczky, 2022; Wieghorst et al., 2022).

Alternatively, there is a lack of studies investigating the long-term effects of cannabinoids in older adults (Abate et al., 2021). Research on the prolonged cognitive effects of heavy cannabis use indicates a negative impact on cognitive functions such as episodic memory, attentional control, and motor inhibition (Croxford, 2003; Kroon et al., 2021). One review identified four RCTs (n = 126) comparing cannabinoids to placebo in treating dementia. Due to the small sample sizes and short follow-up durations, the authors could not determine whether cannabinoids had beneficial or harmful effects. They also noted that cannabinoids primarily influenced the NPS of dementia, with little to no impact on cognitive symptoms (Kuharic et al., 2021).

To conclude, the use of cannabinoids in AD should be approached with caution and based on emerging evidence (Schrot & Hubbard, 2016). There is an urgent need for further studies to explore the short- and long-term effects of cannabinoids in older adults with AD. A significant gap remains in research on drug interactions involving cannabinoids, which is critical to address, given the polypharmacy common among patients with dementia. Understanding these interactions is essential as cannabinoids are explored as a potential therapy for managing NDDs.

Cannabinoids and Parkinson's Disease

To date, there are no large double-blind RCTs assessing the efficacy and safety of cannabinoids in the treatment of PD. Furthermore, available data show mixed and inconclusive results due to the heterogeneity of cannabinoid drugs, administration methods, and lack of standardized protocols (Deuel & Seeberger, 2020). Two recent meta-analyses found insufficient evidence to strongly recommend the use of cannabis for PD (Oikonomou & Jost, 2022; Urbi, Corbett, et al., 2022). These meta-analyses included RCTs and non-randomized studies, with outcomes assessed using standardized tools or self-reported by patients. Despite limitations in the literature, both THC and CBD appear to alleviate non-motor symptoms of PD, particularly anxiety, pain, and sleep disturbances. The findings also suggest potential benefits of CBD in improving quality of life and reducing tremor. Furthermore, a systematic review concluded that non-motor symptoms of PD seem to respond better to cannabinoids than motor symptoms (Figura et al., 2022).

Nonetheless, marijuana use in PD is associated with both short- and long-term effects, including cognitive impairment and dependence, particularly since chronic use is often required for therapeutic effects (Patel et al., 2019). A recent review raised concerns about whether cannabis could exacerbate hallucinations in patients with PD. The authors cited two studies that reported hallucination incidences within the accepted range for PD patients but were unable to draw definitive conclusions due to small sample sizes (Cravanas Jr & Frei, 2020). Consequently, in the absence of high-level evidence or established guidelines, cannabinoids should be considered only as an adjunctive off-label treatment for PD (Costa et al., 2022).

CANNABINOIDS IN THE TREATMENT OF NEUROPSYCHIATRIC SYMPTOMS OF NEURODEGENERATIVE DISORDERS

As mentioned earlier, many NDDs are associated with a wide range of NPS, including alterations in personality and behavioral changes such as irritability, impulsivity, agitation, apathy, aggression, hallucinations, or delusions. These symptoms pose a significant burden on both patients and caregivers (Calsolaro et al., 2021). In clinical practice, various drugs, particularly antipsychotics, are prescribed to manage NPS despite limited strong clinical evidence supporting their effectiveness. Moreover, many of these medications are linked to a high incidence of adverse events and an increased risk of mortality (Calsolaro et al., 2021; Magierski et al., 2020). Emerging evidence suggests that cannabinoids may offer potential benefits in managing NPS. However, significant limitations remain, including the lack of robust, large-scale RCTs and the heterogeneity of cannabinoid products, dosages, and administration routes.

A 2019 systematic review concluded that cannabinoids show promise in managing NPS, particularly in refractory patients, with a reasonable safety profile. Most reported mild adverse events, including sedation, somnolence, and fatigue. However, some studies in the review reported more serious side effects, such as changes in blood pressure, balance issues, and infections, which warrant further investigation in future research. Additionally, long-term safety could not be determined due to the short duration of the studies assessed (Hillen et al., 2019). Another review noted that cannabinoids improved NPS, such as agitation, aggression, impulsivity, nocturnal restlessness, wandering, and poor sleep. Nonetheless, the authors emphasized that these findings should be considered preliminary due to limitations in the included studies (Tampi et al., 2018).

The THC compound in cannabis exhibits sleep-promoting properties, primarily mediated by CB1 receptor agonism, which can influence sleep architecture (Kesner & Lovinger, 2020). While THC may decrease sleep latency, its long-term use could impair sleep quality. In contrast, CBD appears beneficial for treating insomnia, REM sleep behavior disorder, and excessive daytime sleepiness (Babson et al., 2017). A 2019 review by Hodges and Ashpole suggested that cannabinoids may act as chronobiotics, influencing the sleep-wake cycle by modulating the suprachiasmatic nucleus of the hypothalamus, the primary neural site for circadian rhythm integration. This indicates that cannabinoids, at low doses, could help adjust altered circadian rhythms and prevent certain age-related pathologies in older adults (Hodges & Ashpole, 2019). A 2022 systematic review and meta-analysis demonstrated significant improvements in sleep scores and therapeutic effects with cannabinoids, particularly dronabinol, nabilone, and nabiximols (Bilbao & Spanagel, 2022). Similarly, in a survey conducted in two geriatric primary care clinics, 44% of current users reported using marijuana products at least weekly for chronic conditions such as insomnia, with 86% finding these products helpful (Reynolds et al., 2018). More recently, a survey conducted by Kaufmann at a geriatric clinic revealed that 29% of users consumed cannabis for sleep disturbances, with most preferring THC-containing products over CBD-only options (Kaufmann et al., 2023). However, despite these usage patterns in the geriatric population, a 2023 review concluded that the role of cannabinoids in sleep and wakefulness remains inconclusive, as studies report mixed outcomes (Low et al., 2023). For instance, an RCT by Linares found that the acute administration of CBD does not induce sleep in healthy individuals (Linares et al., 2018).

CONCLUSION

In this chapter, we reviewed existing literature evaluating the use of cannabis in geriatric populations diagnosed with NDDs. The medical and therapeutic use of cannabis requires further investigation through rigorous, replicable studies involving larger samples. While cannabis shows therapeutic promise in managing NDDs, challenges such as potential addiction and limited legal access in some regions complicate research efforts. Healthcare professionals and researchers should seize these opportunities to explore the therapeutic benefits and risks of different forms of cannabis and cannabinoids before adopting them as treatment options.

The development of cannabis-derived therapies for NDDs hinges on a deeper understanding of the mechanisms of cannabinoid action in the brain and well-controlled clinical trials to establish efficacy and safety (Hunt et al., 2020). Although cannabis offers potential as a therapeutic tool, its benefits must be carefully weighed against the risks of chronic use, particularly dependency.

Research must also account for the addictive potential of potent and synthetic cannabinoids. Addressing regulatory barriers, stigma, and access challenges is critical for advancing this field while clarifying the complexities of the ECS.

Future research should focus on identifying the most effective cannabinoids, determining their role as stand-alone or adjunctive therapies, and assessing their safety and tolerability. Special attention must be given to the potential risk of psychotic symptoms, as observed in younger populations using cannabis (Weier & Hall, 2017). The scarcity of data and inconsistency in findings remain significant obstacles to drawing robust scientific conclusions about the impact of cannabis on brain health. Most animal studies involve only young rodents, further limiting the generalizability of findings (Cristino et al., 2020). Cross-sectional studies add to these challenges by restricting causal inferences. For example, older adults with preexisting mental or physical health conditions may be more likely to use cannabis, making it unclear whether undesirable health outcomes are caused by cannabis use or underlying conditions. Additional studies are needed to clarify the potential risks and benefits of cannabis use in elderly psychiatric populations, particularly in the context of increased cannabis legalization and marketing. These factors have contributed to cannabis becoming the most commonly used psychoactive substance among older adults (Khoury et al., 2022).

Moving forward, fostering collaboration among researchers, clinicians, policymakers, and patients is imperative to deepen our understanding of cannabis's impact on NDDs. Large-scale clinical trials employing robust methodologies are essential for evaluating the safety, efficacy, and long-term effects of cannabinoid-based therapies. Standardizing formulations, dosages, and delivery methods is crucial for achieving consistent and reliable outcomes.

REFERENCES

2021 Alzheimer's disease facts and figures. (2021). *Alzheimer's & Dementia, 17*(3), 327. https://doi.org/10.1002/alz.12328

Abate, G., Uberti, D., & Tambaro, S. (2021). Potential and limits of cannabinoids in Alzheimer's disease therapy. *Biology (Basel, Switzerland), 10*(6), 542. https://doi.org/10.3390/biology10060542

Abood, M. E., Rizvi, G., Sallapudi, N., & McAllister, S. D. (2001). Activation of the CB1 cannabinoid receptor protects cultured mouse spinal neurons against excitotoxicity. *Neuroscience Letters, 309*(3), 197–201. https://doi.org/10.1016/S0304-3940(01)02065-1

Akinyemi, E., Randhawa, G., Longoria, V., & Zeine, R. (2020). Medical Marijuana effects in movement disorders, focus on Huntington disease: A literature review. *Journal of Pharmacy & Pharmaceutical Sciences, 23*, 389–395. https://doi.org/10.18433/jpps30967

Alzheimer Society of Canada. (2010). Rising tide: The impact of dementia on Canadian society. https://alzheimer.ca/sites/default/files/documents/Rising-tide_Alzheimer-Society.pdf

Arachchige, A. S. P. M. (2023). Marijuana's potential in neurodegenerative diseases: An editorial. *AIMS Neuroscience, 10*(2), 175–177. https://doi.org/10.3934/Neuroscience.2023014

Armstrong, M. J., Miyasaki, J. M., & American Academy of, N. (2012). Evidence-based guideline: Pharmacologic treatment of chorea in Huntington disease: Report of the guideline development Subcommittee of the American Academy of Neurology. *Neurology, 79*(6), 597–603. https://doi.org/10.1212/WNL.0b013e318263c443

Ashizawa, T., & Jankovic, J. (1996). Cervical dystonia as the initial presentation of Huntington's disease. *Movement Disorders: Official Journal of the Movement Disorder Society, 11*(4), 457–459.

Atakan, Z. (2012). Cannabis, a complex plant: Different compounds and different effects on individuals. *Therapeutic Advances in Psychopharmacology, 2*(6), 241. https://doi.org/10.1177/2045125312457586

Babson, K. A., Sottile, J., & Morabito, D. (2017). Cannabis, cannabinoids, and sleep: A review of the literature. *Current Psychiatry Reports, 19*(4), 1–12.

Bahji, A., Meyyappan, A. C., & Hawken, E. R. (2020). Cannabinoids for the neuropsychiatric symptoms of dementia: A systematic review and meta-analysis. *Canadian Journal of Psychiatry, 65*(6), 365–376.

Benninghoff, J., & Perneczky, R. (2022). Anti-dementia medications and anti-Alzheimer's disease drugs: Side effects, contraindications, and interactions. In P. Riederer, G. Laux, T. Nagatsu, W. Le, & C. Riederer (Eds.), *NeuroPsychopharmacotherapy* (pp. 1–10). Springer. https://doi.org/10.1007/978-3-030-62059-2_195

Benoit, M., Dygai, I., Migneco, O., Robert, P., Bertogliati, C., Darcourt, J., Benoliel, J., Aubin-Brunet, V., & Pringuey, D. (1999). Behavioral and psychological symptoms in Alzheimer's disease: Relation between apathy and regional cerebral perfusion. *Dementia and Geriatric Cognitive Disorders*, *10*(6), 511–517.

Bernal-Chico, A., Tepavcevic, V., Manterola, A., Utrilla, C., Matute, C., & Mato, S. (2023). Endocannabinoid signaling in brain diseases: Emerging relevance of glial cells. *Glia*, *71*(1), 103–126.

Bhunia, S., Kolishetti, N., Arias, A. Y., Vashist, A., & Nair, M. (2022). Cannabidiol for neurodegenerative disorders: A comprehensive review. *Frontiers in Pharmacology*, *13*, 989717.

Bilbao, A., & Spanagel, R. (2022). Medical cannabinoids: A pharmacology-based systematic review and meta-analysis for all relevant medical indications. *BMC Medicine*, *20*(1), 259.

Bisogno, T., Delton-Vandenbroucke, I., Milone, A., Lagarde, M., & Di Marzo, V. (1999). Biosynthesis and inactivation of N-arachidonoylethanolamine (Anandamide) and N-docosahexaenoylethanolamine in bovine retina. *Archives of Biochemistry and Biophysics, 370*(2), 300–307. https://doi.org/10.1006/abbi.1999.1410

Bouter, C., Ott, F. W., Günther, D., Weig, L., Schmitz-Peiffer, F., Rozyyeva, M., Beindorff, N., & Bouter, Y. (2023). Chronic exposure to a synthetic cannabinoid alters cerebral brain metabolism and causes long-lasting behavioral deficits in adult mice. *Journal of Neural Transmission*, *130*(8), 1013–1027.

Calsolaro, V., Femminella, G. D., Rogani, S., Esposito, S., Franchi, R., Okoye, C., Rengo, G., & Monzani, F. (2021). Behavioral and psychological symptoms in dementia (BPSD) and the use of antipsychotics. *Pharmaceuticals*, *14*(3), 246.

Cassano, T., Villani, R., Pace, L., Carbone, A., Bukke, V. N., Orkisz, S., Avolio, C., & Serviddio, G. (2020). From *Cannabis sativa* to cannabidiol: Promising therapeutic candidate for the treatment of neurodegenerative diseases. *Frontiers in Pharmacology*, *11*, 124.

Castaneto, M. S., Gorelick, D. A., Desrosiers, N. A., Hartman, R. L., Pirard, S., & Huestis, M. A. (2014). Synthetic cannabinoids: Epidemiology, pharmacodynamics, and clinical implications. *Drug and Alcohol Dependence*, *144*, 12–41.

Cerejeira, J., Lagarto, L., & Mukaetova-Ladinska, E. B. (2012). Behavioral and psychological symptoms of dementia. *Frontiers in Neurology*, *3*, 73.

Coleman, G. R., Bender, A., Hu, K., Sharpe, S. M., Schumann, A. W., Wang, Z., Bagavathiannan, M. V., Boyd, N. S., & Walsh, M. J. (2022). Weed detection to weed recognition: Reviewing 50 years of research to identify constraints and opportunities for large-scale cropping systems. *Weed Technology*, *36*(6), 741–757.

Compton, D. R., Aceto, M. D., Lowe, J., & Martin, B. R. (1996). In vivo characterization of a specific cannabinoid receptor antagonist (SR141716A): inhibition of delta 9-tetrahydrocannabinol-induced responses and apparent agonist activity. *Journal of Pharmacology and Experimental Therapeutics, 277*(2), 586–594.

Connor, J. P., Stjepanović, D., Le Foll, B., Hoch, E., Budney, A. J., & Hall, W. D. (2021). Cannabis use and cannabis use disorder. *Nature Reviews Disease Primers*, *7*(1), 16.

Consroe, P., Laguna, J., Allender, J., Snider, S., Stern, L., Sandyk, R., Kennedy, K., & Schram, K. (1991). Controlled clinical trial of cannabidiol in Huntington's disease. *Pharmacology, Biochemistry and Behavior*, *40*(3), 701–708. https://doi.org/10.1016/0091-3057(91)90386-G

Contino, M., Capparelli, E., Colabufo, N. A., & Bush, A. I. (2017). The CB2 cannabinoid system: A new strategy in neurodegenerative disorder and neuroinflammation. *Frontiers in Neuroscience*, *11*, 259998.

Cooper, Z. D., & Abrams, D. I. (2019). Considering abuse liability and neurocognitive effects of cannabis and cannabis-derived products when assessing analgesic efficacy: A comprehensive review of randomized-controlled studies. *American Journal of Drug and Alcohol Abuse*, *45*(6), 580–595.

Costa, A. C., Joaquim, H. P., Pedrazzi, J. F., Pain, A. d. O., Duque, G., & Aprahamian, I. (2022). Cannabinoids in late life Parkinson's disease and dementia: Biological pathways and clinical challenges. *Brain Sciences*, *12*(12), 1596.

Cowan, C. M., & Raymond, L. A. (2006). Selective neuronal degeneration in Huntington's disease. *Current Topics in Developmental Biology*, *75*, 25–71.

Cravanas Jr, B., & Frei, K. (2020). The effects of Cannabis on hallucinations in Parkinson's disease patients. *Journal of the Neurological Sciences*, *419*, 117206.

Cristino, L., Bisogno, T., & Di Marzo, V. (2020). Cannabinoids and the expanded endocannabinoid system in neurological disorders. *Nature Reviews Neurology*, *16*(1), 9–29.

Croxford, J. L. (2003). Therapeutic potential of cannabinoids in CNS disease. *CNS Drugs*, *17*(3), 179–202. https://doi.org/10.2165/00023210-200317030-00004

Curran, H. V., Freeman, T. P., Mokrysz, C., Lewis, D. A., Morgan, C. J., & Parsons, L. H. (2016). Keep off the grass? Cannabis, cognition and addiction. *Nature Reviews Neuroscience*, *17*(5), 293–306.

Curtis, A., Mitchell, I., Patel, S., Ives, N., & Rickards, H. (2009). A pilot study using nabilone for symptomatic treatment in Huntington's disease. *Movement Disorders*, *24*(15), 2254–2259. https://doi.org/10.1002/mds.22809

Curtis, A., & Rickards, H. (2006). Nabilone could treat chorea and irritability in Huntington's disease. *Journal of Neuropsychiatry and Clinical Neurosciences*, *18*(4), 553. https://doi.org/10.1176/jnp.2006.18.4.553

Daniju, Y., Faulkner, P., Brandt, K., & Allen, P. (2022). Prefrontal cortex and putamen grey matter alterations in cannabis and tobacco users. *Journal of Psychopharmacology*, *36*(12), 1315–1323.

Dash, R., Ali, M. C., Jahan, I., Munni, Y. A., Mitra, S., Hannan, M. A., Timalsina, B., Oktaviani, D. F., Choi, H. J., & Moon, I. S. (2021). Emerging potential of cannabidiol in reversing proteinopathies. *Ageing Research Reviews*, *65*, 101209.

Dearborn, J. T., Nelvagal, H. R., Rensing, N. R., Takahashi, K., Hughes, S. M., Wishart, T. M., Cooper, J. D., Wong, M., & Sands, M. S. (2022). Effects of chronic cannabidiol in a mouse model of naturally occurring neuroinflammation, neurodegeneration, and spontaneous seizures. *Scientific Reports*, *12*(1), 11286.

Deuel, L. M., & Seeberger, L. C. (2020). Complementary therapies in Parkinson disease: A review of acupuncture, Tai Chi, Qi Gong, Yoga, and Cannabis. *Neurotherapeutics*, *17*(4), 1434–1455.

Duncan, R. S., Riordan, S. M., Gernon, M. C., & Koulen, P. (2024). Cannabinoids and endocannabinoids as therapeutics for nervous system disorders: Preclinical models and clinical studies. *Neural Regeneration Research*, *19*(4), 788–799.

Eubanks, L. M., Rogers, C. J., Beuscher, Koob, G. F., Olson, A. J., Dickerson, T. J., & Janda, K. D. (2006). A molecular link between the active component of marijuana and Alzheimer's disease pathology. *Molecular Pharmaceutics*, *3*(6), 773–777.

Factor, S. A., & Friedman, J. H. (1997). The emergin role of clozapine in the treatment of movement disorders. *Movement Disorders: Official Journal of the Movement Disorder Society*, *12*(4), 483–496.

Fernandez, A. C., Coughlin, L., Solway, E. S., Singer, D. C., Kullgren, J. T., Kirch, M., & Malani, P. N. (2024). Prevalence and frequency of cannabis use among adults ages 50–80 in the United States. *Cannabis and Cannabinoid Research*, *9*(1), 59–64.

Fernández-Ruiz, J. (2009). The endocannabinoid system as a target for the treatment of motor dysfunction. *British Journal of Pharmacology, 156*(7), 1029–1040. https://doi.org/10.1111/j.1476-5381.2008.00088.x

Fernández-Ruiz, J., & Gonzáles, S. (2005). Cannabinoid control of motor function at the basal ganglia. *Handbook of Experimental Pharmacology*, (168), 479. https://doi.org/10.1007/3-540-26573-2-16

Fernández-Ruiz, J., Sagredo, O., Pazos, M. R., García, C., Pertwee, R., Mechoulam, R., & Martínez-Orgado, J. (2013). Cannabidiol for neurodegenerative disorders: Important new clinical applications for this phytocannabinoid? *British Journal of Clinical Pharmacology*, *75*(2), 323–333.

Figura, M., Koziorowski, D., & Sławek, J. (2022). Cannabis in Parkinson's disease—The patient's perspective versus clinical trials: A systematic literature review. *Neurologia i Neurochirurgia Polska*, *56*(1), 21–27.

Girgin, M., Isik, S., & Kantarci-Carsibasi, N. (2023). Proposing novel natural compounds against Alzheimer's disease targeting acetylcholinesterase. *Plos One*, *18*(4), e0284994.

Grant, J. E., Chamberlain, S. R., Schreiber, L., & Odlaug, B. L. (2012). Neuropsychological deficits associated with cannabis use in young adults. *Drug and Alcohol Dependence*, *121*(1–2), 159–162.

Han, B. H., & Palamar, J. J. (2020). Trends in cannabis use among older adults in the United States, 2015–2018. *JAMA Internal Medicine*, *180*(4), 609–611.

Harvey, B. S., Ohlsson, K. S., Mååg, J. L., Musgrave, I. F., & Smid, S. D. (2012). Contrasting protective effects of cannabinoids against oxidative stress and amyloid-β evoked neurotoxicity in vitro. *Neurotoxicology*, *33*(1), 138–146.

Hayes, M. T. (2019). Parkinson's disease and parkinsonism. *American Journal of Medicine*, *132*(7), 802–807.

Hazekamp, A., & Fischedick, J. (2012). Cannabis From cultivar to chemovar. *Drug Testing and Analysis*, *4*(7–8), 660–667.

Herkenham, M., Lynn, A. B., de Costa, B. R., & Richfield, E. K. (1991). Neuronal localization of cannabinoid receptors in the basal ganglia of the rat. *Brain Research, 547*(2), 267. https://doi.org/10.1016/0006-8993(91)90970-7

Herrmann, N., Ruthirakuhan, M., Gallagher, D., Verhoeff, N. P. L. G., Kiss, A., Black, S. E., & Lanctôt, K. L. (2019). Randomized placebo-controlled trial of nabilone for agitation in Alzheimer's disease. *American Journal of Geriatric Psychiatry*, *27*(11), 1161–1173. https://doi.org/10.1016/j.jagp.2019.05.002

Heumann, R., Moratalla, R., Herrero, M. T., Chakrabarty, K., Drucker-Colín, R., Garcia-Montes, J. R., Simola, N., & Morelli, M. (2014). Dyskinesia in Parkinson's disease: Mechanisms and current non-pharmacological interventions. *Journal of Neurochemistry*, *130*(4), 472–489.

Hillen, J. B., Soulsby, N., Alderman, C., & Caughey, G. E. (2019). Safety and effectiveness of cannabinoids for the treatment of neuropsychiatric symptoms in dementia: A systematic review. *Therapeutic Advances in Drug Safety, 10*, 2042098619846993.

Hodges, E. L., & Ashpole, N. M. (2019). Aging circadian rhythms and cannabinoids. *Neurobiology of Aging, 79*, 110–118.

Hudson, A., & Hudson, P. (2021). Risk factors for cannabis-related mental health harms in older adults: A review. *Clinical Gerontologist, 44*(1), 3–15.

Hunt, D. A., Keefe, J., Whitehead, T., & Littlefield, A. (2020). Understanding cannabis. *Journal for Nurse Practitioners, 16*(9), 645–649.

Hval, S.-A., Kristiansen, K. M., Lorentzen, B., Fagerstrøm, L., & Tanum, L. (2012). Treatment in nursing home versus hospital for patients with behavioral and psychological symptoms of dementia. *Journal of Gerontopsychology and Geriatric Psychiatry, 25*(2). https://doi.org/10.1024/1662-9647/a000058

Iuvone, T., Esposito, G., De Filippis, D., Scuderi, C., & Steardo, L. (2009). Cannabidiol: A promising drug for neurodegenerative disorders? *CNS Neuroscience & Therapeutics, 15*(1), 65–75. https://doi.org/10.1111/j.1755-5949.2008.00065.x

Janefjord, E., Mååg, J. L., Harvey, B. S., & Smid, S. D. (2014). Cannabinoid effects on β amyloid fibril and aggregate formation, neuronal and microglial-activated neurotoxicity in vitro. *Cellular and Molecular Neurobiology, 34*, 31–42.

Junior, N. C. F., dos- Santos-Pereira, M., Guimarães, F. S., & Del Bel, E. (2020). Cannabidiol and cannabinoid compounds as potential strategies for treating Parkinson's disease and L-DOPA-induced dyskinesia. *Neurotoxicity Research, 37*, 12–29.

Kalia, L. V., & Lang, A. E. (2015). Parkinson's disease. *Lancet, 386*(9996), 896–912.

Kano, M., Ohno-Shosaku, T., Hashimotodani, Y., Uchigashima, M., & Watanabe, M. (2009). Endocannabinoid-mediated control of synaptic transmission. *Physiological Reviews, 89*(1), 309–380. https://doi.org/10.1152/physrev.00019.2008

Karimi-Haghighi, S., Razavi, Y., Iezzi, D., Scheyer, A. F., Manzoni, O., & Haghparast, A. (2022). Cannabidiol and substance use disorder: Dream or reality. *Neuropharmacology, 207*, 108948.

Kaufmann, C. N., Kim, A., Miyoshi, M., & Han, B. H. (2022). Patterns of medical cannabis use among older adults from a cannabis dispensary in New York State. *Cannabis and Cannabinoid Research, 7*(2), 224–230.

Kaufmann, C. N., Malhotra, A., Yang, K. H., Han, B. H., Nafsu, R., Lifset, E. T., Nguyen, K., Sexton, M., & Moore, A. A. (2023). Cannabis use for sleep disturbance among older patients in a geriatrics clinic. *International Journal of Aging and Human Development, 97*(1), 3–17.

Kesner, A. J., & Lovinger, D. M. (2020). Cannabinoids, endocannabinoids and sleep. *Frontiers in Molecular Neuroscience, 13*, 125.

Khoury, R., Maliha, P., & Ibrahim, R. (2022). Cannabis use and misuse in older adults. *Clinics in Geriatric Medicine, 38*(1), 67–83.

Kim, J., Choi, J. Y., Seo, J., & Choi, I. S. (2021). Neuroprotective effect of cannabidiol against hydrogen peroxide in hippocampal neuron culture. *Cannabis and Cannabinoid Research, 6*(1), 40–47.

Kluger, B., Triolo, P., Jones, W., & Jankovic, J. (2015). The therapeutic potential of cannabinoids for movement disorders. *Movement Disorders, 30*(3), 313–327. https://doi.org/10.1002/mds.26142

Koppel, B. S., Brust, J. C. M., Fife, T., Bronstein, J., Youssof, S., Gronseth, G., & Gloss, D. (2014). Systematic review: Efficacy and safety of medical marijuana in selected neurologic disorders: Report of the Guideline Development Subcommittee of the American Academy of Neurology. *Neurology, 82*(17), 1556–1563. https://doi.org/10.1212/WNL.0000000000000363

Kroon, E., Kuhns, L., & Cousijn, J. (2021). The short-term and long-term effects of cannabis on cognition: Recent advances in the field. *Current Opinion in Psychology, 38*, 49–55. https://doi.org/10.1016/j.copsyc.2020.07.005

Kuharic, D. B., Markovic, D., Brkovic, T., Kegalj, M. J., Rubic, Z., Vukasovic, A. V., Jeroncic, A., & Puljak, L. (2021). Cannabinoids for the treatment of dementia. *Cochrane Database of Systematic Reviews, 9*(9), CD012820. https://doi-org.ezproxy.aub.edu.lb/10.1002/14651858.CD012820.pub2.

Lafaye, G., Karila, L., Blecha, L., & Benyamina, A. (2017). Cannabis, cannabinoids, and health. *Dialogues in Clinical Neuroscience, 19*(3), 309–316.

Landucci, E., Mazzantini, C., Lana, D., Davolio, P. L., Giovannini, M. G., & Pellegrini-Giampietro, D. E. (2021). Neuroprotective effects of cannabidiol but not Δ9-tetrahydrocannabinol in rat hippocampal slices exposed to oxygen-glucose deprivation: Studies with cannabis extracts and selected cannabinoids. *International Journal of Molecular Sciences, 22*(18), 9773.

Lastres-Becker, I., Cebeira, M., De Ceballos, M., Zeng, B. Y., Jenner, P., Ramos, J. A., & Fernandez-Ruiz, J. (2001). Increased cannabinoid CB1 receptor binding and activation of GTP-binding proteins in the basal ganglia of patients with Parkinson's syndrome and of MPTP-treated marmosets. *European Journal of Neuroscience, 14*(11), 1827–1832.

Ledent, C., Valverde, O., Cossu, G., Petitet, F., Aubert, J.-F., Beslot, F., Bohme, G. A., Imperato, A., Pedrazzini, T., & Roques, B. P. (1999). Unresponsiveness to cannabinoids and reduced addictive effects of opiates in CB1 receptor knockout mice. *Science, 283*(5400), 401–404.

Linares, I. M., Guimaraes, F. S., Eckeli, A., Crippa, A. C., Zuardi, A. W., Souza, J. D., Hallak, J. E., & Crippa, J. A. (2018). No acute effects of cannabidiol on the sleep-wake cycle of healthy subjects: A randomized, double-blind, placebo-controlled, crossover study. *Frontiers in Pharmacology, 9*, 315.

Liss, J., Seleri Assunção, S., Cummings, J., Atri, A., Geldmacher, D., Candela, S., Devanand, D., Fillit, H., Susman, J., & Mintzer, J. (2021). Practical recommendations for timely, accurate diagnosis of symptomatic Alzheimer's disease (MCI and dementia) in primary care: A review and synthesis. *Journal of Internal Medicine, 290*(2), 310–334.

Lovinger, D. M. (2008). Presynaptic modulation by endocannabinoids. *Handbook of Experimental Pharmacology*, (184), 435. https://doi.org/10.1007/978-3-540-74805-2-14

Løberg, E.-M., Nygård, M., Berle, J. Ø., Johnsen, E., Kroken, R. A., Jørgensen, H. A., & Hugdahl, K. (2012). An fMRI study of neuronal activation in schizophrenia patients with and without previous cannabis use. *Frontiers in Psychiatry, 3*, 94.

López-Sendón Moreno, J. L., García Caldentey, J., Trigo Cubillo, P., Ruiz Romero, C., García Ribas, G., Alonso Arias, M. A. A., García de Yébenes, M. J., Tolón, R. M., Galve-Roperh, I., Sagredo, O., Valdeolivas, S., Resel, E., Ortega-Gutierrez, S., García-Bermejo, M. L., Fernández Ruiz, J., Guzmán, M., & García de Yébenes Prous, J. (2016). A double-blind, randomized, cross-over, placebo-controlled, pilot trial with Sativex in Huntington's disease. *Journal of Neurology, 263*(7), 1390–1400. https://doi.org/10.1007/s00415-016-8145-9

Low, Z. X. B., Lee, X. R., Soga, T., Goh, B. H., Alex, D., & Kumari, Y. (2023). Cannabinoids: Emerging sleep modulator. *Biomedicine & Pharmacotherapy, 165*, 115102.

Magierski, R., Sobow, T., Schwertner, E., & Religa, D. (2020). Pharmacotherapy of behavioral and psychological symptoms of dementia: State of the art and future progress. *Frontiers in Pharmacology, 11*, 481680.

Matsuda, L. A., Lolait, S. J., Brownstein, M. J., Young, A. C., & Bonner, T. I. (1990). Structure of a cannabinoid receptor and functional expression of the cloned cDNA. *Nature (London), 346*(6284), 561–564. https://doi.org/10.1038/346561a0

Menendez-Miranda, I., Garcia-Alvarez, L., Garcia-Portilla, M. P., Gonzalez-Blanco, L., Saiz, P. A., & Bobes, J. (2019). History of lifetime cannabis use is associated with better cognition and worse real-world functioning in schizophrenia spectrum disorders. *European Addiction Research, 25*(3), 111–118.

Mintzer, J. E., Hoernig, K. S., & Mirski, D. F. (1998). Treatment of agitation in patients with dementia. *Clinics in Geriatric Medicine, 14*(1), 147–176.

Munro, S., Thomas, K. L., & Abu-Shaar, M. (1993). Molecular characterization of a peripheral receptor for cannabinoids. *Nature (London), 365*(6441), 61–65. https://doi.org/10.1038/365061a0

Oikonomou, P., & Jost, W. (2022). Randomized controlled trials on the use of cannabis-based medicines in movement disorders: A systematic review. *Journal of Neural Transmission, 129*(10), 1247–1256.

Ożarowski, M., Karpiński, T. M., Zielińska, A., Souto, E. B., & Wielgus, K. (2021). Cannabidiol in neurological and neoplastic diseases: Latest developments on the molecular mechanism of action. *International Journal of Molecular Sciences, 22*(9), 4294.

Pagano, C., Navarra, G., Coppola, L., Avilia, G., Bifulco, M., & Laezza, C. (2022). Cannabinoids: Therapeutic use in clinical practice. *International Journal of Molecular Sciences, 23*(6), 3344. https://doi.org/10.3390/ijms23063344

Patel, R. S., Kamil, S., Shah, M. R., Bhimanadham, N. N., & Imran, S. (2019). Pros and cons of marijuana in treatment of Parkinson's disease. *Cureus, 11*(6), e4813. https://doi.org/10.7759/cureus.4813.

Patel, V., Abu-Hijleh, F., Rigg, N., & Mishra, R. (2023). Cannabidiol protects striatal neurons by attenuating endoplasmic reticulum stress. *Cannabis and Cannabinoid Research, 8*(2), 299–308.

Peres, F. F., Lima, A. C., Hallak, J. E., Crippa, J. A., Silva, R. H., & Abílio, V. C. (2018). Cannabidiol as a promising strategy to treat and prevent movement disorders? *Frontiers in Pharmacology, 9*, 345007.

Perez, M., Benitez, S. U., Cartarozzi, L. P., Del Bel, E., Guimaraes, F. S., & Oliveira, A. L. (2013). Neuroprotection and reduction of glial reaction by cannabidiol treatment after sciatic nerve transection in neonatal rats. *European Journal of Neuroscience, 38*(10), 3424–3434.

Pidgeon, C., & Rickards, H. (2013). The pathophysiology and pharmacological treatment of Huntington disease. *Behavioural Neurology, 26*(4), 245–253. https://doi.org/10.1155/2013/705373

Poewe, W., Seppi, K., Tanner, C. M., Halliday, G. M., Brundin, P., Volkmann, J., Schrag, A.-E., & Lang, A. E. (2017). Parkinson disease. *Nature Reviews Disease Primers*, *3*(1), 1–21.

Quinn, H. R., Matsumoto, I., Callaghan, P. D., Long, L. E., Arnold, J. C., Gunasekaran, N., Thompson, M. R., Dawson, B., Mallet, P. E., & Kashem, M. A. (2008). Adolescent rats find repeated Δ9-THC less aversive than adult rats but display greater residual cognitive deficits and changes in hippocampal protein expression following exposure. *Neuropsychopharmacology*, *33*(5), 1113–1126.

Ramírez, B. G., Blázquez, C., Del Pulgar, T. G., Guzmán, M., & de Ceballos, M. L. (2005). Prevention of Alzheimer's disease pathology by cannabinoids: Neuroprotection mediated by blockade of microglial activation. *Journal of Neuroscience*, *25*(8), 1904–1913.

Reynolds, I. R., Fixen, D. R., Parnes, B. L., Lum, H. D., Shanbhag, P., Church, S., Linnebur, S. A., & Orosz, G. (2018). Characteristics and patterns of marijuana use in community-dwelling older adults. *Journal of the American Geriatrics Society*, *66*(11), 2167–2171.

Rocchetti, M., Crescini, A., Borgwardt, S., Caverzasi, E., Politi, P., Atakan, Z., & Fusar-Poli, P. (2013). Is cannabis neurotoxic for the healthy brain? A meta-analytical review of structural brain alterations in non-psychotic users. *Psychiatry and Clinical Neurosciences*, *67*(7), 483–492.

Rosas, H. D., Salat, D. H., Lee, S. Y., Zaleta, A. K., Hevelone, N., & Hersch, S. M. (2008). Complexity and heterogeneity: What drives the ever-changing brain in Huntington's disease? *Annals of the New York Academy of Sciences*, *1147*(1), 196–205.

Ross, C. A., & Tabrizi, S. J. (2011). Huntington's disease: From molecular pathogenesis to clinical treatment. *Lancet Neurology*, *10*(1), 83–98.

Sagar, K. A., & Gruber, S. A. (2018). Marijuana matters: Reviewing the impact of marijuana on cognition, brain structure and function, & exploring policy implications and barriers to research. *International Review of Psychiatry*, *30*(3), 251–267.

Sagredo, O., Ramos, J. A., Decio, A., Mechoulam, R., & Fernández-Ruiz, J. (2007). Cannabidiol reduced the striatal atrophy caused 3-nitropropionic acid in vivo by mechanisms independent of the activation of cannabinoid, vanilloid TRPV1 and adenosine A2A receptors. *European Journal of Neuroscience*, *26*(4), 843–851.

Sagredo, O., Ruth Pazos, M., Valdeolivas, S., & Fernández-Ruiz, J. (2012). Cannabinoids: Novel medicines for the treatment of Huntington's disease. *Recent Patents on CNS Drug Discovery (Discontinued)*, *7*(1), 41–48. https://www.ingentaconnect.com/content/ben/prn/2012/00000007/00000001/art00005;jsessionid=msx99ek1qj5y.x-ic-live-03

Savage, S. R. M. D. M. S., Romero-Sandoval, A. M. D., Schatman, M. P., Wallace, M. M. D., Fanciullo, G. M. D. M. S., McCarberg, B. M. D., & Ware, M. M. D. M. S. (2016). Cannabis in pain treatment: Clinical and research considerations. *Journal of Pain*, *17*(6), 654–668. https://doi.org/10.1016/j.jpain.2016.02.007

Scheyer, A. F., Laviolette, S. R., Pelissier, A.-L., & Manzoni, O. J. (2023). Cannabis in adolescence: Lasting cognitive alterations and underlying mechanisms. *Cannabis and Cannabinoid Research*, *8*(1), 12–23.

Schrot, R. J., & Hubbard, J. R. (2016). Cannabinoids: Medical implications. *Annals of Medicine*, *48*(3), 128–141.

Sehl, H., Terrett, G., Greenwood, L.-M., Kowalczyk, M., Thomson, H., Poudel, G., Manning, V., & Lorenzetti, V. (2021). Patterns of brain function associated with cannabis cue-reactivity in regular cannabis users: A systematic review of fMRI studies. *Psychopharmacology*, *238*, 2709–2728.

Shen, M., & Thayer, S. A. (1998). Cannabinoid receptor agonists protect cultured rat hippocampal neurons from excitotoxicity. *Molecular Pharmacology, 54*(3), 459–462. https://doi.org/10.1124/mol.54.3.459

Sherman, C., Ruthirakuhan, M., Vieira, D., Lanctôt, K. L., & Herrmann, N. (2018). Cannabinoids for the treatment of neuropsychiatric symptoms, pain and weight loss in dementia. *Current Opinion in Psychiatry*, *31*(2), 140–146. https://doi.org/10.1097/YCO.0000000000000399

Sieradzan, K., Fox, S., Hill, M., Dick, J., Crossman, A., & Brotchie, J. (2001). Cannabinoids reduce levodopa-induced dyskinesia in Parkinson's disease: A pilot study. *Neurology*, *57*(11), 2108–2111.

Singh, K., Bhushan, B., Chanchal, D. K., Sharma, S. K., Rani, K., Yadav, M. K., Porwal, P., Kumar, S., Sharma, A., Virmani, T., Kumar, G., & Noman, A. A. (2023). Emerging therapeutic potential of cannabidiol (CBD) in neurological disorders: A comprehensive review. *Behavioural Neurology*, *2023*, 1–17. https://doi.org/10.1155/2023/8825358

Small, E. (2015). Evolution and classification of *Cannabis sativa* (marijuana, hemp) in relation to human utilization. *Botanical Review*, *81*, 189–294.

Solomon, H. V., Greenstein, A. P., & DeLisi, L. E. (2021). Cannabis use in older adults: A perspective. *Harvard Review of Psychiatry*, *29*(3), 225–233.

Tampi, R. R., Young, J. J., & Tampi, D. J. (2018). Cannabinoids for the treatment of behavioral and psychological symptoms of dementia. *Neurodegenerative Disease Management*, *8*(4), 211–213.

Terry-McElrath, Y. M., Patrick, M. E., O'Malley, P. M., & Johnston, L. D. (2022). Self-reported perceived negative consequences of marijuana use among US young adult users, 2008–2019. *Addictive Behaviors*, *124*, 107098.

Tolosa, E., Garrido, A., Scholz, S. W., & Poewe, W. (2021). Challenges in the diagnosis of Parkinson's disease. *Lancet Neurology*, *20*(5), 385–397.

Tzavara, E. T., Li, D. L., Moutsimilli, L., Bisogno, T., Di Marzo, V., Phebus, L. A., Nomikos, G. G., & Giros, B. (2006). Endocannabinoids activate transient receptor potential vanilloid 1 receptors to reduce hyperdopaminergia-related hyperactivity: Therapeutic implications. *Biological Psychiatry* (1969), *59*(6), 508–515. https://doi.org/10.1016/j.biopsych.2005.08.019

Urbi, B., Corbett, J., Hughes, I., Owusu, M. A., Thorning, S., Broadley, S. A., Sabet, A., & Heshmat, S. (2022). Effects of cannabis in Parkinson's disease: A systematic review and meta-analysis. *Journal of Parkinson's Disease*, *12*(2), 495–508.

Urbi, B., Lee, Y., Hughes, I., Thorning, S., Broadley, S. A., Sabet, A., & Heshmat, S. (2022). Effects of cannabinoids in Parkinson's disease animal models: A systematic review and meta-analysis. *BMJ Open Science*, *6*(1), e100302. https://doi-org.ezproxy.aub.edu.lb/10.1136/bmjos-2022-100302.

Vallée, A., Vallée, J.-N., & Lecarpentier, Y. (2021). Potential role of cannabidiol in Parkinson's disease by targeting the WNT/β-catenin pathway, oxidative stress and inflammation. *Aging (Albany NY)*, *13*(7), 10796.

Van den Elsen, G. A., Ahmed, A. I., Verkes, R.-J., Feuth, T., van der Marck, M. A., & Rikkert, M. G. O. (2015). Tetrahydrocannabinol in behavioral disturbances in dementia: A crossover randomized controlled trial. *American Journal of Geriatric Psychiatry*, *23*(12), 1214–1224. https://doi.org/https://doi.org/10.1016/j.jagp.2015.07.011

Van Den Elsen, G. A., Ahmed, A. I., Verkes, R.-J., Kramers, C., Feuth, T., Rosenberg, P. B., Van Der Marck, M. A., & Olde Rikkert, M. G. (2015). Tetrahydrocannabinol for neuropsychiatric symptoms in dementia: A randomized controlled trial. *Neurology*, *84*(23), 2338–2346. https://doi.org/https://doi.org/10.1212/WNL.0000000000001675

van der Stelt, M., Fox, S. H., Hill, M., Crossman, A. R., Petrosino, S., Di Marzo, V., & Brotchie, J. M. (2005). A role for endocannabinoids in the generation of parkinsonism and levodopa-induced dyskinesia in MPTP-lesioned non-human primate models of Parkinson's disease. *FASEB Journal*, *19*(9), 1140. https://doi.org/10.1096/fj.04-3010fje

van Vliet, S. A., Vanwersch, R. A., Jongsma, M. J., Olivier, B., & Philippens, I. H. (2008). Therapeutic effects of Δ9-THC and modafinil in a marmoset Parkinson model. *European Neuropsychopharmacology*, *18*(5), 383–389.

Volicer, L., Stelly, M., Morris, J., McLaughlin, J., & Volicer, B. J. (1997). Effects of Dronabinol on anorexia and disturbed behavior in patients with Alzheimer's disease. *International Journal of Geriatric Psychiatry*, *12*(9), 913–919. https://doi.org/10.1002/(SICI)1099-1166(199709)12:9<913::AID-GPS663>3.0.CO;2-D

Volkow, N. D., Wang, G.-J., Telang, F., Fowler, J. S., Alexoff, D., Logan, J., Jayne, M., Wong, C., & Tomasi, D. (2014). Decreased dopamine brain reactivity in marijuana abusers is associated with negative emotionality and addiction severity. *Proceedings of the National Academy of Sciences*, *111*(30), E3149–E3156.

Walther, S., & Halpern, M. (2010). Cannabinoids and dementia: A review of clinical and preclinical data. *Pharmaceuticals*, *3*(8), 2689–2708.

Wang, L., Wu, X., Yang, G., Hu, N., Zhao, Z., Zhao, L., & Li, S. (2022). Cannabidiol alleviates the damage to dopaminergic neurons in 1-methyl-4-phenyl-1, 2, 3, 6-tetrahydropyridine-induced Parkinson's disease mice via regulating neuronal apoptosis and neuroinflammation. *Neuroscience*, *498*, 64–72.

Wanke, M., Sandberg, S., Macit, R., & Gülerce, H. (2022). Culture matters! Changes in the global landscape of cannabis. In *Drugs: Education, prevention and policy* (Vol. 29, pp. 317–322). Taylor & Francis. https://doi-org.ezproxy.aub.edu.lb/10.1080/09687637.2022.2091301

Watt, D. (1990). Huntington's disease: A disorder of families. By SE Folstein. (p. 251) Johns Hopkins University Press: Baltimore. 1989. *Psychological Medicine*, *20*(3), 728–731.

Weier, M., & Hall, W. (2017). The use of cannabinoids in treating dementia. *Current Neurology and Neuroscience Reports*, *17*, 1–9.

Weinstein, G., & Sznitman, S. R. (2020). The implications of late-life cannabis use on brain health: A mapping review and implications for future research. *Ageing Research Reviews*, *59*, 101041.

Weintraub, D., Aarsland, D., Chaudhuri, K. R., Dobkin, R. D., Leentjens, A. F., Rodriguez-Violante, M., & Schrag, A. (2022). The neuropsychiatry of Parkinson's disease: Advances and challenges. *Lancet Neurology*, *21*(1), 89–102.

Whiting, P. F., Wolff, R. F., Deshpande, S., Di Nisio, M., Duffy, S., Hernandez, A. V., Keurentjes, J. C., Lang, S., Misso, K., & Ryder, S. (2015). Cannabinoids for medical use: A systematic review and meta-analysis. *JAMA*, *313*(24), 2456–2473.

Wieghorst, A., Roessler, K. K., Hendricks, O., & Andersen, T. E. (2022). The effect of medical cannabis on cognitive functions: A systematic review. *Systematic Reviews*, *11*(1), 210.

Wissel, J., Haydn, T., Müller, J., Brenneis, C., Berger, T., Poewe, W., & Schelosky, L. D. J. J. o. n. (2006). Low dose treatment with the synthetic cannabinoid Nabilone significantly reduces spasticity-related pain: A double-blind placebo-controlled cross-over trial. *Journal of Neurology*, *253*(10), 1337–1341.

Wolfe, D., Corace, K., Butler, C., Rice, D., Skidmore, B., Patel, Y., Thayaparan, P., Michaud, A., Hamel, C., & Smith, A. (2023). Impacts of medical and non-medical cannabis on the health of older adults: Findings from a scoping review of the literature. *Plos One*, *18*(2), e0281826.

Xu, J., Tang, Y., Xie, M., Bie, B., Wu, J., Yang, H., Foss, J. F., Yang, B., Rosenquist, R. W., & Naguib, M. (2016). Activation of cannabinoid receptor 2 attenuates mechanical allodynia and neuroinflammatory responses in a chronic post-ischemic pain model of complex regional pain syndrome type I in rats. *European Journal of Neuroscience*, *44*(12), 3046–3055.

Yanes, J. A., McKinnell, Z. E., Reid, M. A., Busler, J. N., Michel, J. S., Pangelinan, M. M., Sutherland, M. T., Younger, J. W., Gonzalez, R., & Robinson, J. L. (2019). Effects of cannabinoid administration for pain: A meta-analysis and meta-regression. *Experimental and Clinical Psychopharmacology*, *27*(4), 370–382. https://doi.org/10.1037/pha0000281

8 Opioid Use and Neurodegenerative Disorders

Abhishek Ghosh, Ankita Chattopadhyay, and Shinjini Choudhury

INTRODUCTION

Background

Globally, opioids are the second most widely used illicit drugs. Around 60 million of the world's population used opioids in 2022 (UNODC, World Drug Report 2024). Opioids, whether prescription medications such as oxycodone and hydrocodone or illicit drugs like heroin, are highly addictive substances that can cause both physical dependence and severe health consequences. The global burden of opioid use disorder (OUD) is immense, contributing to substantial morbidity and mortality (Volkow & McLellan, 2016). In 2017, nearly 40 million people globally were living with opioid dependence, and over 100,000 died from opioid overdose (Degenhardt et al., 2019). The death toll has mounted sharply, particularly in North America, with United States (US) opioid-related deaths increasing 24-fold since 2010 (UNODC, 2024). US data from 2010 to 2020 revealed nearly 15 non-fatal overdoses for every fatal one (Casillas et al., 2024). Survivors of overdoses may experience cognitive impairments due to brain hypoxia (Mahoney III et al., 2023). Prescription opioid use is more prevalent in high-income countries than in low- and middle-income countries (Jayawardana et al., 2021), with a minority progressing to OUD. Despite opioid agonist maintenance treatment being the most effective public health strategy for OUD, it is available to only 18 of every 100 people who inject drugs (Colledge-Frisby et al., 2023), with likely similar or lower coverage for non-injection opioid use.

The enormous magnitude of the problem, the aging population, increasing non-fatal overdoses and consequent cognitive impairments, increasing use of prescription opioids, and unavailability of evidence-based treatment have led to a substantial untreated population with OUD. This has drawn increased attention to the long-term health impacts of opioid use, particularly its potential role in neurodegenerative disorders (NDDs) (Dublin et al., 2015).

NDDs, such as Alzheimer's disease (AD), Parkinson's disease, and amyotrophic lateral sclerosis (ALS), are characterized by progressive neuronal loss and cognitive decline (Ross & Poirier, 2004). These neurological conditions are among the leading causes of disability and death worldwide, with their prevalence expected to rise due to aging populations (Masters et al., 2015). Understanding the factors contributing to NDD development and progression is critical for developing effective prevention and treatment strategies.

Recent research suggests a potential link between opioid use and an increased risk of NDDs (Tanguturi & Streicher, 2023). Opioids may exacerbate oxidative stress, inflammation, and other pathological processes, contributing to neurodegeneration (Herlinger & Lingford-Hughes, 2022). Chronic, high-dose opioid use may further increase these risks, potentially accelerating cognitive decline and neurodegeneration (Rahman et al., 2021). Given the widespread use of opioids, this association raises significant concerns for public health.

DOI: 10.1201/9781003478485-8

Significance and Aim

The potential link between opioid use and NDDs has significant public health, clinical, and policy implications. Understanding this connection is vital for implementing comprehensive screening of cognitive function in individuals with OUD, particularly those with long-term or high-dose exposure (Kosten & George, 2002). It emphasizes the need for treatment strategies addressing both opioid dependence and neurodegeneration risks. Additionally, public health interventions are necessary to reduce opioid misuse and educate the public on long-term cognitive effects. This chapter explores the epidemiology, pathophysiology, and clinical implications of opioid-related neurodegeneration (Volkow & McLellan, 2016).

EPIDEMIOLOGY OF OPIOID USE IN PATIENTS WITH NEURODEGENERATIVE DISORDERS

The prevalence of opioid use in patients with NDDs, particularly dementia, has been the focus of several studies. A study examining the use of opioids in home-living patients with dementia and nursing home residents (NHRs) in Denmark provided substantial insights into opioid prescription patterns. The study found that NHRs used opioids most frequently, with 41% of NHRs using opioids compared to 27.5% of home-living patients with dementia and 16.9% of home-living patients without dementia. Buprenorphine and fentanyl were particularly common among NHRs and home-living patients with dementia, highlighting the extensive use of potent opioids in these settings. This high prevalence of opioid use, especially among those with dementia, underscores the need for further investigation into the appropriateness and safety of opioid prescribing in these populations (Jensen-Dahm et al., 2015).

A systematic review assessing the prevalence of opioid use in nursing homes over the last decade highlighted significant usage of these drugs among residents, including those with dementia. The review, which analyzed data from studies conducted between 2011 and 2021, found that the annual prevalence of opioid use among NHRs ranged from 6.3% to 50%, with a median prevalence of 22.9%. Among residents with dementia, the prevalence ranged from 10% to 39.6%, suggesting that opioid use remains a common practice in this setting despite concerns about potential safety risks (Sant et al., 2024). Another systematic review compared opioid use in individuals with cognitive impairments to those who are cognitively intact. The analysis revealed that, in many cases, individuals with cognitive impairments, including dementia, were less likely to receive opioids for pain management compared to their cognitively intact counterparts. This undertreatment was evident despite the high prevalence of pain in this population, indicating either caution exercised by the practitioners or a significant gap in the management of pain among those with neurodegenerative conditions (Griffioen et al., 2017).

Longitudinal data also provide insights into trends over time. A study examining opioid use by dementia severity in long-term care NHRs from 2011 to 2017 observed a general decline in opioid use over the years. This decline was consistent across all levels of dementia severity, with the odds of receiving any, prolonged, or high-dose opioids decreasing with increasing dementia severity. For instance, residents with severe dementia had significantly lower odds of receiving any opioid treatment compared to those without dementia. This trend raises concerns about the possibility of suboptimal pain management in patients with severe dementia, who may have difficulties communicating their pain effectively (Mehta et al., 2021).

Geographical variation in opioid use among older adults with dementia has also been observed, highlighting differences in clinical practices and treatment approaches based on location. A nationwide survey conducted in Denmark revealed significant disparities in opioid use among older adult patients with dementia across different municipalities. The study found that 32.5% of older adults with dementia were treated with opioids, compared to 16.9% of those without dementia. Interestingly, for home-living patients with dementia, the use of opioids varied widely, with a four-fold difference between municipalities (ranging from 9.4% to 36.8%) compared to a 1.6-fold difference in older adults without dementia (ranging from 12.7% to 20.2%). Among NHRs, this variation was less pronounced yet still significant (Jensen-Dahm et al., 2019).

MECHANISMS OF INTERACTION BETWEEN OPIOID USE AND NEURODEGENERATION

Neurotoxic Effects

Opioid use, especially heroin, has been linked to accelerated neurodegeneration. A study demonstrated that heroin misuse increases the accumulation of tau proteins in the brain, a hallmark of several NDDs, including AD. This accumulation of hyperphosphorylated tau disrupts normal neuronal function, leading to cognitive impairments and neuronal death. The opioid-induced neurotoxicity is thought to result from chronic exposure, which may disrupt normal tau phosphorylation processes, exacerbating neuronal vulnerability and promoting neurodegenerative changes (Kovacs et al., 2015).

Oxidative Stress and DNA Damage

Oxidative stress is another mechanism by which opioids may contribute to neurodegeneration. Chronic opioid exposure induces oxidative stress, damaging DNA in crucial brain regions like the prefrontal cortex and nucleus accumbens. This DNA damage results in impaired neuronal function and eventual cell death, which can manifest as cognitive decline and increased vulnerability to NDDs. The oxidative stress induced by opioids results from an imbalance between the formation of reactive oxygen species and the brain's antioxidant defenses, causing cellular damage that exacerbates neurodegenerative processes (Y. Wang et al., 2023).

Circadian Rhythm Disruption

Opioid use has been associated with circadian rhythm disruption, which plays a role in neurodegeneration. Circadian rhythms regulate various physiological processes, including sleep-wake cycles, hormone release, and metabolic functions. Chronic opioid use can alter circadian rhythms, leading to disruptions in sleep patterns, hormonal imbalances, and metabolic disturbances. These changes can further contribute to neuronal damage and the progression of NDDs. Disrupted circadian rhythms have been linked to cognitive impairments and an increased risk of developing conditions such as AD (Puig et al., 2023).

Neuroimmune Modulation

Opioids can modulate the neuroimmune system, affecting microglia and astrocytes, the primary immune cells in the central nervous system. Chronic opioid use activates microglia and astrocytes, causing the release of pro-inflammatory cytokines and other neurotoxic substances that contribute to neuronal damage. The activation of the neuroimmune system by opioids is implicated in the development and progression of NDDs. For instance, neuroinflammation and microglial activation are key features of AD, and opioid-induced neuroimmune modulation may exacerbate these pathological processes (Cahill & Taylor, 2017).

Impairment of Neurogenesis and Synaptic Plasticity

Opioids can impair neurogenesis and synaptic plasticity, critical processes for maintaining healthy brain function and cognitive abilities. Neurogenesis, the generation of new neurons, is essential for learning and memory. Opioids can inhibit the proliferation of neural progenitor cells, reducing neurogenesis in regions like the hippocampus, which is crucial for memory formation. Additionally, opioids can alter synaptic plasticity, the capacity of synapses to strengthen or weaken over time in response to activity levels. This impairment in synaptic plasticity can lead to cognitive deficits and exacerbate the progression of NDDs (Cahill & Taylor, 2017).

The interaction between opioid use and neurodegeneration is multifaceted, involving neurotoxic effects, inflammation, oxidative stress, circadian rhythm disruption, impaired neurogenesis, synaptic plasticity, and neuroimmune modulation. Chronic opioid exposure can accelerate neurodegenerative processes, leading to cognitive decline and an elevated risk of developing NDDs. Understanding these mechanisms is fundamental to developing strategies to mitigate the neurodegenerative effects of opioid use and improve outcomes for individuals with opioid dependence.

OPIOID USE AND NEURODEGENERATION IN ANIMAL MODELS

Animal studies using rat models and monkeys have investigated neurodegenerative changes associated with opioid use or dependence. Quintanilla et al. (2023) found that UChB rats with a preference for oral morphine (15 mg/kg/day) exhibited neurodegenerative changes, such as demyelination, axonal degeneration, and decreased glutamate transporter (GLT-1) levels in the prefrontal cortex. Glial neuroinflammation and oxidative stress were also observed in the striatum, similar to changes seen with chronic alcohol and nicotine use. Other studies demonstrated opioid-induced activation of apoptotic pathways, neuroinflammation, and oxidative stress in the prefrontal cortex, neostriatum, and nucleus accumbens (Gibson et al., 2022; Wang et al., 2012; El-Hage et al., 2008). Opioids have been shown to activate toll-like receptor 4 (TLR-4), leading to nuclear factor kappa B (NF-κB) activation and the release of pro-inflammatory mediators like tumor necrosis factor alpha (TNF-α), interleukin 1β (IL-1β), and interleukin-6 (IL-6) in microglia and astrocytes (Wang et al., 2012). Morphine exposure increased CC-chemokine ligand 5 (CCL5) production, triggering pro-inflammatory glial activation and neuronal death (El-Hage et al., 2008). Additionally, opioids induce GLT-1 downregulation in astrocytes, causing excitotoxicity (Mao et al., 2002). Chronic opioid exposure affects oligodendrocytes, causing myelin loss and axonal degeneration (Fan et al., 2018; Quintanilla et al., 2023). In Cynomolgus monkeys, long-term oxycodone self-administration reduced gray matter volume and elevated neurodegenerative biomarkers (Kumar et al., 2021).

CLINICAL EVIDENCE OF THE RELATIONSHIP BETWEEN OPIOID USE AND NEURODEGENERATIVE DISORDERS

The link between long-term opioid use and the risk of dementia has gained attention due to rising opioid prescriptions for chronic pain management. Recent studies suggest a significant association between extended opioid use and the development of dementia.

Dementia Risk with Prescription Opioid Use

A study by Sun et al. (2023) investigated the impact of long-term opioid use on dementia risk in patients with chronic pain. In a propensity score-matched cohort of 41,636 patients, the study compared opioid users (≥180 mg defined daily doses per year) with non-opioid users. The results indicated an adjusted hazard ratio (aHR) of 1.86 for dementia risk among opioid users, with an incidence of 44.09 per 10,000 person-years. This finding highlights the cognitive risks associated with prolonged opioid use, suggesting the need for careful monitoring in chronic pain management.

Similarly, a large-scale cohort study in South Korea involving 1,261,682 patients, including 21,800 regular opioid users, found a 15% increased risk of dementia in opioid users, specifically for AD but not vascular dementia (Oh & Song, 2024). This suggests that public health strategies should focus on early dementia detection in opioid users with chronic non-cancer pain.

Data from the United Kingdom Biobank further supports this association. A prospective cohort study showed an 18% higher dementia risk in regular opioid users compared to non-opioid analgesic users over 15 years (Gao et al., 2024). A nested case-control analysis in the same study demonstrated a dose–response relationship, with more opioid prescriptions linked to an increased dementia risk (Gao et al., 2024). Neuroimaging revealed lower gray matter and hippocampal volumes and higher white matter hyperintensities in regular opioid users, suggesting structural brain changes contributing to dementia risk.

Contrasting these findings, a Finnish nationwide nested case-control study within the medication use and Alzheimer disease study cohort found no significant association between opioid use and AD (Taipale et al., 2018). The study also indicated that neither the duration of opioid use nor high cumulative doses were associated with an increased AD risk. While acute cognitive effects of opioids, such as sedation and delirium, are known, this study suggests these may be reversible and not necessarily lead to AD development. The inconsistency of opioids' impact on cholinergic systems could explain why this study found no association between long-term opioid use and AD.

Further research is needed to explore the impact of opioid use on other forms of dementia, such as vascular dementia, and to investigate the long-term cognitive effects of opioid use in different populations.

OPIOID USE/USE DISORDER AND NEUROCOGNITIVE EFFECTS

Profiles of Cognitive Dysfunction

Neurocognitive impairments are reported in individuals with chronic opioid use. An early study by Hill and Mikhael (1979) on persons with heroin dependence found impairments in tactual performance and tapping tests for memory and location but not in the Category Test, which assesses abstract reasoning. Since the Category Test is linked to frontal lobe function, it was inferred that the frontal lobe might be less impacted by opioid use.

In a comparative study, individuals with chronic heroin use showed deficits in verbal fluency, pattern and spatial recognition, spatial working memory, and attentional set-shifting tasks. Unlike those with amphetamine dependence, who struggled with the extra-dimensional shift task, persons with heroin use had difficulty with the intra-dimensional shift component. They also demonstrated less improvement in sequence generation tasks and more perseverative behavior. Both groups were impaired in visual pattern recognition and sensitive to temporal lobe dysfunction. Persons with heroin dependence showed significant impairment in decision-making speed, executive dysfunction in the visuospatial domain, and general learning deficits (Ornstein et al., 2000).

Chronic opioid use is generally linked to neuropsychological impairments, varying with the duration of use. A meta-analysis found that individuals with chronic heroin use have moderate deficits in verbal fluency, inhibitory control, planning, verbal working memory, and decision-making (Baldacchino et al., 2012).

Persons with opioid use have also been reported to develop toxic leukoencephalopathy, presenting with symptoms such as inattention, personality changes, ataxia, dementia, coma, and even death, although the pathological mechanisms remain unclear (Achamallah et al., 2019). MRI findings indicate subcortical white matter changes, correlating with deep white matter degeneration and axonal damage observed postmortem (Tolomeo et al., 2021).

The type of opioid, mode of administration, and co-use of cannabis can influence cognitive dysfunctions in opioid-dependent individuals compared to healthy controls. Ghosh and colleagues (2023) compared neurocognitive functions in individuals dependent on heroin, pharmaceutical opioids, and natural opioids with controls. The study found that pharmaceutical opioid-dependent individuals exhibited the most severe cognitive impairments, particularly in executive function, abstraction, psychomotor speed, and working memory. Heroin-dependent individuals had significant impairments in verbal and visual working memory but were less affected than pharmaceutical opioid users. Those dependent on natural opioids showed relatively spared cognitive functions but still exhibited some executive dysfunctions.

Additionally, Ghosh et al. (2023) compared inhalational and intravenous heroin use, finding that inhalational users had more significant difficulties with set-shifting abilities and executive functions, while intravenous users showed more errors in verbal working memory.

The co-use of opioids and cannabis was also investigated, with findings indicating that individuals who use both substances exhibit greater neurocognitive impairments than those using opioids alone. This co-use exacerbates deficits in executive function, working memory, and cognitive flexibility (Ghosh et al., 2024), highlighting the compounded negative impact on cognition.

Effects of Opioid Agonist Maintenance Treatment

Deficits in visual memory, working memory, impulsivity, and compulsivity among chronic heroin users have been noted in former users who are abstinent as well as those on methadone maintenance therapy (MMT) (Baldacchino et al., 2015). Interestingly, patients on MMT are often found to have additional cognitive impairments, probably due to general cognitive slowing caused by methadone rather than distinct deficits (Baldacchino et al., 2012). However, buprenorphine-assisted treatment appears to have a less detrimental effect on neuropsychological assessments, with some studies even reporting better decision-making in patients on buprenorphine compared to MMT (Tolomeo et al., 2021). A recent study by Ghosh, Shaktan, Nehra, et al., 2024) investigated cognitive changes in patients with OUD after 6 months of buprenorphine treatment. The study recruited 498 patients and assessed their cognitive functions using tasks like the Wisconsin Card Sorting Test and Trail-Making Tests. Results indicated significant improvement in 25 out of 32 cognitive test parameters. The authors concluded that buprenorphine treatment is associated with cognitive function improvements. Positive cognitive effects of buprenorphine have been demonstrated in a study on EcoHIV-infected mice with established chronic neurocognitive impairment; it was seen that treatment with buprenorphine causes a decrease in brain inflammatory monocytes and a restoration of dendritic damage in the cortex and hippocampus, thus reversing the disease (Murphy et al., 2022). Buprenorphine shows promise as a therapeutic option for alleviating neurocognitive deficits in people with HIV, particularly those with comorbid OUD, but further research is needed to better understand its potential role.

THERAPEUTIC APPROACHES AND POTENTIAL TREATMENTS

Adjunctive Therapies

Memantine, an NMDA receptor antagonist typically used for moderate-to-severe AD, is now being explored for opioid addiction due to its neuroprotective properties. Its neuroprotective effects may mitigate complications arising from chronic neuronal inflammation in opioid addiction. Neurotrophic factors, such as brain-derived neurotrophic factor (BDNF), are critical for neuronal survival and maintenance and are implicated in opioid-induced neuronal degeneration and addiction (Elias et al., 2019).

In a study by Chen et al. (2012), low-dose memantine (0.2–1 mg/kg, subcutaneously) significantly reduced morphine-induced place preference in rats. Morphine treatment for 6 days elevated serum cytokine levels (IL-1β, IL-6) and increased their mRNA expression in the nucleus accumbens and medial prefrontal cortex. BDNF levels decreased in both serum and brain regions associated with addiction. Pretreatment with low-dose memantine reduced morphine-induced increases in cytokines and enhanced BDNF levels in both serum and brain, indicating its anti-inflammatory and neurotrophic effects.

Chang et al. (2015) conducted a double-blind, randomized, placebo-controlled study to assess low-dose memantine's effects on cognitive performance in individuals with opioid dependence undergoing MMT. Participants were randomly assigned to memantine or placebo, with their performance compared to healthy controls over 12 weeks. Both groups showed cognitive improvements, particularly in executive function and sustained attention, but the memantine group exhibited more significant enhancements. Improvements were noted on cognitive tests like the Wisconsin Card Sorting Test and Continuous Performance Test, indicating better executive skills and attention. The memantine group also had a lower dropout rate, suggesting better treatment adherence, and no adverse events were reported. This study suggests that low-dose memantine may offer cognitive benefits and improve clinical outcomes in opioid dependence, warranting further research in this area.

Neurocognitive Rehabilitation

Neurocognitive rehabilitation is being tried as an essential strategy to positively impact the cognitive impairments in attention, working memory, and executive functions, potentially leading to improved relapse rates among individuals with OUD (Rezapour et al., 2019). In a randomized controlled trial assessing the effectiveness of add-on cognitive rehabilitation treatment (CRT) for individuals with OUD participating in an MMT program, the CRT group showed significantly better performance in tests measuring learning, cognitive flexibility, processing speed, working memory, and memory span compared to the control group (Rezapour et al., 2019).

Neurocognitive rehabilitation for substance use disorders is available through various delivery methods, including traditional "paper-and-pencil" testing and computer or web-based technologies in both inpatient and outpatient settings. However, limited research explicitly addresses its effectiveness in preventing neurodegenerative processes (Rezapour et al., 2016).

PUBLIC HEALTH AND POLICY IMPLICATIONS

Prescription Opioid Use in Older Adults with Neurodegenerative Disorders

The use of opioids in older adults, particularly those residing in nursing homes, has become a critical public health issue. Over the past decade, there has been a noticeable increase in the prescription of opioids to older adults, driven in part by the need to manage chronic pain associated with aging and degenerative conditions (Rekatsina et al., 2022). However, this trend has raised significant concerns about the safety and appropriateness of opioid use in this vulnerable population.

Several studies and systematic reviews investigating the prevalence of opioid use in nursing homes reveal that a substantial proportion of residents are prescribed these medications, often for extended periods (Griffioen et al., 2017; Rekatsina et al., 2022). A review of opioid use in nursing homes over the last decade indicates that while opioids are commonly prescribed for pain management, there is often a lack of comprehensive pain assessment and monitoring, leading to potential overuse or misuse (Gustafsson et al., 2021). This is particularly concerning given the heightened sensitivity of elderly individuals to the adverse effects of opioids, including respiratory depression, falls, fractures, and cognitive impairment (Gnjidic et al., 2012).

The safety of opioid use in older adults is further complicated by the high prevalence of polypharmacy in this demographic. Many NHRs take multiple medications for various chronic conditions, increasing the risk of drug interactions and adverse effects (Davies & O'Mahony, 2015). For example, the concomitant use of opioids with benzodiazepines, often prescribed for anxiety or insomnia, can significantly elevate the risk of sedation, respiratory depression, and overdose (Huang et al., 2012). Despite these risks, the use of opioids in nursing homes remains prevalent, highlighting the need for better guidelines and monitoring practices.

Public health policies should focus on improving pain management strategies that are tailored to the needs of older adults to address these concerns. This includes promoting the use of non-opioid analgesics and non-pharmacological interventions as first-line treatments for chronic pain

(American Geriatrics Society Panel on Pharmacological Management of Persistent Pain in Older Persons, 2009). Additionally, there should be stringent monitoring and evaluation of opioid prescriptions in nursing homes, with regular reviews of residents' medication regimens to identify potential risks and ensure that opioid use is appropriate and safe.

Impact of COVID-19

The COVID-19 pandemic has had major effects on public health systems worldwide, including the management of chronic pain and the use of psychotropic and opioid medications in older adult populations. The pandemic has exacerbated existing challenges in healthcare delivery, particularly in assisted-living facilities and nursing homes, where residents are at increased risk of COVID-19 and its complications (Grabowski & Mor, 2020). During the pandemic, there has been a notable shift in the patterns of psychotropic, anticonvulsant, and opioid use among residents of assisted-living facilities. A study examining the use of these medications before and during the COVID-19 pandemic found significant increases in their prescription, likely driven by the heightened anxiety, isolation, and physical pain experienced by residents during lockdowns and restrictions (Maxwell et al., 2022). The increased use of opioids during this period raises concerns about the potential for misuse, dependency, and adverse effects, particularly in settings where healthcare resources were already strained.

Moving forward, public health policies should incorporate lessons learned from the COVID-19 pandemic to enhance the resilience of healthcare systems in nursing homes and assisted living facilities. This includes developing guidelines for the safe use of opioids among older adults, particularly during emergencies, and ensuring that healthcare providers have the resources and training needed to manage complex cases effectively. Additionally, there should be a renewed focus on integrating telemedicine and other innovative solutions into care practices to support ongoing monitoring and consultation for pain management, even when in-person visits are limited.

CONCLUSION

While research has begun to explore the relationship between opioid use and NDDs, significant gaps remain. One major gap is the lack of longitudinal studies that can definitively establish causality between opioid use and NDDs. Many existing studies are cross-sectional or retrospective, limiting their ability to establish a temporal relationship between opioid exposure and neurodegeneration.

Additionally, the focus of current research has been uneven across different NDDs. While there is some understanding of the impact of opioids on Alzheimer's and Parkinson's diseases, other conditions, such as Huntington's disease and ALS, have received less attention. Future studies should explore the effects of opioid use across a broader spectrum of NDDs to provide a more comprehensive understanding.

Another critical gap is the limited research on specific opioids' roles in neurodegeneration. Most studies have generalized opioid use without distinguishing between different types, such as prescription opioids versus illicit opioids like heroin. The pharmacological differences between these substances suggest that they may have distinct effects on neurodegeneration, warranting more targeted research.

Moreover, the role of comorbidities such as chronic pain, psychiatric disorders, and polypharmacy in modulating the relationship between opioid use and NDDs is not well understood. These co-occurring conditions can independently affect neurodegenerative risk, complicating efforts to isolate the effects of opioids. Future research should focus on disentangling these complex interactions to clarify the unique contribution of opioids to neurodegenerative outcomes.

Addressing these gaps requires innovative research methodologies. Advanced neuroimaging methods, such as functional MRI, diffusion tensor imaging, and positron emission tomography, can provide detailed insights into the neurobiological changes associated with opioid use and their potential link to neurodegenerative processes. Longitudinal imaging studies following individuals with OUD over time could be particularly valuable in identifying early biomarkers of neurodegeneration.

Integrating multi-omics approaches—such as genomics, epigenomics, proteomics, and metabolomics—offers a more comprehensive understanding of the molecular mechanisms underlying opioid effects on neurodegeneration. By combining data from different omics layers, researchers can identify novel biomarkers and therapeutic targets that traditional single-approach studies might miss. This systems biology approach could help clarify the role of genetic and environmental factors in modulating neurodegenerative risk associated with opioid use.

Artificial intelligence (AI) and machine learning (ML) also present promising avenues for future research. These technologies can analyze large datasets to identify patterns and associations between opioid use and neurodegenerative outcomes that may not be apparent through conventional methods. AI and ML can also predict individuals at higher risk for NDDs based on their opioid use patterns and other health factors.

Finally, developing advanced animal models that more accurately replicate human neurodegenerative conditions is crucial for preclinical research. Current models often fail to capture the complexity of human NDDs, particularly in the context of chronic opioid exposure. Creating transgenic or humanized models that better mimic human conditions could lead to more relevant findings and accelerate the translation of research into clinical applications.

Acknowledgment: Blessy B. George, MA, MPhil, Scientist-I, Drug Deaddiction and Treatment Centre, Department of Psychiatry, Postgraduate Institute of Medical Education and Research, Chandigarh, India.

REFERENCES

Achamallah, N., Wright, R. S., & Fried, J. (2019). Chasing the wrong dragon: A new presentation of heroin-induced toxic leukoencephalopathy mimicking anoxic brain injury. *Journal of the Intensive Care Society*, *20*(1), 80–85.

American Geriatrics Society Panel on Pharmacological Management of Persistent Pain in Older Persons (2009). Pharmacological management of persistent pain in older persons. *Journal of the American Geriatrics Society, 57*(8), 1331–1346. https://doi.org/10.1111/j.1532-5415.2009.02376.x

Baldacchino, A., Balfour, D. J. K., & Matthews, K. (2015). Impulsivity and opioid drugs: Differential effects of heroin, methadone and prescribed analgesic medication. *Psychological Medicine*, *45*(6), 1167–1179.

Baldacchino, A., Balfour, D. J. K., Passetti, F., Humphris, G., & Matthews, K. (2012). Neuropsychological consequences of chronic opioid use: A quantitative review and meta-analysis. *Neuroscience & Biobehavioral Reviews*, *36*(9), 2056–2068.

Cahill, C. M., & Taylor, A. M. W. (2017). Neuroinflammation—A co-occurring phenomenon linking chronic pain and opioid dependence. *Current Opinion in Behavioral Sciences*, *13*, 171–177.

Casillas, S. M., Pickens, C. M., Tanz, L. J., & Vivolo-Kantor, A. M. (2024). Estimating the ratio of fatal to non-fatal overdoses involving all drugs, all opioids, synthetic opioids, heroin or stimulants, USA, 2010–2020. *Injury Prevention*, *30*(2), 114–124.

Chang, Y.-H., Chen, S.-L., Lee, S.-Y., Chen, P. S., Wang, T.-Y., Lee, I. H., Chen, K. C., Yang, Y. K., Hong, J.-S., & Lu, R.-B. (2015). Low-dose add-on memantine treatment may improve cognitive performance and self-reported health conditions in opioid-dependent patients undergoing methadone-maintenance-therapy. *Scientific Reports*, *5*(1), 9708.

Chen, S.-L., Tao, P.-L., Chu, C.-H., Chen, S.-H., Wu, H.-E., Tseng, L. F., Hong, J.-S., & Lu, R.-B. (2012). Low-dose memantine attenuated morphine addictive behavior through its anti-inflammation and neurotrophic effects in rats. *Journal of Neuroimmune Pharmacology*, *7*, 444–453.

Colledge-Frisby, S., Ottaviano, S., Webb, P., Grebely, J., Wheeler, A., Cunningham, E. B., Hajarizadeh, B., Leung, J., Peacock, A., Vickerman, P., & Others. (2023). Global coverage of interventions to prevent and manage drug-related harms among people who inject drugs: A systematic review. *Lancet Global Health*, *11*(5), e673–e683.

Davies, E. A., & O'mahony, M. S. (2015). Adverse drug reactions in special populations—The elderly. *British Journal of Clinical Pharmacology*, *80*(4), 796–807.

Degenhardt, L., Grebely, J., Stone, J., Hickman, M., Vickerman, P., Marshall, B. D. L., Bruneau, J., Altice, F. L., Henderson, G., Rahimi-Movaghar, A., & others. (2019). Global patterns of opioid use and dependence: Harms to populations, interventions, and future action. *Lancet*, *394*(10208), 1560–1579.

Dublin, S., Walker, R. L., Gray, S. L., Hubbard, R. A., Anderson, M. L., Yu, O., Crane, P. K., & Larson, E. B. (2015). Prescription opioids and risk of dementia or cognitive decline: A prospective cohort study. *Journal of the American Geriatrics Society*, *63*(8), 1519–1526.

El-Hage, N., Bruce-Keller, A. J., Knapp, P. E., & Hauser, K. F. (2008). CCL5/RANTES gene deletion attenuates opioid-induced increases in glial CCL2/MCP-1 immunoreactivity and activation in HIV-1 Tat-exposed mice. *Journal of Neuroimmune Pharmacology*, *3*, 275–285.

Elias, A. M., Pepin, M. J., & Brown, J. N. (2019). Adjunctive memantine for opioid use disorder treatment: A systematic review. *Journal of Substance Abuse Treatment*, *107*, 38–43.

Fan, R., Schrott, L. M., Arnold, T., Snelling, S., Rao, M., Graham, D., Cornelius, A., & Korneeva, N. L. (2018). Chronic oxycodone induces axonal degeneration in rat brain. *BMC Neuroscience*, *19*, 1–20.

Gao, Y., Su, B., Ding, L., Qureshi, D., Hong, S., Wei, J., Zeng, C., Lei, G., & Xie, J. (2024). Association of regular opioid use with incident dementia and neuroimaging markers of brain health in chronic pain patients: Analysis of UK Biobank. *American Journal of Geriatric Psychiatry, 32*(9), 1154–1165.

Ghosh, A., Shaktan, A., Nehra, R., Basu, D., Verma, A., Rana, D. K., Modi, M., & Ahuja, C. K. (2023). Heroin use and neuropsychological impairments: Comparison of intravenous and inhalational use. *Psychopharmacology*, *240*(4), 909–920.

Ghosh, A., Shaktan, A., Nehra, R., Verma, A., Rana, D. K., Ahuja, C. K., Modi, M., Singh, P., & Basu, D. (2024). Neurocognitive functions after 6-month buprenorphine (Naloxone)-based opioid agonist maintenance treatment: A controlled prospective study. *Journal of Clinical Psychopharmacology*, *44*(2), 141–150.

Ghosh, A., Shaktan, A., Verma, A., Nehra, R., Basu, D., Rana, D. K., Ahuja, C. K., Modi, M., & Singh, P. (2023). Neurocognitive functions in opioid dependence: Comparisons between opioid categories. *Journal of Drug Issues*, 00220426231179208.

Ghosh, A., Shaktan, A., Verma, A., Basu, D., Rana, D. K., Nehra, R., Ahuja, C. K., Modi, M., & Singh, P. (2024). Neurocognitive dysfunctions in people with concurrent cannabis use and opioid dependence: A cross-sectional, controlled study. *Journal of Psychoactive Drugs, 57*(1), 71–83.

Gibson, J. M., Chu, T., Zeng, W., Wethall, A. C., Kong, M., Mellen, N., Devlin Phinney, L. A., & Cai, J. (2022). Perinatal methadone exposure attenuates myelination and induces oligodendrocyte apoptosis in neonatal rat brain. *Experimental Biology and Medicine*, *247*(12), 1067–1079.

Gnjidic, D., Hilmer, S. N., Blyth, F. M., Naganathan, V., Cumming, R. G., Handelsman, D. J., McLachlan, A. J., Abernethy, D. R., Banks, E., & Le Couteur, D. G. (2012). High-risk prescribing and incidence of frailty among older community-dwelling men. *Clinical Pharmacology & Therapeutics*, *91*(3), 521–528.

Grabowski, D. C., & Mor, V. (2020). Nursing home care in crisis in the wake of COVID-19. *JAMA*, *324*(1), 23–24.

Griffioen, C., G Willems, E., S Husebo, B., & P Achterberg, W. (2017). Prevalence of the use of opioids for treatment of pain in persons with a cognitive impairment compared with cognitively intact persons: A systematic review. *Current Alzheimer Research*, *14*(5), 512–522.

Gustafsson, M., Lövheim, H., & Sjölander, M. (2021). Pharmacological pain treatment in 2012 and 2017 among older people with major neurocognitive disorder. *Drugs & Aging*, *38*, 1017–1023.

Herlinger, K., & Lingford-Hughes, A. (2022). Opioid use disorder and the brain: A clinical perspective. *Addiction*, *117*(2), 495–505.

Hill, S. Y., & Mikhael, M. A. (1979). Computerized transaxial tomographic and neuropsychological evaluations in chronic alcoholics and heroin abusers. *American Journal of Psychiatry*, *136*(4B), 598–602.

Huang, A. R., Mallet, L., Rochefort, C. M., Eguale, T., Buckeridge, D. L., & Tamblyn, R. (2012). Medication-related falls in the elderly: Causative factors and preventive strategies. *Drugs & Aging*, *29*, 359–376.

Jayawardana, S., Forman, R., Johnston-Webber, C., Campbell, A., Berterame, S., de Joncheere, C., Aitken, M., & Mossialos, E. (2021). Global consumption of prescription opioid analgesics between 2009–2019: A country-level observational study. *EClinicalMedicine*, *42*, 101198.

Jensen-Dahm, C., Gasse, C., Astrup, A., Mortensen, P. B., & Waldemar, G. (2015). Frequent use of opioids in patients with dementia and nursing home residents: A study of the entire elderly population of Denmark. *Alzheimer's & Dementia*, *11*(6), 691–699.

Jensen-Dahm, C., Zakarias, J. K., Gasse, C., & Waldemar, G. (2019). Geographical variation in opioid use in elderly patients with dementia: A nationwide study. *Journal of Alzheimer's Disease*, *70*(4), 1209–1216.

Kosten, T. R., & George, T. P. (2002). The neurobiology of opioid dependence: Implications for treatment. *Science & Practice Perspectives*, *1*(1), 13.

Kovacs, G. G., Horvath, M. C., Majtenyi, K., Lutz, M. I., Hurd, Y. L., & Keller, E. (2015). Heroin abuse exaggerates age-related deposition of hyperphosphorylated tau and p62-positive inclusions. *Neurobiology of Aging*, *36*(11), 3100–3107.

Kumar, A., Kim, S., Su, Y., Sharma, M., Kumar, P., Singh, S., Lee, J., Furdui, C. M., Singh, R., Hsu, F.-C., & others. (2021). Brain cell-derived exosomes in plasma serve as neurodegeneration biomarkers in male Cynomolgus monkeys self-administrating oxycodone. *EBioMedicine*, *63*.

Mahoney III, J. J., Winstanley, E. L., Castillo, F., Luba, R., Marton, J., Alschuler, D. M., Liu, Y., & Comer, S. D. (2023). A pilot study investigating cognitive impairment associated with opioid overdose. *Drug and Alcohol Dependence*, *247*, 109865.

Mao, J., Sung, B., Ji, R.-R., & Lim, G. (2002). Chronic morphine induces downregulation of spinal glutamate transporters: Implications in morphine tolerance and abnormal pain sensitivity. *Journal of Neuroscience*, *22*(18), 8312–8323.

Masters, C. L., Bateman, R., Blennow, K., Rowe, C. C., Sperling, R. A., & Cummings, J. L. (2015). Alzheimer's disease. *Nature Reviews Disease Primers*, *1*(1), 1–18.

Maxwell, C. J., Campitelli, M. A., Cotton, C. A., Hogan, D. B., Iaboni, A., Gruneir, A., Evans, C., & Bronskill, S. E. (2022). Greater opioid use among nursing home residents in Ontario, Canada during the first 2 waves of the COVID-19 pandemic. *Journal of the American Medical Directors Association*, *23*(6), 936–941.

Mehta, H. B., Kuo, Y.-F., Raji, M., Li, S., Westra, J., & Goodwin, J. S. (2021). Time trends in opioid use by dementia severity in long-term care nursing home residents. *Journal of the American Medical Directors Association*, *22*(1), 124–131.

Murphy, A. J., Kelschenbach, J., He, H., Chao, W., Kim, B.-H., Volsky, D. J., & Berman, J. W. (2022). Buprenorphine reverses neurocognitive impairment in EcoHIV infected mice: A potential therapy for HIV-NCI. *Frontiers in Immunology*, *13*, 1004985.

Oh, T. K., & Song, I.-A. (2024). Impact of prescribed opioid use on development of dementia among patients with chronic non-cancer pain. *Scientific Reports*, *14*(1), 3313.

Ornstein, T. J., Iddon, J. L., Baldacchino, A. M., Sahakian, B. J., London, M., Everitt, B. J., & Robbins, T. W. (2000). Profiles of cognitive dysfunction in chronic amphetamine and heroin abusers. *Neuropsychopharmacology*, *23*(2), 113–126.

Puig, S., Xue, X., Salisbury, R., Shelton, M. A., Kim, S.-M., Hildebrand, M. A., Glausier, J. R., Freyberg, Z., Tseng, G. C., Yocum, A. K., & others. (2023). Uncovering circadian rhythm disruptions of synaptic proteome signaling in prefrontal cortex and nucleus accumbens associated with opioid use disorder. *BioRxiv*, 2004–2023.

Quintanilla, M. E., Morales, P., Santapau, D., Ávila, A., Ponce, C., Berrios-Cárcamo, P., Olivares, B., Gallardo, J., Ezquer, M., Herrera-Marschitz, M., & Others. (2023). Chronic voluntary morphine intake is associated with changes in brain structures involved in drug dependence in a rat model of polydrug use. *International Journal of Molecular Sciences*, *24*(23), 17081.

Rahman, S., Rahman, Z. I., Ronan, P. J., Lutfy, K., & Bell, R. L. (2021). Adolescent opioid abuse: Role of glial and neuroimmune mechanisms. *International Review of Neurobiology*, *161*, 147–165.

Rekatsina, M., Paladini, A., Viswanath, O., Urits, I., Myrcik, D., Pergolizzi, J., Breve, F., & Varrassi, G. (2022). Opioids in the elderly patients with cognitive impairment: A narrative review. *Pain and Therapy*, *11*(2), 381–394.

Rezapour, T., DeVito, E. E., Sofuoglu, M., & Ekhtiari, H. (2016). Perspectives on neurocognitive rehabilitation as an adjunct treatment for addictive disorders: From cognitive improvement to relapse prevention. *Progress in Brain Research*, *224*, 345–369.

Rezapour, T., Hatami, J., Farhoudian, A., Sofuoglu, M., Noroozi, A., Daneshmand, R., Samiei, A., & Ekhtiari, H. (2019). Cognitive rehabilitation for individuals with opioid use disorder: A randomized controlled trial. *Neuropsychological Rehabilitation*, *29*(9), 1273–1285. https://doi.org/10.1080/09602011.2018.1439012.

Ross, C. A., & Poirier, M. A. (2004). Protein aggregation and neurodegenerative disease. *Nature Medicine*, *10*(Suppl 7), S10–S17.

Sant, A. M., Portelli, S., Ballard, C., Bezzina-Xuereb, M., Scerri, C., & Sultana, J. (2024). Prevalence of opioid use in nursing homes over the last decade: A systematic literature review. *Journal of Pharmacy Technology*, *40*(3), 123–133.

Sun, M., Chen, W.-M., Wu, S.-Y., & Zhang, J. (2023). Long-term opioid use and dementia risk in patients with chronic pain. *Journal of the American Medical Directors Association*, *24*(9), 1420–1426.

Taipale, H., Hamina, A., Lampela, P., Tanskanen, A., Tiihonen, J., Karttunen, N., Tolppanen, A.-M., & Hartikainen, S. (2018). Is Alzheimer's disease associated with previous opioid use? *Pain Medicine*, *19*(11), 2115–2121.

Tanguturi, P., & Streicher, J. M. (2023). The role of opioid receptors in modulating Alzheimer's disease. *Frontiers in Pharmacology, 14*, 1056402.

Tolomeo, S., Steele, J. D., Ekhtiari, H., & Baldacchino, A. (2021). Chronic heroin use disorder and the brain: Current evidence and future implications. *Progress in Neuro-Psychopharmacology and Biological Psychiatry, 111*, 110148.

United Nations Office on Drugs and Crime. (2024). *World drug report 2024*. United Nations. https://www.unodc.org/unodc/en/data-and-analysis/world-drug-report-2024.html

Volkow, N. D., & McLellan, A. T. (2016). Opioid abuse in chronic pain—Misconceptions and mitigation strategies. *New England Journal of Medicine, 374*(13), 1253–1263.

Wang, X., Loram, L. C., Ramos, K., de Jesus, A. J., Thomas, J., Cheng, K., Reddy, A., Somogyi, A. A., Hutchinson, M. R., Watkins, L. R., & Others. (2012). Morphine activates neuroinflammation in a manner parallel to endotoxin. *Proceedings of the National Academy of Sciences, 109*(16), 6325–6330.

Wang, Y., Singh, A., Li, G., Yue, S., Hertel, K., & Wang, Z.-J. (2023). Opioid induces increased DNA damage in prefrontal cortex and nucleus accumbens. *Pharmacology Biochemistry and Behavior, 224*, 173535.

9 Sedative, Hypnotic, and Anxiolytic Use and Neurodegenerative Disorders

Kirtika Aggarwal, Awanish Mishra, Ramdas Ransing, and Samer El Hayek

INTRODUCTION

Overview of Sedatives, Hypnotics, and Anxiolytics

Sedatives and hypnotics constitute a class of drugs that suppress the central nervous system (CNS) to promote sedation or induce sleep, respectively (Ciraulo, 2014). Sedatives are drugs that calm and subdue excitement without inducing sleep, although they may cause drowsiness. Sedation is characterized by decreased responsiveness to external stimuli and reduced alertness, ideation, and motor activity. Hypnotics are drugs that induce and maintain sleep, similar to normal arousable sleep. Anxiolytics are a class of medications used to treat anxiety symptoms, insomnia, and various anxiety and sleep disorders (Stahl, 2021).

Overview of Neurodegenerative Disorders

NDDs are characterized by progressive neuron structure and function degeneration, leading to various clinical manifestations depending on the affected brain regions (Lamptey et al., 2022). The most well-known NDDs include Alzheimer's disease (AD), Parkinson's disease (PD), Huntington's disease (HD), and amyotrophic lateral sclerosis (ALS) (Lamptey et al., 2022). NDDs constitute a significant public health concern due to their high prevalence and severe impact on individuals and their families. The pathophysiology of NDDs involves a complex interplay of genetic and environmental factors. Abnormal protein aggregation is a common feature in many of these disorders. For instance, amyloid-beta (Aβ) plaques and tau tangles are hallmarks of AD, while alpha-synuclein aggregates are found in PD (Selkoe & Hardy, 2016). Current therapeutic strategies primarily focus on symptom management rather than disease modification.

Importance of Looking at the Interaction between Sedative, Hypnotic, and Anxiolytic Use and Neurodegenerative Disorders

Several modifiable risk factors have been identified for NDDs, including low educational status, excessive alcohol consumption, smoking, depression, anxiety, insomnia, hearing impairment, traumatic brain injury, hypertension, diabetes, obesity, and physical inactivity (Livingston et al., 2020). Sedatives, hypnotics, and anxiolytics are commonly used to treat insomnia, anxiety, and depression, among other symptoms. The evidence also suggests that these drugs have an impact on NDD development and its progression (Bellou et al., 2017). Published literature reveals that benzodiazepine (BZD) prescriptions are received by 10% of women and 6% of men aged 65–80 years (Markota et al., 2016). Moreover, there is a steady increase in both the number and quantity of BZD

 DOI: 10.1201/9781003478485-9

prescriptions in this population group, raising concerns that BZDs may be a significant modifiable risk factor for NDDs, influencing both their development and prognosis. Despite advances, the mechanisms by which these drugs interact with NDDs remain poorly understood, hindering the development of effective treatments and prevention strategies. This chapter's exploration of the published literature aims to identify specific intervention targets to mitigate risks and alternative management options for patients with NDDs who require sedatives, hypnotics, or anxiolytics.

MECHANISMS OF ACTION OF SEDATIVES, HYPNOTICS, AND ANXIOLYTICS

Sedatives, hypnotics, and anxiolytics are broadly classified into two categories based on their mechanisms of action (Table 9.1) (Rang et al., 2011).

Drugs Acting through γ-Aminobutyric Acid Receptor Chloride-Channel Complex

- *BZDs*: BZDs enhance presynaptic and postsynaptic inhibition by acting on a specific BZD receptor, which is part of the γ-aminobutyric acid (GABA)-A receptor chloride-channel complex (Griffin et al., 2013). This BZD receptor boosts GABAergic transmission by increasing the frequency of chloride channels opening in response to GABA, mainly when GABA is present at submaximal concentrations, and by enhancing GABA's ability to bind to its receptor (Griffin et al., 2013; Tripathi, 2018). Notably, BZDs do not directly increase chloride (Cl^-) conductance; they facilitate GABA-induced Cl^- flux without exhibiting intrinsic GABA-mimetic properties.
- *Non-BZD drugs*: Non-BZD hypnotics, also known as Z-drugs, are chemically different from BZDs. They exhibit selective affinity for GABA receptors containing the α1 subunit, resulting primarily in hypnotic and amnesic effects while showing reduced antianxiety, muscle relaxant, and anticonvulsant properties (Bouchette et al., 2024). Non-BZD drugs have a lower abuse potential and are now often preferred over BZDs for the treatment of insomnia due to their shorter duration of action (Bouchette et al., 2024; Tripathi, 2018).
- *Barbiturates*: Barbiturates exert their primary mechanism of action at the GABA-A receptor Cl^- channel complex, enhancing GABAergic inhibition by prolonging the duration of Cl^- channel opening—unlike BZDs, which increase the frequency of Cl^- channel openings (Skibiski et al., 2024). Barbiturates bind to a site on the GABA receptor near, but distinct from, the BZD binding site (located on the α and β subunits), thereby enhancing the receptor's calming effect (Skibiski et al., 2024; Tripathi, 2018).

TABLE 9.1
Classification of Sedatives, Hypnotics, and Anxiolytics

Drugs Acting through GABA-A Receptor Chloride-Channel Complex

- *BZDs*: Lorazepam, diazepam, alprazolam, estazolam, flurazepam, nitrazepam, oxazepam, quazepam, temazepam, triazolam
- *Non-BZD drugs*: Zolpidem, zaleplon, zopiclone, eszopicione
- *Barbiturates*: Amobarbital, butabarbital, pentobarbital, secobarbital

Drugs Acting through Other Mechanisms

- *Melatonin receptor agonists*: Ramelteon, tasimelteon
- *Dual orexin receptor antagonists*: Suvorexant
- *Tricyclic antidepressants*: Doxepin
- *Antihistamines*: Diphenhydramine

Drugs Acting through Other Mechanisms

- *Melatonin receptor agonists*: Melatonin, produced by the pineal gland at night, regulates the sleep-wake cycle and maintains the circadian rhythm. Melatonin receptor agonists activate the melatonin MT1 and MT2 receptors, promoting sleep (Giovannitti & Cooke, 2017).
- *Dual orexin receptor antagonists (DORAs)*: Orexin-A and orexin-B are neuropeptides produced by the hypothalamus that stimulate wakefulness. DORAs suppress wakefulness by blocking the action of orexin neuropeptides (OX1 and OX2 receptors), promoting both rapid eye movement (REM) and non-REM sleep (Markwell & McLellan, 2019).
- *Tricyclic antidepressants*: Doxepin, a sedating tricyclic antidepressant, is used for the treatment of insomnia. It exerts its sedating effects by acting as an antagonist at the histamine H1 receptor (Robertson et al., 2019).

CHARACTERISTICS OF MAJOR NEURODEGENERATIVE DISORDERS

Neurodegeneration, characterized by the gradual and irreversible damage and loss of neurons and axons within the CNS, is the hallmark pathological feature underlying acute and chronic NDDs (Amor et al., 2010). NDDs include AD, PD, ALS, multiple sclerosis, systemic lupus erythematous, neurotropic viral infections, stroke, paraneoplastic disorders, and traumatic brain injury, among others (Amor et al., 2010). Despite varying triggering events, a common feature of NDDs is chronic immune activation. A brief description of the most common NDDs is provided below.

Alzheimer's Disease

AD is the most common cause of dementia among older adults, leading to memory loss, cognitive decline, and, ultimately, the inability to carry out daily activities (Vejandla et al., 2024). The progression of AD is characterized by the deposition of two proteins: Aβ and tau proteins (Bayer & Wirths, 2010). The formation of Aβ-containing plaques in the brain, along with neurofibrillary tangles composed of hyperphosphorylated tau, is the hallmark feature of AD (Butterfield et al., 2002). Hippocampal plaque formation disrupts neural circuits, impairing the conversion of short-term to long-term memory (Spires-Jones & Hyman, 2014). Extensive neuronal loss, faulty synaptic connections, and damage to neurotransmitter systems affect various brain functions in patients with AD. Initially, AD presents with targeted memory deficits, followed by impaired hippocampal and medial temporal lobe functions, such as declarative and episodic memory. As the disease progresses, it impacts executive functions, judgment, and problem-solving skills, leading to additional clinical symptoms (Akhondzadeh & Noroozian, 2002).

Parkinson's Disease

PD primarily affects movement, leading to tremors, rigidity, and bradykinesia, which significantly impair the quality of life (DeMaagd & Philip, 2015). PD is characterized by tremors, muscle stiffness, an unsteady walk, and difficulties with balance and coordination. Risk factors for PD include age, excessive caffeine intake, smoking, and exposure to environmental toxins (Kouli et al., 2018; Priyadarshi et al., 2001). The pathophysiology of PD involves the loss of pigmentation in the locus coeruleus and substantia nigra pars compacta (Bartels & Leenders, 2009). Degeneration of cells in the substantia nigra affects the nigrostriatal pathway, resulting in decreased dopamine concentrations in the striatum and subsequent motor dysfunction (Bartels & Leenders, 2009). Non-motor symptoms in PD arise from cell loss in multiple brain areas, such as the nucleus basalis of Meynert, raphe nuclei, and locus coeruleus, which disrupts diverse neural circuits (Chaudhuri & Schapira, 2009). The progression of PD involves multiple pathological processes, including α-synuclein misfolding, mitochondrial dysfunction, disrupted protein degradation (via the ubiquitin-proteasome and autophagy-lysosome systems), and sustained neuroinflammation (Bartels & Leenders, 2009; Schulz-Schaeffer, 2010).

Other Neurodegenerative Disorders

ALS, also known as motor neuron disease or Lou Gehrig's disease, is an NDD that affects the nerve cells and spinal cord, leading to muscle weakness and paralysis (Morris, 2015). Mutations in genes such as C9orf72, TARDBP, SOD1, and FUS, which regulate DNA repair, homeostasis, mitochondrial function, and glial cell activity, contribute to the degenerative changes observed in patients with ALS (Hardiman et al., 2017). In this regard, genetic mutations play a crucial role in several NDDs. For example, HD is caused by a mutation in the HTT gene, leading to an abnormal repetition of the CAG nucleotide sequence (Walker, 2007). Other NDDs, such as multiple sclerosis, systemic lupus erythematosus, neurotropic viral infections, stroke, paraneoplastic disorders, and traumatic brain injury, can also cause neurodegeneration in various brain regions, resulting in motor, non-motor, and neuropsychiatric symptoms (Kobeissy & Moshourab, 2015).

SEDATIVES AND NEURODEGENERATION: IMPACT ON COGNITION, MOTOR SYMPTOMS, AND SLEEP

The use of sedatives, particularly BZDs and related drugs, has been studied for its potential association with an increased risk of dementia. Several epidemiological studies suggest a link between long-term use of these medications and an elevated risk of neurodegeneration (Ferreira et al., 2022). The impact of sedatives, hypnotics, and anxiolytics on brain function in older adults and neurodegeneration can be categorized into three main areas.

Cognitive Effects

One of the primary concerns is that BZDs, due to their sedative properties and effects on the CNS, may contribute to cognitive impairment over time. Ettcheto et al. (2020) provide a comprehensive summary of clinical studies evaluating the relationship between this category of medications and adults with and without AD, mainly focusing on the relationship between their use and loss of cognitive function (Ettcheto et al., 2020). Margolis et al. (2021) found that exposure to anticholinergics and sedatives in adults aged 65 years and above who were at risk of AD was associated with significant subjective cognitive decline. A study by Billioti de Gage et al. (2014) found that BZD use was associated with an increased risk of AD, with the risk growing as the duration of BZD use increased. Specifically, the study reported that individuals who used BZDs for more than 6 months had a significantly higher risk of developing AD compared to non-users (Billioti de Gage et al., 2014). A retrospective cohort study of 268,170 subjects aged 50 years and above evaluated the association between the use of sedatives and the risk of subsequent AD. Subjects exposed to sedatives showed a significantly higher risk of AD than those who were not. The risk was particularly elevated among subjects exposed to BZDs or zolpidem, with the increased risk occurring regardless of the dose or half-life of the BZDs (Lee et al., 2018). Along the same lines, long-term BZD use in older adults was associated with an increased risk of cognitive decline (Paterniti et al., 2002). However, these findings have not been consistent throughout the literature. A recent systematic review and meta-analysis of 35 studies examining the association between sedative-hypnotic medications and the risk of dementia found no overall association with an increased risk of dementia, except for BZDs. However, this observed association was no longer significant after a sensitivity analysis, notably when studies that did not adjust for age were excluded and after adjustment for protopathic bias (AlDawsari et al., 2022). A narrative review by Yoon and colleagues (2024) concluded that the relationship between hypnotics and dementia remains complex and controversial, with conflicting and inconclusive evidence (Yoon et al., 2024) highlighting the need for future research.

Motor Function Effects

Sedative, hypnotic, and anxiolytic use is also known to affect coordination, increase the risk of falls, and exacerbate Parkinson's symptoms (Woolcott, 2009). As NDDs progress, patients may become more susceptible to the adverse effects of these medications. A cohort study of 2,961 subjects found that the incidence of PD was higher among zolpidem users compared to BZD users, with the risk increasing in proportion to the dose of zolpidem. The authors concluded that zolpidem use might unmask preclinical PD (H.-C. Huang et al., 2015). Alternatively, two studies found that BZD use in patients with PD did not alter motor function (Gaztanaga et al., 2021) or impact antiparkinsonian drug treatment (Van De Vijver et al., 2002).

Sleep Disturbances

The impact of sedative use on NDDs is influenced by various confounding factors, including sleep disorders, underlying psychiatric conditions, physical comorbidities, and lifestyle factors (Ju et al., 2014). Psychiatric comorbidities are associated with faster cognitive decline and reduced quality of life, and using sedatives to manage these conditions can create a negative feedback loop, worsening both mental and cognitive symptoms (AlDawsari et al., 2022; Lee et al., 2018). Similarly, patients with NDDs often experience sleep disturbances, which lead to increased sedative use. Paradoxically, sedatives can impair sleep quality (Butterfield et al., 2002). This, in turn, can further accelerate NDD progression by promoting Aβ plaque deposition and altering brain structure and function, resulting in memory impairment, reduced cognitive flexibility, and other deficits. This creates a vicious cycle that exacerbates both the sleep disturbance and its treatment (AlDawsari et al., 2022; Lee et al., 2018).

As cognitive decline advances in patients with NDDs, neuropsychiatric symptoms tend to worsen, leading to increased use of sedatives despite more pronounced side effects, such as confusion, dizziness, and an increased risk of falls (Ju et al., 2014). Therefore, it is crucial for clinicians to carefully manage and monitor medication use in patients with or at risk of these conditions. Minimizing the use of sedatives, particularly in older adults, and exploring alternative treatments for neuropsychiatric symptoms are essential steps in reducing the potential adverse effects on cognitive health.

MECHANISMS OF SEDATIVES, HYPNOTICS, AND ANXIOLYTICS IN THE PROGRESSION OF NEURODEGENERATIVE DISORDERS

The mechanisms underlying the associations between sedative use and NDDs are not fully understood, but several hypotheses have been proposed.

Alteration of GABAergic Neurotransmission

BZDs may contribute to neurodegeneration by affecting GABAergic neurotransmission, which is crucial for maintaining neural plasticity and cognitive function. By enhancing GABA, the brain's chief inhibitory neurotransmitter, chronic BZD use can lead to changes in the GABAergic system, including downregulation of GABA receptors and reduced responsiveness. These changes impair synaptic plasticity (Lader, 2011), clinically manifesting as cognitive deficits, including memory problems, reduced attention, and slower processing speed (Barker et al., 2004).

Disruption of Sleep Architecture

Sedatives can significantly affect sleep patterns, mainly by altering the architecture of sleep stages. For instance, BZDs increase non-REM sleep while reducing REM sleep, which is essential for memory consolidation and cognitive function (Wilson & Nutt, 2008). Disruption of REM sleep and chronic sleep disturbances can lead to impaired cognitive performance and increased deposition of Aβ plaques, a hallmark of AD (Ju et al., 2014).

Potential Neurotoxicity

There is evidence suggesting that BZDs may have direct neurotoxic effects (Billioti de Gage et al., 2014). Neuroimaging studies have shown that long-term BZD use is associated with reduced CNS volume, particularly in the hippocampus, and structural changes in the brain, which can contribute to cognitive impairment (Livingston et al., 2017). Additionally, BZDs can induce oxidative stress and apoptosis in neuronal cells, further promoting neurodegeneration (Baldwin et al., 2013).

Interactions with Other Neurotransmitter Systems

Sedatives also interact with other neurotransmitter systems, including the cholinergic system, which is essential for cognitive function. BZDs can inhibit acetylcholine release, leading to impaired cholinergic transmission. This interaction is particularly concerning because acetylcholine plays a crucial role in attention, learning, and memory. Impairment of cholinergic function can exacerbate cognitive deficits and contribute to the development of neurocognitive disorders (Paterniti et al., 2002).

Further into the discussion, Ettchello et al. (2020) comprehensively assess preclinical studies on the molecular pathways through which BZDs and related drugs could be involved in AD pathogenesis (Ettcheto et al., 2020). The authors discuss how insomnia increases cognitive impairment through microglial activation and oxidative stress. Similarly, BZDs can heighten this risk through several mechanisms, including reducing brain-derived neurotrophic factor (BDNF) and increasing β-amyloid precursor protein mRNA levels and tau phosphorylation.

MISUSE, DEPENDENCE, AND WITHDRAWAL OF SEDATIVES, HYPNOTICS, AND ANXIOLYTICS IN OLDER ADULTS

Long-term sedative use, particularly BZDs, often leads to dependence, making discontinuation difficult due to withdrawal symptoms. This is particularly problematic in older adults and those with NDDs. A cross-sectional study of 340 patients with a mean age of 72 years found long-term use of BZDs and Z-drugs to be significantly associated with unemployment and problematic medication use (Mokhar et al., 2018). Another study reported a high prevalence of BZD and Z-drug use among nursing home residents with dementia, reaching 39.2%. Furthermore, many prescriptions did not follow guidelines regarding indications, dosing, or treatment duration (Rijksen et al., 2021). However, trends in medication use have been changing. Vasudev et al. (2015) noted a decline in BZD use among individuals aged 65 and above, potentially being substituted by sedative and non-sedative antidepressants (Vasudev et al., 2015).

For individuals with mild or early dependence, minimal interventions (e.g., advisory letters or general practitioner advice) may be sufficient (Cowen et al., 2012). However, for those with established dependence, a gradual reduction over several weeks is recommended. The BZD dose should be reduced by about one-eighth every 2 weeks (Cowen et al., 2012; Lingford-Hughes et al., 2012). Individuals taking short half-life BZDs are more prone to withdrawal symptoms. Therefore, switching to a long half-life BZD before initiating a gradual taper may be helpful (Cowen et al., 2012; Lingford-Hughes et al., 2012).

The Screening, Brief Intervention, and Referral to Treatment (SBIRT) model can be used to manage sedative, hypnotic, and anxiolytic use disorders in patients with NDDs (Babor et al., 2007). This model emphasizes universal screening in healthcare settings, followed by brief interventions for those with or at risk of developing such disorders. Universal screening includes a pre-screening questionnaire (e.g., "In the last year, have you used prescription or other drugs more than you meant to?") (Schonfeld et al., 2015). A positive pre-screening result leads to the administration of the Alcohol, Smoking, and Substance Involvement Screening Test (ASSIST), which stratifies substance use into mild, moderate, and high categories (Schonfeld et al., 2015). Red flags for sedative misuse include long-term use, rebound anxiety and insomnia upon withdrawal, a strong desire to use BZDs, driving under the influence of BZDs, concurrent use of BZDs with other hypnotics, and continuing to use BZDs despite a physician's recommendation to discontinue (Markota et al., 2016). Screening also includes assessments for depression, anxiety, post-traumatic stress disorder, and cognitive impairment. Management involves brief pharmacological and non-pharmacological interventions and treatment of any underlying mental or medical condition (M.D., 2024).

CLINICAL IMPLICATIONS

Risks and Benefits of Prescribing in Older Adults

Older adults exhibit increased sensitivity to sedatives, hypnotics, and anxiolytics, as well as reduced pharmacokinetic clearance of these medications (Glass et al., 2005). The use of BZDs in this population is associated with a higher risk of paradoxical reactions, such as increased agitation, aggression, or irritability (Mancuso et al., 2004), and medical complications like sedation, falls (Richardson et al., 2021), delirium, pneumonia, and respiratory depression (K.-H. Huang et al., 2021). Many older adults also have multiple comorbidities, necessitating the use of medications like anti-diabetic and anti-hypertensive agents. As such, healthcare providers must carefully consider potential drug–drug interactions between these medications and sedatives when co-prescribed.

In patients with NDDs, the use of sedatives, hypnotics, and anxiolytics may worsen the prognosis, although evidence supporting this claim remains inconclusive and debated (Billioti de Gage et al., 2014; Yoon et al., 2024). Long-term sedative use can lead to physical dependence and tolerance, requiring escalating doses to achieve the same therapeutic effects (Baldwin et al., 2013). This creates a cycle of increasing medication use, heightening the risk of adverse effects and further cognitive decline. Furthermore, withdrawal from these medications in such patients may result in complex symptoms, including severe anxiety, agitation, and worsening neuropsychiatric symptoms.

However, the careful and appropriate use of sedatives, hypnotics, and anxiolytics can be beneficial for older adults and patients with NDDs, particularly those with neuropsychiatric symptoms. These medications are intended for short-term use to manage conditions such as acute anxiety and insomnia. Additionally, anxiolytics like BZDs can be effective in alleviating anxiety or agitation without inducing delirium in individuals approaching the end of life (Salt et al., 2017). In cases of agitated delirium, whether in patients with NDDs or those requiring end-of-life care, antipsychotic agents are typically used, often supplemented with BZDs to provide additional sedation and anxiolytic effects (National Clinical Guideline Centre [UK], 2015).

General Principles for Prescribing in Older Adults

Non-pharmacological interventions (Table 9.2) are often preferred to manage comorbid neuropsychiatric disorders in patients with NDDs, either alone or in combination with sedative, hypnotic, and anxiolytic medications.

When medications are prescribed, the following recommendations should be considered:

1. Use sedatives, hypnotics, and anxiolytics only for appropriate and limited indications that cannot be managed with non-pharmacological interventions. Consider safer pharmacological interventions (e.g., melatonin for sleep). Avoid long-term use of sedatives, hypnotics, and anxiolytics; instead, consider short-term use (2–4 weeks) with a gradual tapering strategy to prevent withdrawal (Baldwin et al., 2013; Crocco et al., 2017).
2. Start low and go slow, keeping in mind that older adults, particularly those with NDDs, are susceptible to this group of medications (Subramanyam et al., 2018).

3. Choose short-acting BZDs (e.g., lorazepam) with no active metabolites to minimize prolonged sedation, as long-acting BZDs (e.g., diazepam) may accumulate due to slower metabolism in older adults (McIntosh et al., 2011).
4. Engage in regular reassessments, including checking vital signs and conducting periodic cognitive assessments to monitor for any deterioration of executive functions or memory (Barker et al., 2004; McIntosh et al., 2011).
5. Monitor adverse effects (such as confusion, increased risk of falls, and paradoxical reactions) and be mindful of potential drug–drug interactions, particularly with concurrent use of opioids or antipsychotics, which can exacerbate sedative effects (McIntosh et al., 2011).
6. Engage family members and caregivers in the treatment plan, encouraging them to monitor the patient for changes in behavior, mood, and physical function (Subramanyam et al., 2018).

In conclusion, balancing risks and benefits when prescribing sedatives, hypnotics, and anxiolytics is critical for older adults with NDDs. The main goal should always be to minimize harm while effectively addressing acute needs. By following the above best practices, clinicians can optimize the safety and efficacy of these medications while considering the needs of this vulnerable population.

CONCLUSION

Chronic sedative use, combined with environmental and genetic factors, may increase the risk of developing NDDs and influence their progression. However, the complex mechanisms and bidirectional relationship between sedative, hypnotic, and anxiolytic use and NDDs require further investigation. Rigorous studies focusing on causality and the potential role of sedatives in preventing or treating NDDs, particularly in at-risk populations, are essential. Future research should utilize robust methodologies and large sample sizes to deepen our understanding of this relationship.

TABLE 9.2
Recommended Non-Pharmacological Interventions for Patients with NDDs and Concurrent Psychiatric Disorders

Lifestyle modifications	• Structured daily activities. • *Physical exercise*: Regular, simple walk to brisk exercise for a few minutes to hours. If wheelchair-bound, then upper body exercises such as simple stretches. • *Sleep hygiene and counseling*: Educating on sleep hygiene reduces erroneous sleep-related expectations. • *Nutrition*: Monitoring through a simple nutrition chart to check appetite and intake of various nutrients.
Therapy	• *Relaxation therapy*: Classical Jacobson's progressive muscle relaxation technique, guided imagery. • Cognitive behavioral therapy. • Mindfulness. • Systemic desensitization for phobias or unspecified fears. • Eye movement desensitization and reprocessing for post-traumatic stress disorder. • Exposure and response prevention for obsessive and compulsive disorder.
Miscellaneous	• Yoga. • Art, dance, and music therapy in individual or group settings. • Cognitive rehabilitation (e.g., memory games). • Social activities or networking. • *Other therapies*: Complementary and alternative therapies (e.g., touch, reflexology, massage, reiki); limited evidence.

Sources: Chen et al., 2021, p. 35; Subramanyam et al., 2018; Teixeira et al., 2012.

Clinically, non-pharmacological interventions should be prioritized when possible, and sedatives should be prescribed only for short-term, well-defined indications. Close monitoring for adverse effects is critical, with pharmacological interventions tailored to the individual's age, comorbidities, and cognitive status to minimize risks and optimize outcomes.

REFERENCES

AlDawsari, A., Bushell, T. J., Abutheraa, N., Sakata, S., Al Hussain, S., & Kurdi, A. (2022). Use of sedative-hypnotic medications and risk of dementia: A systematic review and meta-analysis. *British Journal of Clinical Pharmacology*, *88*(4), 1567–1589. https://doi.org/10.1111/bcp.15113

Akhondzadeh, S., & Noroozian, M. (2002). Alzheimer's disease: Pathophysiology and pharmacotherapy. *IDrugs: The Investigational Drugs Journal*, *5*(11), 1062–1069.

Amor, S., Puentes, F., Baker, D., & van der Valk, P. (2010). Inflammation in neurodegenerative diseases. *Immunology*, *129*(2), 154–169. https://doi.org/10.1111/j.1365-2567.2009.03225.x

Babor, T. F., McRee, B. G., Kassebaum, P. A., Grimaldi, P. L., Ahmed, K., & Bray, J. (2007). Screening, Brief Intervention, and Referral to Treatment (SBIRT): Toward a public health approach to the management of substance abuse. *Substance Abuse*, *28*(3), 7–30. https://doi.org/10.1300/J465v28n03_03

Baldwin, D. S., Aitchison, K., Bateson, A., Curran, H. V., Davies, S., Leonard, B., Nutt, D. J., Stephens, D. N., & Wilson, S. (2013). Benzodiazepines: Risks and benefits. A reconsideration. *Journal of Psychopharmacology (Oxford, England)*, *27*(11), 967–971. https://doi.org/10.1177/0269881113503509

Barker, M. J., Greenwood, K. M., Jackson, M., & Crowe, S. F. (2004). Cognitive effects of long-term benzodiazepine use: A meta-analysis. *CNS Drugs*, *18*(1), 37–48. https://doi.org/10.2165/00023210-200418010-00004

Bartels, A. L., & Leenders, K. L. (2009). Parkinson's disease: The syndrome, the pathogenesis and pathophysiology. *Cortex; A Journal Devoted to the Study of the Nervous System and Behavior*, *45*(8), 915–921. https://doi.org/10.1016/j.cortex.2008.11.010

Bayer, T. A., & Wirths, O. (2010). Intracellular accumulation of amyloid-beta—A predictor for synaptic dysfunction and neuron loss in Alzheimer's disease. *Frontiers in Aging Neuroscience*, *2*, 8. https://doi.org/10.3389/fnagi.2010.00008

Bellou, V., Belbasis, L., Tzoulaki, I., Middleton, L. T., Ioannidis, J. P. A., & Evangelou, E. (2017). Systematic evaluation of the associations between environmental risk factors and dementia: An umbrella review of systematic reviews and meta-analyses. *Alzheimer's & Dementia: The Journal of the Alzheimer's Association*, *13*(4), 406–418. https://doi.org/10.1016/j.jalz.2016.07.152

Billioti de Gage, S., Moride, Y., Ducruet, T., Kurth, T., Verdoux, H., Tournier, M., Pariente, A., & Bégaud, B. (2014). Benzodiazepine use and risk of Alzheimer's disease: Case-control study. *BMJ (Clinical Research Ed.)*, *349*, g5205. https://doi.org/10.1136/bmj.g5205

Bouchette, D., Akhondi, H., Patel, P., & Quick, J. (2024). *Zolpidem*. In StatPearls. StatPearls Publishing. http://www.ncbi.nlm.nih.gov/books/NBK442008/

Butterfield, D. A., Griffin, S., Munch, G., & Pasinetti, G. M. (2002). Amyloid beta-peptide and amyloid pathology are central to the oxidative stress and inflammatory cascades under which Alzheimer's disease brain exists. *Journal of Alzheimer's Disease: JAD*, *4*(3), 193–201. https://doi.org/10.3233/jad-2002-4309

Chaudhuri, K. R., & Schapira, A. H. V. (2009). Non-motor symptoms of Parkinson's disease: Dopaminergic pathophysiology and treatment. *Lancet. Neurology*, *8*(5), 464–474. https://doi.org/10.1016/S1474-4422(09)70068-7

Chen, Y., Li, X., Pan, B., Wang, B.-, Jing, G., Liu, Q., Li, Y., Bing, Z., Yang, K., Han, X., & Ge, L. (2021). Non-pharmacological interventions for older adults with depressive symptoms: A network meta-analysis of 35 randomized controlled trials. *Aging & Mental Health*, *25*(5), 773–786. https://doi.org/10.1080/13607863.2019.1704219

Ciraulo, D. A. (2014). Sedatives, hypnotics, and anxiolytics. In H. R. Kranzler, D. A. Ciraulo, & L. R. Zindel (Eds.), *Clinical manual of addiction psychopharmacology* (2nd ed., pp. 199–260). American Psychiatric Publishing.

Cowen, P., Harrison, P., & Burns, T. (2012). *Shorter Oxford textbook of psychiatry*. Oxford University Press.

Crocco, E. A., Jaramillo, S., Cruz-Ortiz, C., & Camfield, K. (2017). Pharmacological management of anxiety disorders in the elderly. *Current Treatment Options in Psychiatry*, *4*(1), 33–46. https://doi.org/10.1007/s40501-017-0102-4

DeMaagd, G., & Philip, A. (2015). Parkinson's disease and its management: Part 1: Disease entity, risk factors, pathophysiology, clinical presentation, and diagnosis. *P & T: A Peer-Reviewed Journal for Formulary Management, 40*(8), 504–532.

Ettcheto, M., Olloquequi, J., Sánchez-López, E., Busquets, O., Cano, A., Manzine, P. R., Beas-Zarate, C., Castro-Torres, R. D., García, M. L., Bulló, M., Auladell, C., Folch, J., & Camins, A. (2020). Benzodiazepines and related drugs as a risk factor in Alzheimer's disease dementia. *Frontiers in Aging Neuroscience, 11*, 344. https://doi.org/10.3389/fnagi.2019.00344

Ferreira, P., Ferreira, A. R., Barreto, B., & Fernandes, L. (2022). Is there a link between the use of benzodiazepines and related drugs and dementia? A systematic review of reviews. *European Geriatric Medicine, 13*(1), 19–32. https://doi.org/10.1007/s41999-021-00553-w

Gaztanaga, W., Sarno, M., Margolesky, J., Luca, C., Singer, C., Moore, H., Jagid, J., & Levin, B. (2021). Do benzodiazepines impair motor and nonmotor symptoms in a sample of Parkinson's disease patients? *Cureus, 13*(2), e13220. https://doi.org/10.7759/cureus.13220

Giovannitti, J. A., & Cooke, M. R. (2017). Sedative-hypnotics, antianxiety drugs, and centrally acting muscle relaxants. In *Pharmacology and therapeutics for dentistry* (pp. 156–175). Elsevier. https://doi.org/10.1016/B978-0-323-39307-2.00011-4

Glass, J., Lanctôt, K. L., Herrmann, N., Sproule, B. A., & Busto, U. E. (2005). Sedative hypnotics in older people with insomnia: Meta-analysis of risks and benefits. *BMJ (Clinical Research Ed.), 331*(7526), 1169. https://doi.org/10.1136/bmj.38623.768588.47

Griffin, C. E., Kaye, A. M., Bueno, F. R., & Kaye, A. D. (2013). Benzodiazepine pharmacology and central nervous system-mediated effects. *Ochsner Journal, 13*(2), 214–223.

Hardiman, O., Al-Chalabi, A., Chio, A., Corr, E. M., Logroscino, G., Robberecht, W., Shaw, P. J., Simmons, Z., & van den Berg, L. H. (2017). Amyotrophic lateral sclerosis. *Nature Reviews. Disease Primers, 3*, 17085. https://doi.org/10.1038/nrdp.2017.85

Huang, H.-C., Tsai, C.-H., Muo, C.-H., Lin, K.-H., Lu, M.-K., Sung, F.-C., & Kao, C.-H. (2015). Risk of Parkinson's disease following zolpidem use: A retrospective, population-based cohort study. *Journal of Clinical Psychiatry, 76*(1), e104–110. https://doi.org/10.4088/JCP.13m08790

Huang, K.-H., Tai, C.-J., Kuan, Y.-H., Chang, Y.-C., Tsai, T.-H., & Lee, C.-Y. (2021). Pneumonia risk associated with the use of individual benzodiazepines and benzodiazepine related drugs among the elderly with Parkinson's disease. *International Journal of Environmental Research and Public Health, 18*(17), 9410. https://doi.org/10.3390/ijerph18179410

Ju, Y.-E. S., Lucey, B. P., & Holtzman, D. M. (2014). Sleep and Alzheimer disease pathology—A bidirectional relationship. *Nature Reviews: Neurology, 10*(2), 115–119. https://doi.org/10.1038/nrneurol.2013.269

Kobeissy, F., & Moshourab, R. A. (2015). Autoantibodies in CNS trauma and neuropsychiatric disorders: A new generation of biomarkers. In F. H. Kobeissy (Ed.), *Brain neurotrauma: Molecular, neuropsychological, and rehabilitation aspects*. CRC Press/Taylor & Francis. http://www.ncbi.nlm.nih.gov/books/NBK299208/

Kouli, A., Torsney, K. M., & Kuan, W.-L. (2018). Parkinson's disease: Etiology, neuropathology, and pathogenesis. In T. B. Stoker & J. C. Greenland (Eds.), *Parkinson's disease: Pathogenesis and clinical aspects*. Codon Publications. http://www.ncbi.nlm.nih.gov/books/NBK536722/

Lader, M. (2011). Benzodiazepines revisited—Will we ever learn? *Addiction (Abingdon, England), 106*(12), 2086–2109. https://doi.org/10.1111/j.1360-0443.2011.03563.x

Lamptey, R. N. L., Chaulagain, B., Trivedi, R., Gothwal, A., Layek, B., & Singh, J. (2022). A review of the common neurodegenerative disorders: Current therapeutic approaches and the potential role of nanotherapeutics. *International Journal of Molecular Sciences, 23*(3), 1851. https://doi.org/10.3390/ijms23031851

Lee, J., Jung, S. J., Choi, J., Shin, A., & Lee, Y. J. (2018). Use of sedative-hypnotics and the risk of Alzheimer's dementia: A retrospective cohort study. *PLoS One, 13*(9), e0204413. https://doi.org/10.1371/journal.pone.0204413

Lingford-Hughes, A. R., Welch, S., Peters, L., Nutt, D. J., & British Association for Psychopharmacology, Expert Reviewers Group. (2012). BAP updated guidelines: Evidence-based guidelines for the pharmacological management of substance abuse, harmful use, addiction and comorbidity: Recommendations from BAP. *Journal of Psychopharmacology (Oxford, England), 26*(7), 899–952. https://doi.org/10.1177/0269881112444324

Livingston, G., Huntley, J., Sommerlad, A., Ames, D., Ballard, C., Banerjee, S., Brayne, C., Burns, A., Cohen-Mansfield, J., Cooper, C., Costafreda, S. G., Dias, A., Fox, N., Gitlin, L. N., Howard, R., Kales, H. C., Kivimäki, M., Larson, E. B., Ogunniyi, A., … & Mukadam, N. (2020). Dementia prevention, intervention, and care: 2020 report of the Lancet Commission. *Lancet, 396*(10248), 413–446. https://doi.org/10.1016/S0140-6736(20)30367-6

Livingston, G., Sommerlad, A., Orgeta, V., Costafreda, S. G., Huntley, J., Ames, D., Ballard, C., Banerjee, S., Burns, A., Cohen-Mansfield, J., Cooper, C., Fox, N., Gitlin, L. N., Howard, R., Kales, H. C., Larson, E. B., Ritchie, K., Rockwood, K., Sampson, E. L., … & Mukadam, N. (2017). Dementia prevention, intervention, and care. *Lancet (London, England)*, *390*(10113), 2673–2734. https://doi.org/10.1016/S0140-6736(17)31363-6

Mancuso, C. E., Tanzi, M. G., & Gabay, M. (2004). Paradoxical reactions to benzodiazepines: Literature review and treatment options. *Pharmacotherapy*, *24*(9), 1177–1185. https://doi.org/10.1592/phco.24.13.1177.38089

Margolis, S. A., Kelly, D. A., Daiello, L. A., Davis, J., Tremont, G., Pillemer, S., Denby, C., & Ott, B. R. (2021). Anticholinergic/sedative drug burden and subjective cognitive decline in older adults at risk of Alzheimer's disease. *Journals of Gerontology: Series A*, *76*(6), 1037–1043. https://doi.org/10.1093/gerona/glaa222

Markota, M., Rummans, T. A., Bostwick, J. M., & Lapid, M. I. (2016). Benzodiazepine use in older adults: Dangers, management, and alternative therapies. *Mayo Clinic Proceedings*, *91*(11), 1632–1639. https://doi.org/10.1016/j.mayocp.2016.07.024

Markwell, P., & McLellan, S. L. F. (2019). Jet lag. In *Travel medicine* (pp. 417–422). Elsevier. https://doi.org/10.1016/B978-0-323-54696-6.00045-8

McIntosh, B., Clark, M., & Spry, C. (2011). Summary of findings. In *Benzodiazepines in older adults: A review of clinical effectiveness, cost-effectiveness, and guidelines* [Internet]. Canadian Agency for Drugs and Technologies in Health. https://www.ncbi.nlm.nih.gov/books/NBK174566/

M.D, A. W. (2024). *Substance use in older adults*. American Psychiatric Pub.

Mokhar, A., Tillenburg, N., Dirmaier, J., Kuhn, S., Härter, M., & Verthein, U. (2018). Potentially inappropriate use of benzodiazepines and Z-drugs in the older population-analysis of associations between long-term use and patient-related factors. *PeerJ*, *6*, e4614. https://doi.org/10.7717/peerj.4614

Morris, J. (2015). Amyotrophic lateral sclerosis (ALS) and related motor neuron diseases: An overview. *Neurodiagnostic Journal*, *55*(3), 180–194. https://doi.org/10.1080/21646821.2015.1075181

National Clinical Guideline Centre (UK). (2015). *Care of dying adults in the last days of life*. National Institute for Health and Care Excellence (NICE). http://www.ncbi.nlm.nih.gov/books/NBK338140/

Paterniti, S., Dufouil, C., & Alpérovitch, A. (2002). Long-term benzodiazepine use and cognitive decline in the elderly: The Epidemiology of Vascular Aging Study. *Journal of Clinical Psychopharmacology*, *22*(3), 285–293. https://doi.org/10.1097/00004714-200206000-00009

Priyadarshi, A., Khuder, S. A., Schaub, E. A., & Priyadarshi, S. S. (2001). Environmental risk factors and Parkinson's disease: A metaanalysis. *Environmental Research*, *86*(2), 122–127. https://doi.org/10.1006/enrs.2001.4264

Rang, H. P., Dale, M. M., Ritter, J. M., Flower, R. J., & Henderson, G. (2011). *Rang & Dale's pharmacology*. Elsevier Health Sciences.

Richardson, K., Savva, G. M., Boyd, P. J., Aldus, C., Maidment, I., Pakpahan, E., Loke, Y. K., Arthur, A., Steel, N., Ballard, C., Howard, R., & Fox, C. (2021). Non-benzodiazepine hypnotic use for sleep disturbance in people aged over 55 years living with dementia: A series of cohort studies. *Health Technology Assessment (Winchester, England)*, *25*(1), 1–202. https://doi.org/10.3310/hta25010

Rijksen, D. O. C., Zuidema, S. U., & De Haas, E. C. (2021). Use of benzodiazepines and Z-drugs in nursing home residents with dementia: Prevalence and appropriateness. *Journal of Alzheimer's Disease Reports*, *5*(1), 871–879. https://doi.org/10.3233/ADR-210041

Robertson, I., Cheung, A., & Fan, X. (2019). Insomnia in patients with schizophrenia: Current understanding and treatment options. *Progress in Neuro-Psychopharmacology and Biological Psychiatry*, *92*, 235–242. https://doi.org/10.1016/j.pnpbp.2019.01.016

Salt, S., Mulvaney, C. A., & Preston, N. J. (2017). Drug therapy for symptoms associated with anxiety in adult palliative care patients. *Cochrane Database of Systematic Reviews*, *5*(5), CD004596. https://doi.org/10.1002/14651858.CD004596.pub3

Schonfeld, L., Hazlett, R. W., Hedgecock, D. K., Duchene, D. M., Burns, L. V., & Gum, A. M. (2015). Screening, brief intervention, and referral to treatment for older adults with substance misuse. *American Journal of Public Health*, *105*(1), 205–211. https://doi.org/10.2105/AJPH.2013.301859

Schulz-Schaeffer, W. J. (2010). The synaptic pathology of alpha-synuclein aggregation in dementia with Lewy bodies, Parkinson's disease and Parkinson's disease dementia. *Acta Neuropathologica*, *120*(2), 131–143. https://doi.org/10.1007/s00401-010-0711-0

Selkoe, D. J., & Hardy, J. (2016). The amyloid hypothesis of Alzheimer's disease at 25 years. *EMBO Molecular Medicine*, *8*(6), 595–608. https://doi.org/10.15252/emmm.201606210

Skibiski, J., Patel, P., & Abdijadid, S. (2024). *Barbiturates*. In StatPearls. StatPearls Publishing. http://www.ncbi.nlm.nih.gov/books/NBK539731/

Spires-Jones, T. L., & Hyman, B. T. (2014). The intersection of amyloid beta and tau at synapses in Alzheimer's disease. *Neuron*, *82*(4), 756–771. https://doi.org/10.1016/j.neuron.2014.05.004

Stahl, S. M. (2021). *Stahl's essential psychopharmacology: Neuroscientific basis and practical applications*. Cambridge university press.

Subramanyam, A. A., Kedare, J., Singh, O. P., & Pinto, C. (2018). Clinical practice guidelines for geriatric anxiety disorders. *Indian Journal of Psychiatry*, *60*(Suppl 3), S371–S382. https://doi.org/10.4103/0019-5545.224476

Teixeira, C. V. L., Gobbi, L. T. B., Corazza, D. I., Stella, F., Costa, J. L. R., & Gobbi, S. (2012). Non-pharmacological interventions on cognitive functions in older people with mild cognitive impairment (MCI). *Archives of Gerontology and Geriatrics*, *54*(1), 175–180. https://doi.org/10.1016/j.archger.2011.02.014

Tripathi, K. D. (2018). *Essentials of medical pharmacology*. Jaypee Brothers Medical Publishers.

Van De Vijver, D. A. M. C., Roos, R. A. C., Jansen, P. A. F., Porsius, A. J., & De Boer, A. (2002). Influence of benzodiazepines on antiparkinsonian drug treatment in levodopa users: Benzodiazepines in levodopa users. *Acta Neurologica Scandinavica*, *105*(1), 8–12. https://doi.org/10.1034/j.1600-0404.2002.00132.x

Vasudev, A., Shariff, S. Z., Liu, K., Burhan, A. M., Herrmann, N., Leonard, S., & Mamdani, M. (2015). Trends in psychotropic dispensing among older adults with dementia living in long-term care facilities: 2004–2013. *American Journal of Geriatric Psychiatry: Official Journal of the American Association for Geriatric Psychiatry*, *23*(12), 1259–1269. https://doi.org/10.1016/j.jagp.2015.07.001

Vejandla, B., Savani, S., Appalaneni, R., Veeravalli, R. S., & Gude, S. S. (2024). Alzheimer's disease: The past, present, and future of a globally progressive disease. *Cureus*, *16*(1), e51705. https://doi.org/10.7759/cureus.51705

Walker, F. O. (2007). Huntington's disease. *Lancet (London, England)*, *369*(9557), 218–228. https://doi.org/10.1016/S0140-6736(07)60111-1

Wilson, S., & Nutt, D. (2008). Drug treatment of chronic insomnia—Dawn at the end of a long night? *Journal of Psychopharmacology*, *22*(7), 703–706. https://doi.org/10.1177/0269881108096605

Woolcott, J. C. (2009). Meta-analysis of the impact of 9 medication classes on falls in elderly persons. *Archives of Internal Medicine*, *169*(21), 1952. https://doi.org/10.1001/archinternmed.2009.357

Yoon, S.-H., Kim, Y.-C., Seo, H. J., Hong, S. C., Kim, T. W., Jeong, J.-H., & Um, Y. H. (2024). Association between hypnotics and dementia: A mini narrative review. *Psychiatry Investigation*, *21*(5), 457–463. https://doi.org/10.30773/pi.2023.0383

10 Stimulants and Neurodegenerative Disorders

Sarah El Halabi, Ismail Deniz, Bita Vahdani, Odel Zadeh, and Sanya Virani

INTRODUCTION

Neurodegenerative disorders (NDDs) are characterized by the progressive degeneration of the structure and function of the nervous system (Lamptey et al., 2022). These conditions often involve the gradual loss of neurons, leading to a decline in cognitive function, movement, and overall health (Lamptey et al., 2022). The most prevalent NDDs include Alzheimer's disease (AD), Parkinson's disease (PD), amyotrophic lateral sclerosis (ALS), Huntington's disease (HD), and frontotemporal dementia (FTD) (Lamptey et al., 2022). AD occurs secondary to the accumulation of amyloid-beta and tau proteins, which form plaques that disrupt the hippocampus (Lamptey et al., 2022). Both genetic and non-genetic factors cause PD. Its pathophysiology involves atrophy of the frontal cortex, ventricular enlargement, and significant cell loss, all of which contribute to decreased dopamine levels in the brain (Lamptey et al., 2022). ALS involves the degeneration of motor neurons, protein aggregation, gliosis, and oxidative stress, which are some factors in the disease's progression (Lamptey et al., 2022). HD is an autosomal dominant disorder caused by the elongation of CAG repeats (Ajitkumar & De Jesus, 2023), while abnormal protein aggregates in the frontal and temporal lobes lead to FTD (Khan & De Jesus, 2023).

Dementia, a syndrome characterized by cognitive decline beyond what is expected from normal aging, affects multiple higher-level brain functions, including memory, attention, executive function, language, and social cognition (Emmady, 2022). These impairments severely hinder an individual's ability to function independently. According to the World Health Organization, around 55 million people globally were living with dementia as of 2020, with projections indicating a sharp rise to 78 million by 2030 and 139 million by 2050. From 1990 to 2019, the global incidence and prevalence of dementia, as well as the number of deaths attributed to it, increased significantly, with cases expected to reach 152.8 million by 2050 (WHO fact sheet, 2021). Over this period, the burden of AD and other dementias grew from 9.66 million to 25.28 million disability-adjusted life years (DALYs) (Li, 2022).

The increasing prevalence of NDDs has prompted more research on their causes and potential treatments. One intriguing area of study explores the potential therapeutic applications of amphetamines and related stimulants, which have been used to manage conditions such as attention-deficit hyperactivity disorder, narcolepsy, and depression (Sassi, 2020). Stimulants, such as amphetamines, cocaine, and methamphetamine, are widely known for their cognitive-enhancing effects and are among the most commonly abused substances in the United States. While amphetamines were discovered in the early 20th century and are now tightly regulated due to their abuse potential, they have shown promise in treating age-related cognitive and mood disorders, particularly by addressing dopamine deficiencies. For example, reduced dopamine levels, often observed in older adults, are associated with impaired cognitive and motor performance. In this regard, amphetamines may help by increasing the availability of dopamine and norepinephrine (Faraone, 2018).

This chapter provides an overview of stimulants, detailing the mechanisms of action of key substances, and explores the relationship between stimulant use and NDDs, particularly in the context

 DOI: 10.1201/9781003478485-10

of comorbidity. It covers the definition and epidemiology of stimulant use disorder, including its subtypes and pathophysiology, and examines how stimulant use influences the pathophysiology of NDDs. The chapter further discusses diagnostic considerations and current treatment approaches for managing comorbid stimulant use and NDDs, focusing on integrated care models, pharmacological interventions, behavioral therapies, and lifestyle modifications. Finally, it highlights the need for ongoing research to refine diagnostic tools and treatment strategies to improve patient outcomes in this complex clinical scenario.

STIMULANT USE DISORDERS: DEFINITION AND EPIDEMIOLOGY

Substance use disorders affect approximately 40 million Americans and are a significant cause of mortality in this population (Drossel et al., 2023). These disorders involve the excessive use of substances such as nicotine, alcohol, and illicit drugs, including cannabis, sedatives, hypnotics, anxiolytics, inhalants, opioids, hallucinogens, and stimulants, all of which negatively impact an individual's quality of life (Jahan & Burges, 2023). Each substance uniquely affects the central nervous system, producing varying effects such as increased energy, euphoria, and depressant effects (Jahan & Burges, 2023). Stimulants, in particular, have a distinct impact. The *Diagnostic and Statistical Manual of Mental Disorders,* Fifth Edition, Text Revision (DSM-5-TR) defines stimulants as "psychoactive substances that increase activity in the brain," leading to euphoria or heightened focus (American Psychiatric Association, 2022). The criteria for stimulant use disorder are outlined in Table 10.1 below. Globally, amphetamine use disorder affects 96 people per 100,000, while cocaine use disorder affects 64 per 100,000 (Farrell et al., 2019). The highest prevalence of amphetamine use disorder is seen in Australia and high-income North American countries, while cocaine dependence is most prevalent in high-income North American countries (Farrell et al., 2019).

While the diagnostic criteria for stimulant use disorder resemble those of other substance use disorders, stimulants themselves are unique substances that produce a specific set of symptoms, including euphoria, decreased need for sleep, and loss of appetite. The DSM-5-TR discusses several

TABLE 10.1
Criteria for Stimulant Use Disorder

A pattern of amphetamine-type substance, cocaine, or other stimulant use leading to clinically significant impairment or distress, as manifested by at least two of the following occurring within a 12-month period:

1. Stimulant often taken in larger amounts or over a longer period than intended.
2. Persistent desire or unsuccessful efforts to cut down or control stimulant use.
3. Considerable time spent obtaining the stimulant, using it, or recovering from its effects.
4. Craving or strong desire to use the stimulant.
5. Recurrent stimulant use resulting in failure to fulfill major obligations at work, school, or home.
6. Continued use despite persistent or recurrent social or interpersonal problems caused or exacerbated by the effects of the stimulant.
7. Important social, occupational, or recreational activities abandoned or reduced because of stimulant use.
8. Recurrent stimulant use in physically hazardous situations.
9. Continued use despite knowledge of persistent or recurrent physical or psychological problems likely caused or exacerbated by the stimulant.
10. Tolerance, defined by either:
    a. A need for markedly increased amounts to achieve intoxication or the desired effect.
    b. A markedly diminished effect with continued use of the same amount of the stimulant.
11. Withdrawal, as manifested by either of the following:
    a. The characteristic withdrawal syndrome for the stimulant.
    b. The stimulant (or a closely related substance) is taken to relieve or avoid withdrawal symptoms.

Source: Adapted from the DSM-5-TR.

stimulants in detail, including amphetamine-based substances, cocaine, methylphenidate, caffeine, and nicotine. These substances are commonly ingested orally, intravenously injected, or inhaled, with the latter two methods often used recreationally. Regardless of the route of administration, stimulant use disorders can develop rapidly, depending on both the frequency and the quantity of use (American Psychiatric Association, 2022).

Research on stimulant use disorders has shown that individuals between the ages of 18 and 25 are the most likely to use stimulants, including nicotine, caffeine, cocaine, and amphetamines. The National Survey on Drug Use and Health (NSDUH) reported that, within the 26 and older age group, the prevalence of prescribed stimulant misuse was 1.5%, compared to 4.8% among young adults aged 18 to 25 (Substance Abuse and Mental Health Services Administration, 2021). Stimulant use disorder rates are also higher in males and individuals who identify as Hispanic or non-Hispanic White (American Psychiatric Association, 2022). In adolescents and young adults, stimulant misuse is often motivated by desires for weight control or improved academic performance through enhanced focus (American Psychiatric Association, 2022). Individuals with underlying psychiatric disorders, such as bipolar disorder, schizophrenia, antisocial personality disorder, and other substance use disorders, are also at increased risk of developing stimulant use disorders (American Psychiatric Association, 2022). The lack of data on stimulant use disorders in individuals aged 65 and older presents a challenge in understanding the relationship between stimulant use and NDDs, which are common in this age group.

STIMULANT USE DISORDERS: SUBTYPES AND PATHOPHYSIOLOGY

Stimulant use disorder is highly prevalent, which is unsurprising given the neurobiology of addiction. Both animal and human studies indicate that addiction involves not only the brain's reward pathway—specifically, dopaminergic projections from the ventral tegmental area (VTA) of the midbrain to the nucleus accumbens—but also the interaction of multiple other neural circuits (Volkow & Morales, 2015). These circuits include those governing conditioning and memory, executive control, motivation/drive, interoception, and aversion/avoidance. This model suggests that addiction amplifies the expected value of the drug within the reward, motivation, and memory circuits, ultimately overriding control circuits. This process creates a positive feedback loop sustained by the heightened activation of the motivation/drive and memory circuits, promoting continued drug use.

Although the term "stimulants" in the literature broadly refers to any psychoactive substances with stimulant effects, this chapter will focus on those associated with neurodegenerative illnesses, such as amphetamines, methamphetamines, and cocaine.

Amphetamines

Amphetamines increase the concentration of monoamines in the synaptic cleft through various mechanisms. These include enhancing monoamine release by inhibiting and reversing the action of monoamine transporters dopamine transporter (DAT), norepinephrine transporter (NET), and serotonin transporter (SERT), as well as promoting release via the vesicular monoamine transporter type-2 (VMAT-2) (Cadet et al., 2007). Amphetamines also weakly inhibit monoamine oxidase (MAO) in mitochondria. As a result, they increase the availability of dopamine, norepinephrine, and serotonin in presynaptic neurons, facilitating their release into the synaptic cleft and amplifying their effects on post-synaptic receptors (Ferrucci et al., 2019).

Methamphetamines

Methamphetamine is a derivative of amphetamine, both of which belong to the class of compounds known as phenethylamines. Structurally, methamphetamine differs from amphetamine by the addition of a methyl group attached to the nitrogen atom. This slight structural difference significantly alters its pharmacokinetics and pharmacodynamics, making methamphetamine more potent and longer-lasting in its effects compared to amphetamine. Consequently, methamphetamine is more neurotoxic, particularly to dopaminergic and serotonergic neurons, and has a longer duration of action due to its greater potency and more effective inhibition of MAOs (Cadet et al., 2007).

Methamphetamine exhibits profoundly addictive stimulant properties and causes substantial dysregulation of neural circuitry (Park et al., 2013). It readily crosses the blood–brain barrier (BBB), provoking rapid and intense euphoria (Murnane et al., 2013). Immediate effects often include methamphetamine-induced restlessness, insomnia, and irritability. Continued use can result in adverse effects such as dyskinesia and other movement-related disorders. Chronic methamphetamine use contributes to damage in dopaminergic nerve terminals, leading to motor deficits, cognitive decline, anxiety, depression, hallucinations, violent behaviors, and delusions (Murnane et al., 2012; Andersen et al., 2013; McGregor et al., 2005). Prolonged exposure triggers several neurotoxic mechanisms, including oxidative stress, endoplasmic reticulum stress, mitochondrial dysfunction, impaired axonal transport, autophagy, and programmed cell death. These mechanisms ultimately contribute to neurodegeneration (Galimberti & Scarpini, 2018; Pregeljc et al., 2020).

Cocaine

Acute cocaine administration significantly impacts neuronal activity and cerebral blood flow. It binds to monoamine transporters for dopamine, norepinephrine, and serotonin in the synaptic cleft, inhibiting their reuptake and enhancing post-synaptic activity. The resulting elevated dopamine levels activate dopaminergic G-protein-coupled receptors (GPCRs) in the mesocorticolimbic dopamine pathway, which underlies the drug's reinforcing properties (Baik, 2013).

Additionally, the transcriptional regulator ΔFosB, stimulated by D1-receptor activation, modulates the synthesis of AMPA glutamate receptors. Following cocaine's effects, ΔFosB increases the levels of the glutamate receptor subunit GluR1 in the ventral tegmental area, potentially contributing to addiction development (Kalivas & Volkow, 2005).

Neuroimaging

In addition to these molecular insights, neuroimaging provides a valuable perspective on the effects of chronic stimulant use disorder. Individuals with chronic stimulant use disorder often show reductions in gray matter volume (GMV) in brain regions critical for cognitive control, decision-making, and emotional regulation (Ersche et al., 2011). A meta-analysis of voxel-based morphology studies in stimulant dependence revealed deficits in the insula, thalamus, anterior cingulate cortex, and ventral prefrontal cortex, with no regions identified as having increased GMV.

Evidence also points to changes in the basal ganglia, particularly the striatum, which is involved in reward processing. While some studies report reduced volume, others indicate increased volume, highlighting the need for further research to reconcile these discrepancies (Ersche et al., 2013; Ersche et al., 2024). As most of these studies are cross-sectional, it remains uncertain whether differences in GMV predispose individuals to stimulant use disorder or result from chronic drug exposure. Nevertheless, these findings suggest that GMV alterations may serve as a biomarker for stimulant use disorder.

NEURODEGENERATIVE DISORDERS: PATHOPHYSIOLOGY AND CORRELATION WITH STIMULANT USE DISORDERS

NDDs are progressive disorders characterized by irreversible decline, continuing until the individual's death. Diagnosis relies on clinical history and physical examinations, although these diseases share genetic and pathological characteristics that aid in their classification (Galimberti & Scarpini, 2018).

A significant subset of NDDs exhibits protein aggregates, which can be grouped based on pathological aberrations. For instance, ubiquitinated neuronal nuclear inclusions manifest in polyglutaminopathies, including HD. Parenchymal or vascular accumulation of amyloid-beta (Aβ) is associated with aging and is also present in AD, particularly in individuals with Down syndrome. The presence of Lewy bodies is a hallmark of PD, with or without dementia, and affects various classes of neurons. Beyond these aggregates, neuronal degeneration is another defining feature. While degeneration occurs in different brain regions, histological findings show large neostriatal neurons degenerating in progressive supranuclear palsy and AD. Microscopic examination of affected cells often reveals neuronal gliosis and cell loss—key unifying features of neurodegenerative illnesses (Galimberti & Scarpini, 2018).

Extensive literature explores the connections between stimulant use disorder and PD. For example, a computational study in *Drosophila* delineated three dopamine transport pathways: dopamine transporters, exocytosis, and vesicular dopamine transporters. Amphetamines, which block vesicular transporters of dopamine (unlike cocaine or substances that only block dopamine transporters), were implicated in PD-related neurodegeneration (Pregeljc et al., 2020). Methamphetamines, long associated with PD, are frequently used in lab studies to induce PD in animal models. This is attributed to their lipophilic nature, which allows them to cross the BBB. Once inside the brain, methamphetamines are thought to cause widespread damage, including dopamine depletion, oxidative stress, activation of astrocytes and microglial cells, impaired axonal transport, autophagy, and apoptosis—all mechanisms of neurodegeneration (Shukla & Vincent, 2020). Chronic cocaine use has also been linked to increased alpha-synuclein expression in the striatum, a hallmark of PD (Qin et al., 2005).

Regarding clinical studies, a retrospective study in Utah examined statewide medical records from 1996 to 2011 and observed an increased risk of PD and PD/PT in methamphetamine/amphetamine users, while cocaine users did not exhibit this elevated risk (Curtin et al., 2015). Similarly, a retrospective cohort study using statewide inpatient records from California hospitals compared methamphetamine/amphetamine users to cocaine users and a control group with appendicitis. The methamphetamine/amphetamine cohort demonstrated an increased risk of PD compared to both the appendicitis and cocaine groups. In contrast, the cocaine group did not show a higher hazard of PD compared to the appendicitis group (Callaghan et al., 2012).

The deleterious effects of methamphetamines extend beyond PD, as they have also been implicated in the pathogenesis of AD. This association is supported by evidence of epigenetic changes (Shrestha et al., 2022), microRNA alterations (Limanaqi et al., 2018), and the indirect promotion of beta-amyloid secretion (Roberts et al., 2017). Amphetamines similarly appear to elevate AD risk. A study by Alrafiah et al. suggested that amphetamine users are at increased risk of developing AD at an earlier age due to inflammatory biomarker changes, such as ACT, PEDF, and MIP-4 (Alrafiah et al., 2022). Prescription amphetamine misuse has also been linked to heightened dementia risk. A national study in Taiwan found that amphetamine use disorder is associated with a nearly five-fold increase in the likelihood of developing dementia, including AD and other forms of dementia (Tzeng et al., 2020).

Conversely, coffee—a widely used stimulant—has shown protective effects against AD. The CADIE trial suggested that midlife coffee consumption is associated with a lower risk of dementia and AD in later life compared to little or no coffee consumption (Eskelinen et al., 2009). Additionally, a study by Sassi et al. reported the effective use of methylphenidate in treating apathy, a significant comorbidity in dementia patients. The findings encourage further randomized controlled trials to explore methylphenidate's utility in this population (Sassi et al., 2020).

Concerning less-common NDDs, cocaine has been implicated in the pathogenesis of ALS and HD through the dissociation of SIG-1R and disruption of Sig-1R pathways. Moreover, cocaine use disorder has been correlated with both AD and PD, further underscoring its neurodegenerative potential (Cai et al., 2017).

DIAGNOSIS AND TREATMENT APPROACHES FOR COMORBID STIMULANT USE AND NEURODEGENERATIVE DISORDERS

The concurrent occurrence of stimulant use disorder and NDDs presents a significant clinical challenge, necessitating thorough diagnostic evaluations and comprehensive treatment strategies. While stimulant use, including substances such as cocaine and methamphetamine, is associated with both immediate and long-term neurological consequences, NDDs like AD and PD are characterized by progressive cognitive decline and motor dysfunction. Understanding the complex relationship between stimulant use disorder and NDDs is essential for developing practical diagnostic and therapeutic approaches.

Diagnostic Considerations

Accurately diagnosing the comorbidity of stimulant use disorder and NDDs requires a comprehensive assessment that includes clinical manifestations, substance use history, and neurocognitive functioning. The DSM-5-TR provides diagnostic criteria for substance use disorders, while standardized neurological evaluations, including neuroimaging and neuropsychological assessments, help identify NDDs. Employing screening, brief intervention, and referral to treatment (SBIRT) methodologies allows for a thorough understanding of the patient's symptoms and severity. Additionally, addressing substance-related concerns, including their effects on family dynamics, relationships, employment, and legal issues, is essential for effective intervention (TIP 26, 2020).

Integrated Care Models

Integrated care models that concurrently address substance use disorder and NDDs offer a holistic approach. This collaborative model involves coordination among addiction specialists, neurologists, psychiatrists, geriatricians, nurses, social workers, and other healthcare professionals to develop personalized treatment plans. Integrated care may include pharmacotherapy, psychotherapy, cognitive rehabilitation, and lifestyle modifications tailored to each patient's needs.

Pharmacological Interventions

Pharmacotherapy plays a vital role in addressing the challenges of comorbid stimulant use disorder and NDDs. Currently, no US Food and Drug Administration (FDA)-approved pharmacological interventions exist specifically for treating cocaine or amphetamine use disorder (Karila et al., 2010; SAMHSA, 2021). Moreover, there are no approved therapies tailored for older adults experiencing stimulant use disorders alongside cognitive impairment.

A systematic review by Ronsley et al. (2020) evaluated various pharmacological interventions for stimulant use disorder, including antidepressants, disulfiram, dopamine agonists, antipsychotics, anticonvulsants, topiramate, psychostimulants, opioid agonists, and N-acetylcysteine (NAC). The findings revealed that antidepressants, dopamine agonists, antipsychotics, and anticonvulsants generally lacked significant efficacy. Disulfiram demonstrated mixed results, with some benefits for individuals dependent on cocaine. Topiramate showed promise in promoting cocaine abstinence, while psychostimulants also yielded positive outcomes for maintaining abstinence, though further research is necessary to validate their effectiveness (Ronsley et al., 2020). Additionally, specific medications have shown potential in managing stimulant use disorder. For example, bupropion has demonstrated efficacy in reducing cravings and preventing relapse in cocaine use disorder (Chan B., 2019), while naltrexone has shown effectiveness for methamphetamine use disorder (Jayaram-Lindström N, 2008). A combination of both compounds was also found to be effective in individuals with methamphetamine dependence (Trivedi et al., 2021).

In the context of NDDs, cholinesterase inhibitors such as donepezil, rivastigmine, and galantamine are commonly used in mild to moderate AD to enhance acetylcholine synthesis and address cortical cholinergic dysfunction, offering modest symptomatic relief. Memantine, an NMDA receptor antagonist, is thought to be neuroprotective and is often combined with cholinesterase inhibitors in moderate to severe AD cases (Folch, 2018). Recent advancements include the approval of monoclonal antibodies targeting amyloid-beta. Aducanumab (Padda, 2024) and lecanemab target aggregated forms of amyloid-beta for mild AD. Lecanemab, in particular, has demonstrated reductions in amyloid markers and moderate slowing of cognitive and functional decline over 18 months. However, it is associated with adverse events (Van Dyck, 2023).

For other types of dementia, no FDA-approved treatments currently exist. Symptomatic treatments focus on alleviating neurobehavioral, cognitive, and neuropsychiatric symptoms in conditions such as FTD, dementia with Lewy bodies, and vascular dementia (Arvanitakis, 2019). Dopaminergic agents are also used to address motor and cognitive symptoms across various forms of dementia, though careful consideration is required to avoid drug interactions and adverse effects in individuals with comorbid conditions.

Clinicians must also evaluate the effects of amphetamine and methylphenidate on cognitive impairment in patients with dementia. Methylphenidate works by inhibiting norepinephrine and dopamine reuptake into presynaptic neurons while increasing their release into the extraneuronal space, potentially improving behavioral and neurocognitive symptoms. Dopamine neuron hypofunction, a hallmark of apathy and cognitive impairment in dementia, may be alleviated through methylphenidate's mechanism of action, which restores dopamine neuron function (Kishi, 2020).

Behavioral Therapies

Behavioral therapies such as cognitive-behavioral therapy (CBT), contingency management (CM), community reinforcement approach, and motivational interviewing are highly effective in managing stimulant use disorder. These approaches address maladaptive behaviors and reinforce adaptive coping strategies (SAMHSA, 2021).

A comprehensive meta-analysis involving nearly 7,000 individuals across 12 psychosocial interventions for cocaine and amphetamine use disorder identified the combination of CM and the community reinforcement approach as the most effective and well-accepted treatment, both in the short term and over extended periods (De Crescenzo, 2018). When adapted to include cognitive rehabilitation techniques, these therapies can also address the specific challenges faced by individuals with comorbid NDDs.

Family education and supportive interventions also play pivotal roles by enhancing social support networks and fostering greater functional independence. These strategies improve treatment outcomes and overall well-being, making them valuable components of a holistic care plan.

Lifestyle Modifications

Lifestyle modifications, including regular physical exercise, a balanced diet, management of co-existing medical conditions such as diabetes, hypertension, and heart disease, as well as stress management techniques, are essential components in treating both stimulant use disorder (Greer et al., 2012) and neurodegenerative disorders (Santiago & Potashkin, 2023). Physical activity mitigates the neurotoxic effects of stimulant use and fosters neuroplasticity, while a nutrient-rich diet high in antioxidants may help slow neurodegenerative processes (Pickersgill, 2022; Cutuli, 2023). Additionally, mindfulness-based practices and relaxation techniques reduce anxiety and improve mood, offering significant benefits for individuals with comorbid conditions.

The diagnosis and treatment of comorbid stimulant use disorder and NDDs demand a multidisciplinary approach that simultaneously addresses substance-related and neurological symptoms. Integrated care models, pharmacological interventions, behavioral therapies, and lifestyle modifications are critical components of effective treatment strategies. Further research is essential to elucidate the underlying mechanisms driving these comorbidities and to develop targeted interventions that cater to the unique needs of affected individuals.

CONCLUSION

This chapter has examined the nature and epidemiology of stimulant use disorders while exploring the intricate relationship between stimulant use and select NDDs. The existing literature predominantly emphasizes the connection between stimulant use and PD, as well as the adverse effects of methamphetamines. Although discussions on other NDDs and stimulant use disorders are more limited, this chapter has sought to address these gaps by presenting insights into the diagnosis and treatment of NDDs. The strategies discussed include integrated approaches combining lifestyle modifications, psychotherapy, and pharmacological interventions tailored to individual needs. However, the irreversible nature of neurodegenerative illnesses underscores prevention as the most effective course of action. Further research is warranted to deepen our understanding of the interactions between various stimulants, such as cocaine and prescription amphetamines, and NDDs, as well as to identify innovative pharmacological solutions for their management.

REFERENCES

Alrafiah, A. R., Abu-Illah, M. M., Magadmi, R. M., Aqeel, A., Najmi, A., & Jaddoh, S. (2022). Increased inflammatory markers at AMPH-addicts are related to neurodegenerative conditions: Alzheimer's disease. *Applied Sciences*, *12*(11), 5536.

Ajitkumar, A., & De Jesus, O. *Huntington disease.* [Updated 2023 Aug 23]. In StatPearls [Internet]. Treasure Island (FL): StatPearls Publishing; 2024 Jan. https://www.ncbi.nlm.nih.gov/books/NBK559166/

American Psychiatric Association. (2022). Bipolar and related disorders. In *Diagnostic and statistical manual of mental disorders* (5th ed., text rev.). https://doi.org/10.1176/appi.books.9780890425787.x03_Bipolar_and_Related_Disorders

Andersen, M. L., Diaz, M. P., Murnane, K. S., & Howell, L. L. (2013). Effects of methamphetamine self-administration on actigraphy-based sleep parameters in rhesus monkeys. *Psychopharmacology*, *227*, 101–107.

Arvanitakis, Z., Shah, R. C., & Bennett, D. A. (2019). Diagnosis and management of dementia: Review. *JAMA*, *322*(16), 1589–1599. https://doi.org/10.1001/jama.2019.4782. PMID: 31638686; PMCID: PMC7462122.

Baik, J. H. (2013). Dopamine signaling in reward-related behaviors. *Frontiers in Neural Circuits*, *7*, 152.

Cadet, J. L., Krasnova, I. N., Jayanthi, S., & Lyles, J. (2007). Neurotoxicity of substituted amphetamines: Molecular and cellular mechanisms. *Neurotoxicity Research*, *11*(3-4), 183–202. https://doi.org/10.1007/BF03033567

Cai, Y., Yang, L., Niu, F., Liao, K., & Buch, S. (2017). Role of sigma-1 receptor in cocaine abuse and neurodegenerative disease. *Advances in Experimental Medicine and Biology*, *964*, 163–175. https://doi.org/10.1007/978-3-319-50174-1_12

Callaghan, R. C., Cunningham, J. K., Sykes, J., & Kish, S. J. (2012). Increased risk of Parkinson's disease in individuals hospitalized with conditions related to the use of methamphetamine or other amphetamine-type drugs. *Drug and Alcohol Dependence*, *120*(1–3), 35–40. https://doi.org/10.1016/j.drugalcdep.2011.06.013

Chan, B., Kondo, K., Freeman, M., Ayers, C., Montgomery, J., & Kansagara, D. (2019). Pharmacotherapy for cocaine use disorder—A systematic review and meta-analysis. *Journal of General Internal Medicine*, *34*, 2858–2873.

Curtin, K., Fleckenstein, A. E., Robison, R. J., Crookston, M. J., Smith, K. R., & Hanson, G. R. (2015). Methamphetamine/amphetamine abuse and risk of Parkinson's disease in Utah: A population-based assessment. *Drug and Alcohol Dependence*, *146*, 30–38. https://doi.org/10.1016/j.drugalcdep.2014.10.027

Cutuli, D., Decandia, D., Giacovazzo, G., & Coccurello, R. (2023). Physical exercise as disease-modifying alternative against Alzheimer's disease: A gut–muscle–brain partnership. *International Journal of Molecular Sciences*, *24*(19), 14686.

De Crescenzo, F., Ciabattini, M., D'Alò, G. L., De Giorgi, R., Del Giovane, C., Cassar, C., ... & Cipriani, A. (2018). Comparative efficacy and acceptability of psychosocial interventions for individuals with cocaine and amphetamine addiction: A systematic review and network meta-analysis. *PLoS Medicine, 15*(12), e1002715.

Drossel, G., Brucar, L. R., Rawls, E., Hendrickson, T. J., & Zilverstand, A. (2023, April 18). Subtypes in addiction and their neurobehavioral profiles across three functional domains. *Nature News*. https://www.nature.com/articles/s41398-023-02426-1

Emmady, P. D., Schoo, C., & Tadi, P. *Major Neurocognitive Disorder (Dementia)* [Updated 2022 Nov 19].

Eskelinen, M. H., Ngandu, T., Tuomilehto, J., Soininen, H., & Kivipelto, M. (2009). Midlife coffee and tea drinking and the risk of late-life dementia: A population-based CAIDE study. *Journal of Alzheimer's Disease*: *JAD, 16*(1), 85–91. https://doi.org/10.3233/JAD-2009-0920

Ersche, K. D., Barnes, A., Jones, P. S., Morein-Zamir, S., Robbins, T. W., & Bullmore, E. T. (2011). Abnormal structure of frontostriatal brain systems is associated with aspects of impulsivity and compulsivity in cocaine dependence. *Brain: A Journal of Neurology, 134*(Pt 7), 2013–2024. https://doi.org/10.1093/brain/awr138

Ersche, K. D., Williams, G. B., Robbins, T. W., & Bullmore, E. T. (2013). Meta-analysis of structural brain abnormalities associated with stimulant drug dependence and neuroimaging of addiction vulnerability and resilience. *Current Opinion in Neurobiology, 23*(4), 615–624.

Ersche, K. D., Stochl, J., Brühl, A. B., & Gurnell, M. (2024). Evidence of hypothalamic-pituitary-adrenal and-gonadal dysfunction in cocaine-addicted men. *European Addiction Research, 30*(2), 114–120.

Faraone, S. V. (2018, April). The pharmacology of amphetamine and methylphenidate: Relevance to the neurobiology of attention-deficit/hyperactivity disorder and other psychiatric comorbidities. *Neuroscience & Biobehavioral Reviews 87,* 255–270. https://doi.org/10.1016/j.neubiorev.2018.02.001.

Farrell, M., Martin, N. K., Stockings, E., Bórquez, A., Cepeda, J. A., Degenhardt, L., Ali, R., Tran, L. T., Rehm, J., Torrens, M., Shoptaw, S., & McKetin, R. (2019). Responding to global stimulant use: Challenges and opportunities. *Lancet (London, England), 394*(10209), 1652–1667. https://doi.org/10.1016/S0140-6736(19)32230-5

Ferrucci, M., Limanaqi, F., Ryskalin, L., Biagioni, F., Busceti, C. L., & Fornai, F. (2019). The effects of amphetamine and methamphetamine on the release of norepinephrine, dopamine and acetylcholine from the brainstem reticular formation. *Frontiers in Neuroanatomy, 13*, 48. https://doi.org/10.3389/fnana.2019.00048

Folch, J., Busquets, O., Ettcheto, M., Sánchez-López, E., Castro-Torres, R. D., Verdaguer, E., Garcia, M. L., Olloquequi, J., Casadesús, G., Beas-Zarate, C., Pelegri, C., Vilaplana, J., Auladell, C., & Camins, A. (2018). Memantine for the treatment of dementia: A review on its current and future applications. *Journal of Alzheimer's Disease 62*(3), 1223–1240. https://doi.org/10.3233/JAD-170672.

Galimberti, D., & Scarpini, E. (2018). *Neurogenerative diseases: Clinical considerations, molecular genetics and biomarkers*. Oxford Press.

Greer, T. L., Ring, K. M., Warden, D., Grannemann, B. D., Church, T. S., Somoza, E., Blair, S. N., Szapocznik, J., Stoutenberg, M., Rethorst, C., Walker, R., Morris, D. W., Kosinski, A. S., Kyle, T., Marcus, B., Crowell, B., Oden, N., Nunes, E., & Trivedi, M. H. (2012). Rationale for using exercise in the treatment of stimulant use disorders. *Journal of Global Drug Policy and Practice, 6*(1), http://ctndisseminationlibrary.org/display/825.htm.

Jahan, A. R., & Burgess, D. M. *Substance use disorder*. [Updated 2023 Jul 21]. In: StatPearls [Internet]. Treasure Island (FL): StatPearls Publishing; 2024 Jan. https://www.ncbi.nlm.nih.gov/books/NBK570642/

Jayaram-Lindström, N., Hammarberg, A., Beck, O., & Franck, J. (2008). Naltrexone for the treatment of amphetamine dependence: a randomized, placebo-controlled trial. *American Journal of Psychiatry, 165*(11), 1442–1448.

Kalivas, P. W., & Volkow, N. D. (2005). The neural basis of addiction: A pathology of motivation and choice. *American Journal of Psychiatry, 162*(8), 1403–1413.

Karila, L., Weinstein, A., Aubin, H. J., Benyamina, A., Reynaud, M., & Batki, S. L. (2010). Pharmacological approaches to methamphetamine dependence: A focused review. *British Journal of Clinical Pharmacology, 69*(6), 578–592.

Khan, I., & De Jesus, O. *Frontotemporal lobe dementia*. [Updated 2023 Aug 23]. In: StatPearls [Internet]. Treasure Island (FL): StatPearls Publishing; 2024 Jan. https://www.ncbi.nlm.nih.gov/books/NBK559286/

Kishi, T., Sakuma, K., & Iwata, N. (2020). Efficacy and safety of psychostimulants for Alzheimer's disease: A systematic review and meta-analysis. *Pharmacopsychiatry, 53*(03), 109–114.

Lamptey, R. N. L., Chaulagain, B., Trivedi, R., Gothwal, A., Layek, B., & Singh, J. (2022). A review of the common neurodegenerative disorders: Current therapeutic approaches and the potential role of nanotherapeutics. *International Journal of Molecular Sciences*, *23*(3), 1851. https://doi.org/10.3390/ijms23031851

Limanaqi, F., Gambardella, S., Biagioni, F., Busceti, C. L., & Fornai, F. (2018). Epigenetic effects induced by methamphetamine and methamphetamine-dependent oxidative stress. *Oxidative Medicine and Cellular Longevity*, 2018, 4982453. https://doi.org/10.1155/2018/4982453.

Li, X., Feng, X., Sun, X., Hou, N., Han, F., & Liu Y. (2022, October 10). Global, regional, and national burden of Alzheimer's disease and other dementias, 1990–2019. *Frontiers in Aging Neuroscience14,* 937486. https://doi.org/10.3389/fnagi.2022.937486.

McGregor, C., Srisurapanont, M., Jittiwutikarn, J., Laobhripatr, S., Wongtan, T., & White, J. M. (2005). The nature, time course and severity of methamphetamine withdrawal. *Addiction*, *100*(9), 1320–1329.

Murnane, K. S., Andersen, M. L., Rice, K. C., & Howell, L. L. (2013). Selective serotonin 2A receptor antagonism attenuates the effects of amphetamine on arousal and dopamine overflow in non-human primates. *Journal of Sleep Research*, *22*, 581–588.

Murnane, K. S., Perrine, S. A., Finton, B. J., Galloway, M. P., Howell, L. L., & Fantegrossi, W. E. (2012). Effects of exposure to amphetamine derivatives on passive avoidance performance and the central levels of monoamines and their metabolites in mice: Correlations between behavior and neurochemistry. *Psychopharmacology*, *220*, 495–508.

Padda, I. S., & Parmar, M. Aducanumab. [Updated 2024 Feb 26]. In: StatPearls [Internet]. Treasure Island (FL): StatPearls Publishing; 2024 Jan-. https://www.ncbi.nlm.nih.gov/books/NBK573062/

Park, M., Kim, H. J., Lim, B., Wylegala, A., & Toborek, M. Methamphetamine-induced occludin endocytosis is mediated by the Arp2/3 complex-regulated actin rearrangement. *Journal of Biological Chemistry*, *288*, 33324–33334.

Pickersgill, J. W., Turco, C. V., Ramdeo, K., Rehsi, R. S., Foglia, S. D., & Nelson, A. J. (2022). The combined influences of exercise, diet and sleep on neuroplasticity. *Frontiers in Psychology*, *13*, 831819.

Pregeljc, D., Teodorescu-Perijoc, D., Vianello, R., Umek, N., & Mavri, J. (2020). How important is the use of cocaine and amphetamines in the development of Parkinson disease? A computational study. *Neurotoxicity Research*, *37*(3), 724–731. https://doi.org/10.1007/s12640-019-00149-0

Qin Yujing1, Ouyang, Qinjie1, Pablo, John1, Mash, & Deborah, C.1 2 CA. (2005, September 8). Cocaine abuse elevates alpha-synuclein and dopamine transporter levels in the human striatum. *NeuroReport*, *16*(13), 1489–1493. https://doi.org/10.1097/01.wnr.0000175617.39054.ba

Roberts, H. L., Schneider, B. L., & Brown, D. R. (2017). α-Synuclein increases β-amyloid secretion by promoting β-/γ-secretase processing of APP. *PloS One*, *12*(2), e0171925

Ronsley, C., Nolan, S., Knight, R., Hayashi, K., Klimas, J., Walley, A., ... & Fairbairn, N. (2020). Treatment of stimulant use disorder: A systematic review of reviews. *PloS One*, *15*(6), e0234809.

Santiago, J. A., & Potashkin, J. A. (2023). Physical activity and lifestyle modifications in the treatment of neurodegenerative diseases. *Frontiers in Aging Neuroscience*, *15*, 1185671.

Sassi, K. L. M., Rocha, N. P., Colpo, G. D., John, V., & Teixeira, A. L. (2020). Amphetamine use in the elderly: A systematic review of the literature. *Current Neuropharmacology* *18*(2), 126–135. https://doi.org/10.2174/1570159X17666191010093021.

Shrestha, P., Katila, N., Lee, S., Seo, J. H., Jeong, J. H., & Yook, S. (2022). Methamphetamine induced neurotoxic diseases, molecular mechanism, and current treatment strategies. *Biomedicine & Pharmacotherapy = Biomedecine & Pharmacotherapie*, *154*, 113591. https://doi.org/10.1016/j.biopha.2022.113591

Shukla, M., & Vincent, B. (2020). The multi-faceted impact of methamphetamine on Alzheimer's disease: From a triggering role to a possible therapeutic use. *Ageing Research Reviews*, *60*, 101062.

Substance Abuse and Mental Health Services Administration. (2021). Key substance use and mental health indicators in the United States: Results from the 2020 National Survey on Drug Use and Health (HHS Publication No. PEP21-07-01-003, NSDUH Series H-56). Rockville, MD: Center for Behavioral Health Statistics and Quality, Substance Abuse and Mental Health Services Administration. https://www.samhsa.gov/data/

Trivedi, M. H., Walker, R., Ling, W., Dela Cruz, A., Sharma, G., Carmody, T., ... & Shoptaw, S. (2021). Bupropion and naltrexone in methamphetamine use disorder. *New England Journal of Medicine*, *384*(2), 140–153.

Tzeng, N. S., Chien, W. C., Chung, C. H., Chang, H. A., Kao, Y. C., & Liu, Y. P. (2020). Association between amphetamine-related disorders and dementia-a nationwide cohort study in Taiwan. *Annals of Clinical and Translational Neurology*, *7*(8), 1284–1295. https://doi.org/10.1002/acn3.51113

Van Dyck, C. H., Swanson, C. J., Aisen, P., Bateman, R. J., Chen, C., Gee, M., Kanekiyo, M., Li, D., Reyderman, L., Cohen, S., Froelich, L., Katayama, S., Sabbagh, M., Vellas, B., Watson, D., Dhadda, S., Irizarry, M., Kramer, L. D., & Iwatsubo, T. (2023, January 5). Lecanemab in early Alzheimer's disease. *New England Journal of Medicine, 388*(1), 9–21. https://doi.org/10.1056/NEJMoa2212948.

Volkow, N. D., & Morales, M. (2015). The brain on drugs: From reward to addiction. *Cell, 162*(4), 712–725. https://doi.org/10.1016/j.cell.2015.07.046

World Health Organization. Fact sheets of dementia [Internet]. Geneva, Switzerland: World Health Organization; 2021 [cited 2022 Apr 13]. https:// www.who.int/news-room/fact-sheets/detail/dementia

11 Tobacco and Neurodegenerative Disorders

Celal Yesilkaya and Serkan Turan

INTRODUCTION

Tobacco harbors a complex amalgamation of over 5,000 chemical compounds, among which more than 30 have been conclusively identified as human carcinogens (Hajdusianek et al., 2021; Suwannatad, 2022). Constituents such as acetaldehyde, ammonia, hydrogen cyanide, and formaldehyde, prevalent in tobacco smoke, inflict deleterious effects by compromising the integrity of the ciliary epithelium (Tareq Mussa et al., 2018). Notably, these substances are also found in electronic cigarettes, raising concerns about their safety (Gülşen & Uslu, 2020). The World Health Organization (WHO) estimates that around 1.3 billion individuals worldwide engage in daily tobacco consumption (WHO, 2020), with initiation of smoking and other tobacco practices commonly occurring during adolescence (Naik & Cucullo, 2015). A multinational longitudinal study involving youth aged 13 to 15 revealed a prevalence rate of 17.9% among boys and 11.5% among girls for any tobacco product consumption within the preceding 30 days (Ma et al. 2021). A 30-year cohort study in the United States indicated that the smoking rate among individuals over the age of 55 was reported to be 11.8% in 2018, compared to 15.9% in 1998 (Hunt et al., 2023).

During smoking, only 15% of the emitted smoke is absorbed, while the remaining 85% is released into the environment, significantly impacting the surrounding air quality (Suwannatad, 2022). The devastating toll of tobacco use is starkly evident in epidemiological data, with an estimated 100 million deaths attributed to tobacco-related causes in the 20th century alone and current statistics indicating approximately 6 million deaths annually worldwide (Adhikari et al., 2009; Jha, 2009). Leading the roster of smoking-associated fatalities in the US are cardiovascular disease (CVD), chronic obstructive pulmonary disease (COPD), and various malignancies (US Department of Health and Human Services, 2014). A study delineated the extensive economic burden posed by active and passive smoking, tallying approximately $289 billion in additional costs between 2009 and 2012, encompassing medical care expenses and productivity losses (Adhikari et al. 2009).

Many of the health afflictions linked with tobacco use exhibit potential for regression following sustained smoking cessation, underscoring the modifiable nature of smoking as a risk factor for adverse health outcomes. Moreover, immune dysregulation, rheumatoid arthritis, diabetes, ocular disorders, and inflammatory bowel diseases are among the spectrum of conditions believed to bear associations with tobacco exposure (US Department of Health and Human Services, 2014).

Past investigations scrutinizing the nexus between smoking and cerebrovascular ailments have established correlations with silent cerebral infarction, stroke, cerebral small vessel disease (CSVD), and cerebral aneurysms (Chalouhi et al., 2012; Naik & Cucullo, 2015). Additionally, burgeoning scientific evidence underscores the potential links between tobacco consumption and hippocampal volume diminution, alongside cognitive impairments in the learning and memory domains, positing these neurocognitive deficits as potential precursors to Alzheimer's disease (AD) (Almeida et al., 2008; Durazzo et al., 2012, 2013). Numerous meta-analyses and longitudinal inquiries corroborate a significantly heightened risk for AD and related dementias among individuals with a history of smoking (Cataldo et al., 2010).

Nicotine primarily exerts its pharmacological effects by activating nicotinic (nAchR) and muscarinic (mAchR) acetylcholine receptors, with both nicotine and acetylcholine exhibiting robust

DOI: 10.1201/9781003478485-11

affinities for these receptors. The central nervous system (CNS) harbors 42 subtypes of these receptors, each characterized by varying affinities (Akaike et al., 2018; Shimohama & Kawamata, 2018). Nicotine's stimulation of nAchRs within dopaminergic neurons originating from the ventral tegmental area (VTA) and projecting to the nucleus accumbens precipitates disproportionate dopamine release within the latter, a pivotal component of the reward circuitry implicated in the addiction process (Tiwari et al., 2020). Meta-analyses have documented reductions in dopamine transporters (DAT) among tobacco users, while D2 dopamine receptor function remains ostensibly unaffected (Ashok et al., 2019; Palermo & Ceravolo, 2019).

This chapter discusses the mechanisms of tobacco's effects on the blood–brain barrier (BBB) and the CNS, its effects on the pathogenesis of AD and Parkinson's disease (PD), the treatments planned based on these effects, and the potential roles of other tobacco components in prognosis.

MECHANISMS OF NEURODEGENERATIVE PROCESSES WITH TOBACCO

Blood–Brain Barrier

The BBB constitutes a dynamic interface primarily comprised of endothelial cells lining the luminal surfaces of cerebral vessels alongside juxtaposed astrocytic end-foot projections and supportive pericytes, collectively contributing to the safeguarding and selective permeability of the BBB (Science et al., 2002). Unlike their counterparts in peripheral circulation, the endothelial cells forming the BBB lack fenestrations, and their transmembrane carrier distribution exhibits distinct patterns. Integral to the stringent regulation of molecular flux across the BBB are connecting elements such as zonula occludens, occludin, and claudins, which effectively impede the transit of hydrophilic substances while facilitating the transport of essential nutrients via hydrophilic carrier systems (Naik & Cucullo, 2012). This intricate balance prevents detrimental external agents from permeating the brain parenchyma while facilitating the uptake of vital nutrients critical for neural function. Paramount in determining the impermeability of the BBB are tight junctions, orchestrating the formidable barrier function essential for maintaining CNS homeostasis (Bazzoni & Dejana, 2004).

Effect of Tobacco Smoking on the Blood–Brain Barrier and Central Nervous System

The detrimental impact of tobacco use on neurological and neurovascular function is intricately linked to the heightened levels of reactive oxidative species (ROS), which incite oxidative and inflammatory damage within the CNS (Bennett et al., 2009; Cojocaru et al., 2013; Khanna et al., 2013). Notably, exposure to cigarette smoke has been correlated with the down-regulation of critical elements comprising the BBB, including zona occludens, occludin, and cadherin, with nicotine itself exhibiting an affinity for these junctional proteins (Deane & Zlokovic, 2007; Naik & Cucullo, 2015; Lockman et al., 2005).

Elevated leukocyte counts, mainly neutrophils, are an additional inflammatory hallmark in smokers, with these cells implicated in the direct induction of endothelial damage through the secretion of free radicals, elastase, and collagenase (Naik & Cucullo, 2012). Platelet activation is another frequent occurrence in smoking-related studies, further exacerbating endothelial dysfunction. The heightened release and activity of matrix metalloproteinase-2 and -9 represent additional pathways through which tobacco use compromises BBB integrity, primarily by degrading basal lamina components and facilitating immune cell trafficking into the CNS (Hossain et al., 2009; Strazielle et al., 2003). A study suggested that endothelium-derived CD105 and CD49e release represents another inflammatory change that plays a role in impairing immune system function. CD105 is associated with angiogenesis and extravasation, while CD49e deficiency has been shown to result in impaired BBB integrity in mice. Furthermore, smokers show increased monocyte-derived CD36 expression alongside decreased levels of interleukin (IL)-33, IL-10, IL-8, and brain-derived neurotrophic factors (Garza et al., 2023).

Augmented nitric oxide (NO) level is an additional mediator of tobacco-induced BBB compromise, exerting multiple effects on vascular tone regulation, leukocyte-endothelial adhesion, and platelet aggregation (Wong et al., 2004). While NO initially functions protectively by inducing vasodilation in response to acute ischemic insults, sustained ischemic conditions trigger an upsurge in inducible NO synthase (iNOS) activation, leading to peroxynitrite and superoxide anion radical formation, which creates a pro-inflammatory and pro-thrombotic environment (Peluffo et al., 2009).

In the context of the CNS, characterized by high oxygen concentrations and low levels of antioxidant enzymes, neurons are particularly susceptible to the deleterious effects of ROS-induced oxidative stress (Olmez & Ozyurt, 2012). Epidemiological and in vitro investigations consistently underscore the neurologically and neurovascularly deleterious consequences of tobacco use, further highlighting the imperative for targeted interventions aimed at mitigating these effects (Naik & Cucullo, 2015).

Another pathway explaining the relationship between tobacco and AD is the disruption in the oral microbiota. One study suggested that both active and passive smoking are associated with a decline in healthy bacteria and an increase in pathogenic ones in the oral mucosa. This imbalance results in the direct invasion of pathogenic bacteria into the CNS, along with increased amyloid-beta (Aß) deposition and activation of the immune response, leading to both an earlier onset of AD and a worse prognosis (Dai et al., 2024).

The above-mentioned mechanisms mediating the effects of tobacco smoking on the BBB and CNS are summarized in Table 11.1.

ALZHEIMER'S DISEASE

AD stands as the foremost cause of dementia, a debilitating disorder characterized by a gradual loss of cognitive and social abilities (Veljkovic et al., 2018). Late-onset AD, typically manifesting after the age of 65, constitutes the predominant form, encompassing nearly 95% of cases globally (Prince et al., 2013). Presently, an estimated 44 million individuals worldwide grapple with AD or related dementias, with projections suggesting a staggering rise to approximately 100 million cases by 2050 (Veljkovic et al., 2018; Brookmeyer et al., 2007).

Early-stage AD commonly manifests through challenges in recent memory recall, with the disease typified by an insidious trajectory featuring protracted asymptomatic preclinical phases preceding overt dementia symptoms, which may span several decades (Sperling et al., 2011). As pathological hallmarks associated with AD accrue, individuals often transition to a stage of mild cognitive impairment (MCI), characterized by significant memory deficits, signifying an elevated risk of progression to full-blown AD (Drago et al., 2011). Advanced disease stages herald a myriad of behavioral disturbances, encompassing speech impairment, spatial disorientation, emotional

TABLE 11.1
Summary of Tobacco's Impact on the BBB and CNS

ROS-mediated oxidative stress.
Down-regulation of main components of the BBB, such as zonula occludens, occludin, and cadherin.
Increase in neutrophil-mediated inflammatory response.
Disruptive effect of matrix metalloproteinase-2 and -9 on the basal lamina and inflammatory cell migration to the CNS.
Increased NO levels in response to increased iNOS activity secondary to ischemia, resulting in progression to pro-inflammatory and pro-thrombotic pathways.
Direct pathogen invasion and immune response caused by disruption in oral microbiota.

lability, compromised self-care, and behavioral aggression, ultimately culminating in a profound decline in both familial and societal functioning, leading to eventual demise (Veljkovic et al., 2018).

Pathologically, AD is characterized by the degeneration and demise of brain cells, precipitating a relentless decline in memory and cognitive functions. Notable neuropathological features include the deposition of extracellular amyloid and neuritic plaques alongside intracellular neurofibrillary tangles (Durazzo et al., 2014). Amyloid plaques arise from the aggregation of insoluble Aß peptides, while neuritic plaques comprise insoluble Aß peptides, degenerated neurites, and hyperphosphorylated tau proteins (Nelson et al., 2012). Notably, Aß deposition and plaque formation are believed to commence up to two decades before AD symptom onset, with cerebral Aß accumulation influenced by age and APOE-ε4 genotype (Durazzo et al., 2014; Morris et al., 2010). Concomitantly, reactive microgliosis and extensive neuronal, white matter, and synaptic losses ensue, further exacerbating neurodegenerative processes (Veljkovic et al., 2018).

Current paradigms postulate a multifactorial etiology for AD pathogenesis, implicating aberrations in Aß metabolism as the initial trigger, followed by cortical tau pathology, neuroinflammation, synaptic deterioration, and neuronal demise (Mondragón-Rodríguez et al., 2010). Notably, cerebrospinal fluid (CSF) levels of ß-amyloid 42 (Aß42) hold diagnostic specificity for AD, with elevated Aß42 levels observed in AD and MCI cases, suggesting an intricate interplay between Aß dysregulation, oxidative stress, neuroinflammation, and neurodegeneration in AD pathogenesis (Olsson et al., 2016; Kametani & Hasegawa, 2018).

The relationship between tobacco use and AD risk remains contentious, with divergent findings in the literature. While some studies assert an association between tobacco use and heightened AD risk, others propose inconclusive or even contradictory outcomes (Daviglus et al., 2011; Cataldo et al., 2010). Notably, meta-analyses have underscored the significance of smoking as a potent risk factor for AD development, with a particularly pronounced risk observed among middle-aged smokers and APOE-ε4 carriers (Rusanen et al., 2010). Moreover, cumulative evidence suggests a dose-dependent relationship between smoking duration and AD risk, with prolonged exposure correlating with heightened disease susceptibility (Durazzo et al., 2014; Reitz et al., 2007; Tyas et al., 2003). Gender disparities have also been noted, with male smokers exhibiting an elevated risk of AD compared to their female counterparts (Launer et al., 1999). Notably, smoking has been implicated in hastening disease onset and exacerbating neurodegeneration, with heavy smoking linked to a more than 100% increased risk of dementia and AD following prolonged exposure (Rusanen et al., 2011). Moreover, tobacco use has been associated with accelerated cognitive decline, gray matter loss, and heightened neuronal atrophy in patients with AD, underscoring the profound neurotoxic effects of smoking (Hajdusianek et al., 2021; Suwannatad, 2022).

Neuronal nAChRs play a pivotal role in AD pathophysiology, with diminished nAChR expression documented in cholinergic neurons and synapses in AD brains, alongside the frequent occurrence of neuritic and Aß plaques within nAChR-rich regions (Oz et al., 2013). Chronic nicotine exposure has been shown to exacerbate tau pathology in AD mouse models, further implicating nAChR dysregulation in disease progression (Oddo et al., 2005). Paradoxically, nicotine exhibits dual effects on AD pathogenesis, with studies suggesting both neuroprotective and neurotoxic properties. While nicotine has been implicated in mitigating Aß- and N-methyl-D-aspartate (NMDA)-mediated toxicity through antioxidative mechanisms, it may also induce oxidative stress and structural brain damage upon chronic exposure (Dong et al., 2020; Saad et al., 2020). Notably, the neuroprotective effects of nicotine have been attributed to α7-nAChR activation, with α7-nAChR agonists posited as potential therapeutic agents against neurodegenerative processes (Nullens et al., 2016; Dong et al., 2020). A review suggested that the dextrorotatory enantiomer (R)-nicotine has the potential as a less toxic pharmacotherapeutic intervention (Pogocki et al., 2007).

Considering its role in the development of AD, interventions aimed at treating tobacco addiction may also contribute to decreasing the risk of the disease. For this purpose, pharmacological interventions targeting nAchRs have been developed, and studies suggest that variations in cytochrome P450 2A6 (CYP2A6), catechol-O-methyltransferase (COMT), dopamine D2 (DRD2), and DRD4 receptor genes are determinants of response to nicotine replacement therapies and other pharmacological agents, e.g., bupropion (Almahasneh et al., 2024).

In sum, the intricate interplay between tobacco use and AD pathogenesis underscores the multifaceted nature of disease etiology, necessitating the comprehensive elucidation of the underlying mechanisms to inform targeted therapeutic interventions and mitigate disease burden.

PARKINSON'S DISEASE

Parkinson's disease stands as the predominant movement disorder, characterized by a progressive loss of dopaminergic neurons within the substantia nigra and subsequent dopamine depletion in the striatum and other brain regions, culminating in motor dysfunction (Dexter & Jenner, 2013; Suwannatad, 2022). Despite the heterogeneity of its etiology, PD afflicts an estimated 10 million individuals worldwide, with familial and sporadic forms accounting for its diverse clinical presentations (Hajdusianek et al., 2021; Ball et al., 2019). While genetic mutations underpin familial PD cases, sporadic instances, comprising the majority, lack clearly defined causative factors, though environmental influences such as traumatic brain injury, pesticide exposure, and dietary habits have been implicated (Kenborg et al., 2015;Nicholatos et al., 2018; Spataro et al., 2015). Notably, tobacco and caffeine consumption have emerged as protective factors against PD, contrasting with other detrimental lifestyle factors (Ascherio & Schwarzschild, 2016).

PD pathology is marked by the accumulation of Lewy bodies, intracellular inclusions primarily composed of α-synuclein (α-Syn) within surviving dopaminergic neurons, along with neuroinflammation and oxidative stress (Carvajal-Oliveros et al., 2021). Oxidative stress, in particular, precipitates ribonucleic acid (RNA) oxidation, exacerbates synuclein aggregation, and compromises dopamine release via ATP-sensitive potassium channel activation, exacerbating dopaminergic neuron demise (Kuzkina et al., 2019).

Current PD management strategies primarily address symptomatic relief, underscoring the need for disease-modifying therapies targeting dopaminergic neuron preservation and protein accumulation mitigation (Connolly & Lang, 2014). Notably, nicotine, a key component of tobacco, has garnered interest for its purported neuroprotective properties in PD, with preclinical studies demonstrating an attenuation of motor symptoms and dopaminergic neuron loss via $\alpha 4\beta 2$ and/or $\alpha 7$-nAchR activation (Carvajal-Oliveros et al., 2021). Nicotine-induced dopamine release and subsequent reduction in cytoplasmic dopamine and iron concentrations contribute to oxidative stress mitigation, further supporting its therapeutic potential (Carvajal-Oliveros et al., 2021).

Contrary to its well-documented adverse health effects, tobacco use exhibits a consistent inverse association with PD development and progression, with a more substantial protective effect observed in current smokers and a dose–response relationship between smoking duration/intensity and PD risk (Chen et al., 2010; Gallo et al., 2019; Ma et al., 2017; Quik & Kulak, 2002). Exposure to secondhand smoke has also been implicated in PD risk reduction (Gallo et al., 2019). While the precise mechanisms underlying this inverse relationship remain elusive, theories attributing it to the modulation of dopamine-related reward circuits and reverse causality have been proposed, yet robust evidence refutes these conjectures (Ritz et al., 2014; Gallo et al., 2019). Notably, the protective effect of tobacco against PD transcends demographic and genetic variables, further underscoring its potential as a therapeutic avenue in PD management.

THE ROLE OF NON-NICOTINE CONSTITUENTS OF TOBACCO IN NEURODEGENERATIVE DISORDERS

The inverse relationship between tobacco use and PD has been speculated to involve substances beyond nicotine present in tobacco. For instance, in mouse models, 2,3,6-trimethyl-1,4-naphthoquinone (TMN) found in tobacco has demonstrated the ability to inhibit both monoamine oxidase (MAO) A and B, thereby reducing endogenous dopamine degradation and alleviating neurodegeneration by mitigating oxidative stress (Gallo et al., 2019; Quik et al., 2012). Notably, MAO inhibitors, such as selegiline and rasagiline, have a long history in depression treatment, and MAO-B inhibitors are now used in PD therapy (Rascol et al., 2011). However, the potential for dangerous interactions with dietary substances like tyramine—found in certain foods such as cheese—and other medications raises concerns about the use of selective MAO inhibitors, highlighting the need for new alternatives (Sathyanarayana Rao & Yeragani, 2009; Veljkovic et al., 2018).

Tobacco alkaloids such as anabasine and anatabine, structurally similar to nicotine, also merit attention as potential contributors to the observed inverse correlation with PD. Anatabine, for instance, has exhibited the capacity to reduce Aß expression in AD cell models and reduce Aß1-40 and Aß1-42 levels in the brains of AD mice, suggesting neuroprotective properties (Paris et al., 2011). Anabasine, functioning as a partial agonist for α4β2 nAchRs and a full agonist for the α7 subtype, holds promise in memory enhancement, as evidenced in rat experiments (Veljkovic et al., 2018; Hunter et al., 1994).

Additionally, cotinine, the primary metabolite of nicotine in mammals, has emerged as a potential therapeutic agent in conditions like AD, schizophrenia, and depression, exhibiting cytoprotective and neuroprotective effects in vitro (Burgess et al., 2011; Echeverria et al., 2016). These findings underscore the multifaceted nature of tobacco constituents and their potential roles in neuroprotection and disease modulation beyond nicotine's well-established pharmacological actions.

CONCLUSION

Tobacco use plays a role in the course of conditions such as CVD, COPD, various malignancies, and autoimmune diseases, as well as stroke, CSVD, aneurysms, and neurodegenerative disorders. BBB damage, neuroinflammatory processes, ROS-mediated damage, and pro-inflammatory and prothrombotic processes mediate the effects of tobacco on neurodegenerative disorders. Understanding these mechanisms is necessary for developing targeted pharmacotherapeutic interventions and public health strategies to mitigate the harmful effects of tobacco on the nervous system and overall health.

REFERENCES

Adhikari, B., Kahende, J., Malarcher, A., Pechacek, T., & Tong, V. (2009). Smoking-attributable mortality, years of potential life lost, and productivity losses—United States, 2000–2004. *JAMA*, *301*(6), 593. https://doi.org/10.1001/JAMA.301.6.593

Akaike, A., Shimohama, S., & Misu, Y. (2018). *Nicotinic acetylcholine receptor signaling in neuroprotection* (1st ed.). Springer. https://doi.org/10.1007/978-981-10-8488-1

Almahasneh, F., Gerges, R. H., Abu-El-Rub, E., & Khasawneh, R. R. (2024). Nicotine abuse and neurodegeneration: Novel pharmacogenetic targets to aid quitting and reduce the risk of dementia. *CNS & Neurological Disorders—Drug Targets*, *23*(1), 2–8.

Almeida, O. P., Garrido, G. J., Lautenschlager, N. T., Hulse, G. K., Jamrozik, K., & Flicker, L. (2008). Smoking is associated with reduced cortical regional gray matter density in brain regions associated with incipient Alzheimer disease. *American Journal of Geriatric Psychiatry*, *16*(1), 92–98. https://doi.org/10.1097/JGP.0B013E318157CAD2

Ascherio, A., & Schwarzschild, M. A. (2016). The epidemiology of Parkinson's disease: Risk factors and prevention. *Lancet Neurology*, *15*(12), 1257–1272. https://doi.org/10.1016/S1474-4422(16)30230-7

Ashok, A. H., Mizuno, Y., & Howes, O. D. (2019). Tobacco smoking and dopaminergic function in humans: A meta-analysis of molecular imaging studies. *Psychopharmacology*, *236*(4), 1119–1129. https://doi.org/10.1007/S00213-019-05196-1/FIGURES/2

Ball, N., Teo, W. P., Chandra, S., & Chapman, J. (2019). Parkinson's disease and the environment. *Frontiers in Neurology*, *10*, 421551. https://doi.org/10.3389/FNEUR.2019.00218/BIBTEX

Bazzoni, G., & Dejana, E. (2004). Endothelial cell-to-cell junctions: Molecular organization and role in vascular homeostasis. *Physiological Reviews*, *84*(3), 869–901. https://doi.org/10.1152/PHYSREV.00035.2003/ASSET/IMAGES/LARGE/Z9J0030403181006.JPEG

Bennett, S., Grant, M. M., & Aldred, S. (2009). Oxidative stress in vascular dementia and Alzheimer's disease: A common pathology. *Journal of Alzheimer's Disease*, *17*(2), 245–257. https://doi.org/10.3233/JAD-2009-1041

Brookmeyer, R., Johnson, E., Ziegler-Graham, K., & Arrighi, H. M. (2007). Forecasting the global burden of Alzheimer's disease. *Alzheimer's & Dementia*, *3*(3), 186–191. https://doi.org/10.1016/J.JALZ.2007.04.381

Burgess, S., Zeitlin, R., & Echeverria, V. (2011). Cotinine inhibits amyloid-β peptide neurotoxicity and oligomerization. *Journal of Clinical Toxicology*, *6*. https://doi.org/10.4172/2161-0495.S6-003

Carvajal-Oliveros, A., Domínguez-Baleón, C., Zárate, R. V., Campusano, J. M., Narváez-Padilla, V., & Reynaud, E. (2021). Nicotine suppresses Parkinson's disease-like phenotypes induced by Synphilin-1 overexpression in *Drosophila melanogaster* by increasing tyrosine hydroxylase and dopamine levels. *Scientific Reports*, *11*(1), 1–13. https://doi.org/10.1038/s41598-021-88910-4

Cataldo, J. K., Prochaska, J. J., & Glantz, S. A. (2010). Cigarette smoking is a risk factor for Alzheimer's disease: An analysis controlling for tobacco industry affiliation. *Journal of Alzheimer's Disease*, *19*(2), 465–480. https://doi.org/10.3233/JAD-2010-1240

Chalouhi, N., Ali, M. S., Starke, R. M., Jabbour, P. M., Tjoumakaris, S. I., Gonzalez, L. F., Rosenwasser, R. H., Koch, W. J., & Dumont, A. S. (2012). Cigarette smoke and inflammation: Role in cerebral aneurysm formation and rupture. *Mediators of Inflammation*, *2012*, Article 271582. https://doi.org/10.1155/2012/271582

Chen, H., Huang, X., Guo, X., Mailman, R. B., Park, Y., Kamel, F., Umbach, D. M., Xu, Q., Hollenbeck, A., Schatzkin, A., & Blair, A. (2010). Smoking duration, intensity, and risk of Parkinson disease. *Neurology*, *74*(11), 878–884. https://doi.org/10.1212/WNL.0B013E3181D55F38/ASSET/B01E1170-58B0-4E5B-AB51-39B04A4926AF/ASSETS/GRAPHIC/PAP.JPEG

Cojocaru, I. M., Cojocaru, M., Sapira, V., & Ionescu, A. (2013). Evaluation of oxidative stress in patients with acute ischemic stroke. *Romanian Journal of Internal Medicine*, *51*(2), 97–106. https://europepmc.org/article/med/24294813

Connolly, B. S., & Lang, A. E. (2014). Pharmacological treatment of Parkinson disease: A review. *JAMA*, *311*(16), 1670–1683. https://doi.org/10.1001/JAMA.2014.3654

Dai, X., Liang, R., Dai, M., Li, X., & Zhao, W. (2024). Smoking impacts Alzheimer's disease progression through oral microbiota modulation. *Molecular Neurobiology*, 1–26.

Daviglus, M. L., Plassman, B. L., Pirzada, A., Bell, C. C., Bowen, P. E., Burke, J. R., Connolly, E. S., Dunbar-Jacob, J. M., Granieri, E. C., McGarry, K., Patel, D., Trevisan, M., & Williams, J. W. (2011). Risk factors and preventive interventions for Alzheimer disease: State of the science. *Archives of Neurology*, *68*(9), 1185–1190. https://doi.org/10.1001/ARCHNEUROL.2011.100

Deane, R., & Zlokovic, B. V. (2007). Role of the blood-brain barrier in the pathogenesis of Alzheimer's disease. *Current Alzheimer Research*, *4*(2), 191–197. https://doi.org/10.2174/156720507780362245

Dexter, D. T., & Jenner, P. (2013). Parkinson disease: From pathology to molecular disease mechanisms. *Free Radical Biology and Medicine*, *62*, 132–144. https://doi.org/10.1016/J.FREERADBIOMED.2013.01.018

Dong, Y., Bi, W., Zheng, K., Zhu, E., Wang, S., Xiong, Y., Chang, J., Jiang, J., Liu, B., Lu, Z., & Cheng, Y. (2020). Nicotine prevents oxidative stress-induced hippocampal neuronal injury through α7-nAChR/Erk1/2 signaling pathway. *Frontiers in Molecular Neuroscience*, *13*, Article 557647. https://doi.org/10.3389/FNMOL.2020.557647/BIBTEX

Drago, V., Babiloni, C., Bartrés-Faz, D., Caroli, A., Bosch, B., Hensch, T., Didic, M., Klafki, H. W., Pievani, M., Jovicich, J., Venturi, L., Spitzer, P., Vecchio, F., Schoenknecht, P., Wiltfang, J., Redolfi, A., Forloni, G., Blin, O., Irving, E., … & Frisoni, G. B. (2011). Disease tracking markers for Alzheimer's disease at the prodromal (MCI) stage. *Journal of Alzheimer's Disease*, *26*(s3), 159–199. https://doi.org/10.3233/JAD-2011-0043

Durazzo, T. C., Insel, P. S., & Weiner, M. W. (2012). Greater regional brain atrophy rate in healthy elderly subjects with a history of cigarette smoking. *Alzheimer's & Dementia*, *8*(6), 513–519. https://doi.org/10.1016/J.JALZ.2011.10.006

Durazzo, T. C., Mattsson, N., & Weiner, M. W. (2014). Smoking and increased Alzheimer's disease risk: A review of potential mechanisms. *Alzheimer's & Dementia*, *10*(3 Suppl.), S122–S145. https://doi.org/10.1016/J.JALZ.2014.04.009

Durazzo, T. C., Meyerhoff, D. J., & Nixon, S. J. (2013). Interactive effects of chronic cigarette smoking and age on hippocampal volumes. *Drug and Alcohol Dependence*, *133*(2), 704–711. https://doi.org/10.1016/J.DRUGALCDEP.2013.08.020

Echeverria, V., Alex Grizzell, J., & E. Barreto, G. (2016). Neuroinflammation: a therapeutic target of cotinine for the treatment of psychiatric disorders? *Current Pharmaceutical Design*, *22*(10), 1324–1333.

Gallo, V., Vineis, P., Cancellieri, M., Chiodini, P., Barker, R. A., Brayne, C., Pearce, N., Vermeulen, R., Panico, S., Bueno-De-Mesquita, B., Vanacore, N., Forsgren, L., Ramat, S., Ardanaz, E., Arriola, L., Peterson, J., Hansson, O., Gavrila, D., Sacerdote, C., … & Riboli, E. (2019). Exploring causality of the association between smoking and Parkinson's disease. *International Journal of Epidemiology*, *48*(3), 912–925. https://doi.org/10.1093/IJE/DYY230

Garza, A. P., Morton, L., Pállinger, É., Buzás, E. I., Schreiber, S., Schott, B. H., & Dunay, I. R. (2023). Initial and ongoing tobacco smoking elicits vascular damage and distinct inflammatory response linked to neurodegeneration. *Brain, Behavior, & Immunity-Health*, *28*, Article 100597.

Gülşen, A., & Uslu, B. (2020). Health hazards and complications associated with electronic cigarettes: A review. *Turkish Thoracic Journal*, *21*(3), 201. https://doi.org/10.5152/TURKTHORACJ.2019.180203

Hajdusianek, W., Żórawik, A., Waliszewska-Prosół, M., Poręba, R., & Gać, P. (2021). Tobacco and nervous system development and function—New findings 2015–2020. *Brain Sciences*, *11*(6), Article 797. https://doi.org/10.3390/BRAINSCI11060797

Hossain, M., Sathe, T., Fazio, V., Mazzone, P., Weksler, B., Janigro, D., Rapp, E., & Cucullo, L. (2009). Tobacco smoke: A critical etiological factor for vascular impairment at the blood–brain barrier. *Brain Research*, *1287*, 192–205. https://doi.org/10.1016/J.BRAINRES.2009.06.033

Hunt, L. J., Covinsky, K. E., Cenzer, I., Espejo, E., Boscardin, W. J., Leutwyler, H., … & Cataldo, J. (2023). The epidemiology of smoking in older adults: A national cohort study. *Journal of General Internal Medicine*, *38*(7), 1697–1704.

Hunter, B. E., de Fiebre, C. M., Papke, R. L., Kem, W. R., & Meyer, E. M. (1994). A novel nicotinic agonist facilitates induction of long-term potentiation in the rat hippocampus. *Neuroscience Letters*, *168*(1–2), 130–134. https://doi.org/10.1016/0304-3940(94)90433-2

Jha, P. (2009). Avoidable global cancer deaths and total deaths from smoking. *Nature Reviews Cancer*, *9*(9), 655–664. https://doi.org/10.1038/nrc2703

Kametani, F., & Hasegawa, M. (2018). Reconsideration of amyloid hypothesis and tau hypothesis in Alzheimer's disease. *Frontiers in Neuroscience*, *12*, Article 328460. https://doi.org/10.3389/FNINS.2018.00025/BIBTEX

Kenborg, L., Lassen, C. F., Ritz, B., Andersen, K. K., Christensen, J., Schernhammer, E. S., Hansen, J., Wermuth, L., Rod, N. H., & Olsen, J. H. (2015). Lifestyle, family history, and risk of idiopathic Parkinson disease: A large Danish case-control study. *American Journal of Epidemiology*, *181*(10), 808–816. https://doi.org/10.1093/AJE/KWU332

Khanna, A., Guo, M., Mehra, M., & Royal, W. (2013). Inflammation and oxidative stress induced by cigarette smoke in Lewis rat brains. *Journal of Neuroimmunology*, *254*(1–2), 69–75. https://doi.org/10.1016/J.JNEUROIM.2012.09.006

Kuzkina, A., Schulmeyer, L., Monoranu, C. M., Volkmann, J., Sommer, C., & Doppler, K. (2019). The aggregation state of α-synuclein deposits in dermal nerve fibers of patients with Parkinson's disease resembles that in the brain. *Parkinsonism & Related Disorders*, *64*, 66–72. https://doi.org/10.1016/J.PARKRELDIS.2019.03.003

Launer, L. J., Andersen, K., Dewey, M. E., Letenneur, L., Ott, A., Amaducci, L. A., Brayne, C., Copeland, J. R. M., Dartigues, J. F., Kragh-Sorensen, P., Lobo, A., Martinez-Lage, J. M., Stijnen, T., & Hofman, A. (1999). Rates and risk factors for dementia and Alzheimer's disease: Results from EURODEM pooled analyses. *Neurology*, *52*(1), 78–84. https://doi.org/10.1212/WNL.52.1.78

Lockman, P. R., Van Der Schyf, C. J., Abbruscato, T. J., & Allen, D. D. (2005). Chronic nicotine exposure alters blood–brain barrier permeability and diminishes brain uptake of methyllycaconitine. *Journal of Neurochemistry*, *94*(1), 37–44. https://doi.org/10.1111/J.1471-4159.2005.03162.X

Ma, C., Liu, Y., Neumann, S., & Gao, X. (2017). Nicotine from cigarette smoking and diet and Parkinson disease: A review. *Translational Neurodegeneration*, *6*(1), Article 9. https://doi.org/10.1186/S40035-017-0090-8

Ma, C., Xi, B., Li, Z., Wu, H., Zhao, M., Liang, Y., & Bovet, P. (2021). Prevalence and trends in tobacco use among adolescents aged 13–15 years in 143 countries, 1999–2018: Findings from the Global Youth Tobacco Surveys. *Lancet Child and Adolescent Health*, *5*(4), 245–255. https://doi.org/10.1016/S2352-4642(20)30390-4

Mondragón-Rodríguez, S., Basurto-Islas, G., Lee, H. G., Perry, G., Zhu, X., Castellani, R. J., & Smith, M. A. (2010). Causes versus effects: The increasing complexities of Alzheimer's disease pathogenesis. *Expert Review of Neurotherapeutics*, *10*(5), 683–691. https://doi.org/10.1586/ERN.10.27

Morris, J. C., Roe, C. M., Xiong, C., Fagan, A. M., Goate, A. M., Holtzman, D. M., & Mintun, M. A. (2010). APOE predicts amyloid-beta but not tau Alzheimer pathology in cognitively normal aging. *Annals of Neurology*, *67*(1), 122–131. https://doi.org/10.1002/ANA.21843

Mussa, M. T., Kamal, M. M., Abdullah-Al-Mahmud, M., Hai, M. A., Begum, S., & Das, S. K. (2018). In vivo effects of tobacco smoking injury to trachea in a mice model. *Alexandria Journal of Veterinary Sciences*, *58*(1), 24–30. https://doi.org/10.5455/ajvs.301175

Naik, P., & Cucullo, L. (2012). In vitro blood–brain barrier models: Current and perspective technologies. *Journal of Pharmaceutical Sciences*, *101*(4), 1337–1354. https://doi.org/10.1002/JPS.23022

Naik, P., & Cucullo, L. (2015). Pathobiology of tobacco smoking and neurovascular disorders: Untied strings and alternative products. *Fluids and Barriers of the CNS*, *12*(1), Article 22. https://doi.org/10.1186/S12987-015-0022-X

Nelson, P. T., Alafuzoff, I., Bigio, E. H., Bouras, C., Braak, H., Cairns, N. J., Castellani, R. J., Crain, B. J., Davies, P., Tredici, K. Del, Duyckaerts, C., Frosch, M. P., Haroutunian, V., Hof, P. R., Hulette, C. M., Hyman, B. T., Iwatsubo, T., Jellinger, K. A., Jicha, G. A., … & Beach, T. G. (2012). Correlation of Alzheimer disease neuropathologic changes with cognitive status: A review of the literature. *Journal of Neuropathology & Experimental Neurology*, *71*(5), 362–381. https://doi.org/10.1097/NEN.0B013E31825018F7

Nicholatos, J. W., Francisco, A. B., Bender, C. A., Yeh, T., Lugay, F. J., Salazar, J. E., Glorioso, C., & Libert, S. (2018). Nicotine promotes neuron survival and partially protects from Parkinson's disease by suppressing SIRT6. *Acta Neuropathologica Communications*, *6*(1), Article 120. https://doi.org/10.1186/S40478-018-0625-Y/FIGURES/7

Nullens, S., Staessens, M., Peleman, C., Schrijvers, D. M., Malhotra-Kumar, S., Francque, S., Matteoli, G., Boeckxstaens, G. E., De Man, J. G., & De Winter, B. Y. (2016). Effect of GTS-21, an alpha7 nicotinic acetylcholine receptor agonist, on CLP-induced inflammatory, gastrointestinal motility, and colonic permeability changes in mice. *Shock*, *45*(4), 450–459. https://doi.org/10.1097/SHK.0000000000000519

Oddo, S., Caccamo, A., Green, K. N., Liang, K., Tran, L., Chen, Y., Leslie, F. M., & LaFerla, F. M. (2005). Chronic nicotine administration exacerbates tau pathology in a transgenic model of Alzheimer's disease. *Proceedings of the National Academy of Sciences of the United States of America*, *102*(8), 3046–3051. https://doi.org/10.1073/PNAS.0408500102/SUPPL_FILE/08500FIG7.JPG

Olmez, I., & Ozyurt, H. (2012). Reactive oxygen species and ischemic cerebrovascular disease. *Neurochemistry International*, *60*(2), 208–212. https://doi.org/10.1016/J.NEUINT.2011.11.009

Olsson, B., Lautner, R., Andreasson, U., Öhrfelt, A., Portelius, E., Bjerke, M., Hölttä, M., Rosén, C., Olsson, C., Strobel, G., Wu, E., Dakin, K., Petzold, M., Blennow, K., & Zetterberg, H. (2016). CSF and blood biomarkers for the diagnosis of Alzheimer's disease: A systematic review and meta-analysis. *Lancet Neurology*, *15*(7), 673–684. https://doi.org/10.1016/S1474-4422(16)00070-3

Oz, M., E. Lorke, D., S. Yang, K.-H., & Petroianu, G. (2013). On the interaction of ß-amyloid peptides and α7-nicotinic acetylcholine receptors in Alzheimer's disease. *Current Alzheimer Research*, *10*(6), 618–630.

Palermo, G., & Ceravolo, R. (2019). Molecular imaging of the dopamine transporter. *Cells*, *8*(8), Article 872. https://doi.org/10.3390/CELLS8080872

Paris, D., Beaulieu-Abdelahad, D., Bachmeier, C., Reed, J., Ait-Ghezala, G., Bishop, A., Chao, J., Mathura, V., Crawford, F., & Mullan, M. (2011). Anatabine lowers Alzheimer's Aβ production in vitro and in vivo. *European Journal of Pharmacology*, *670*(2–3), 384–391. https://doi.org/10.1016/J.EJPHAR.2011.09.019

Peluffo, G., Calcerrada, P., Piacenza, L., Pizzano, N., & Radi, R. (2009). Superoxide-mediated inactivation of nitric oxide and peroxynitrite formation by tobacco smoke in vascular endothelium: Studies in cultured cells and smokers. *American Journal of Physiology—Heart and Circulatory Physiology*, *296*(6), 1781–1792. https://doi.org/10.1152/AJPHEART.00930.2008

Prince, M., Bryce, R., Albanese, E., Wimo, A., Ribeiro, W., & Ferri, C. P. (2013). The global prevalence of dementia: A systematic review and meta-analysis. *Alzheimer's & Dementia*, *9*(1), 63-75.e2. https://doi.org/10.1016/J.JALZ.2012.11.007

Pogocki, D., Ruman, T., Danilczuk, M., Celuch, M., & Wałajtys-Rode, E. (2007). Application of nicotine enantiomers, derivatives and analogues in therapy of neurodegenerative disorders. *European Journal of Pharmacology*, *563*(1-3), 18-39.

Quik, M., & Kulak, J. M. (2002). Nicotine and nicotinic receptors: Relevance to Parkinson's disease. *NeuroToxicology*, *23*(4–5), 581–594. https://doi.org/10.1016/S0161-813X(02)00036-0

Quik, M., Perez, X. A., & Bordia, T. (2012). Nicotine as a potential neuroprotective agent for Parkinson's disease. *Movement Disorders*, *27*(8), 947–957. https://doi.org/10.1002/MDS.25028

Rasmussen Eid, H., Rosness, T. A., Bosnes, O., Salvesen, Ø., Knutli, M., & Stordal, E. (2019). Smoking and obesity as risk factors in frontotemporal dementia and Alzheimer's disease: The HUNT Study. *Dementia and Geriatric Cognitive Disorders Extra*, *9*(1), 1–10.

Rascol, O., Fitzer-Attas, C. J., Hauser, R., Jankovic, J., Lang, A., Langston, J. W., Melamed, E., Poewe, W., Stocchi, F., Tolosa, E., Eyal, E., Weiss, Y. M., & Olanow, C. W. (2011). A double-blind, delayed-start trial of rasagiline in Parkinson's disease (the ADAGIO study): Prespecified and post-hoc analyses of the need for additional therapies, changes in UPDRS scores, and non-motor outcomes. *Lancet Neurology, 10*(5), 415–423. https://doi.org/10.1016/S1474-4422(11)70073-4

Reitz, C., Den Heijer, T., Van Duijn, C., Hofman, A., & Breteler, M. M. B. (2007). Relation between smoking and risk of dementia and Alzheimer disease: The Rotterdam Study. *Neurology, 69*(10), 998–1005. https://doi.org/10.1212/01.WNL.0000271395.29695.9A

Ritz, B., Lee, P. C., Lassen, C. F., & Arah, O. A. (2014). Parkinson disease and smoking revisited: Ease of quitting is an early sign of the disease. *Neurology, 83*(16), 1396–1402. https://doi.org/10.1212/WNL.0000000000000879/SUPPL_FILE/TABLE-FIGURE-APPENDIX.DOC

Rusanen, M., Kivipelto, M., Quesenberry, C. P., Zhou, J., & Whitmer, R. A. (2011). Heavy smoking in midlife and long-term risk of Alzheimer disease and vascular dementia. *Archives of Internal Medicine, 171*(4), 333–339. https://doi.org/10.1001/ARCHINTERNMED.2010.393

Rusanen, M., Rovio, S., Ngandu, T., Nissinen, A., Tuomilehto, J., Soininen, H., & Kivipelto, M. (2010). Midlife smoking, apolipoprotein E and risk of dementia and Alzheimer's disease: A population-based cardiovascular risk factors, aging and dementia study. *Dementia and Geriatric Cognitive Disorders, 30*(3), 277–284. https://doi.org/10.1159/000320484

Saad, A. Ben, Rjeibi, I., Brahmi, N., Elaloui, E., & Zouari, N. (2020). Nicotine-induced oxidative stress, testis injury, AChE inhibition and brain damage alleviated by *Mentha spicata. Inflammopharmacology, 28*(4), 939–948. https://doi.org/10.1007/S10787-019-00650-0/FIGURES/6

Sathyanarayana Rao, T. S., & Yeragani, V. K. (2009). Hypertensive crisis and cheese. *Indian Journal of Psychiatry, 51*(1), 65–66. https://doi.org/10.4103/0019-5545.44910

Science, B., Morris, J. F., Christian, H. C., Epton, M. J., & Chapman, L. P. (2002). Astrocyte–endothelial interactions and blood–brain barrier permeability. *Journal of Anatomy, 200*(5), 523–534. https://doi.org/10.1046/J.1469-7580.2002.00047_13.X

Shimohama, S., & Kawamata, J. (2018). Roles of nicotinic acetylcholine receptors in the pathology and treatment of Alzheimer's and Parkinson's diseases. In *Nicotinic acetylcholine receptor signaling in neuroprotection* (pp. 137–158). https://doi.org/10.1007/978-981-10-8488-1_8

Spataro, N., Calafell, F., Cervera-Carles, L., Casals, F., Pagonabarraga, J., Pascual-Sedano, B., Campolongo, A., Kulisevsky, J., Lleó, A., Navarro, A., Clarimón, J., & Bosch, E. (2015). Mendelian genes for Parkinson's disease contribute to the sporadic forms of the disease. *Human Molecular Genetics, 24*(7), 2023–2034. https://doi.org/10.1093/HMG/DDU616

Sperling, R. A., Aisen, P. S., Beckett, L. A., Bennett, D. A., Craft, S., Fagan, A. M., Iwatsubo, T., Jack, C. R., Kaye, J., Montine, T. J., Park, D. C., Reiman, E. M., Rowe, C. C., Siemers, E., Stern, Y., Yaffe, K., Carrillo, M. C., Thies, B., Morrison-Bogorad, M., … & Phelps, C. H. (2011). Toward defining the preclinical stages of Alzheimer's disease: Recommendations from the National Institute on Aging-Alzheimer's Association workgroups on diagnostic guidelines for Alzheimer's disease. *Alzheimer's & Dementia, 7*(3), 280–292. https://doi.org/10.1016/J.JALZ.2011.03.003

Strazielle, N., Khuth, S. T., Murat, A., Chalon, A., Giraudon, P., Belin, M. F., & Ghersi-Egea, J. F. (2003). Pro-inflammatory cytokines modulate matrix metalloproteinase secretion and organic anion transport at the blood-cerebrospinal fluid barrier. *Journal of Neuropathology & Experimental Neurology, 62*(12), 1254–1264. https://doi.org/10.1093/JNEN/62.12.1254

Suwannatad, P. (2022). Impacts of nicotine on Neurodegenerative disorders. *International Journal of Science and Healthcare Research, 7*(2), 118. https://doi.org/10.52403/ijshr.20220418

Tiwari, R. K., Sharma, V., Pandey, R. K., & Shukla, S. S. (2020). Nicotine addiction: Neurobiology and mechanism. *Journal of Pharmacopuncture, 23*(1), 1. https://doi.org/10.3831/KPI.2020.23.001

Tyas, S. L., White, L. R., Petrovitch, H., Ross, G. W., Foley, D. J., Heimovitz, H. K., & Launer, L. J. (2003). Mid-life smoking and late-life dementia: The Honolulu-Asia aging study. *Neurobiology of Aging, 24*(4), 589–596. https://doi.org/10.1016/S0197-4580(02)00156-2

U.S. Department of Health and Human Services. (2014). The health consequences of smoking—50 years of progress: A report of the Surgeon General.

Veljkovic, E., Xia, W., Phillips, B., Wong, E. T., Ho, J., Oviedo, A., Hoeng, J., & Peitsch, M. (2018). *Nicotine and other tobacco compounds in neurodegenerative and psychiatric: Overview of epidemiological data on smoking and preclinical and clinical data on nicotine diseases.* Elsevier Academic Press.

Wong, D., Dorovini-Zis, K., & Vincent, S. R. (2004). Cytokines, nitric oxide, and cGMP modulate the permeability of an in vitro model of the human blood–brain barrier. *Experimental Neurology, 190*(2), 446–455. https://doi.org/10.1016/J.EXPNEUROL.2004.08.008

World Health Organization. (2020). Tobacco fact sheet key facts.

12 Behavioral Addictions in Neurodegenerative Disorders

Shyam Sundar Kanagarajan and Kavitha Parthiban

INTRODUCTION

Behavioral addictions are disorders characterized by repetitive engagement in certain behaviors, which become compulsive and continue despite harmful consequences, often associated with socio-occupational impairment. They include but are not limited to gambling disorder, gaming disorder, sex addiction, social media addiction, smartphone addiction, internet addiction, food addiction, work addiction, and compulsive buying. The neurobiological and psychological factors underlying behavioral addictions are analogous to those of substance use disorders (SUDs). They also share some features of obsessive-compulsive and related disorders and impulse control disorders. Similar to SUDs, behavioral addictions involve components of both impulsivity and compulsivity (Robbins & Clark, 2015). Their clinical features include salience, mood modification, tolerance, withdrawal symptoms, conflict, and relapse (Griffiths, 1996).

Behavioral addictions are not recognized in current classificatory systems as a separate group of disorders. However, gambling disorder (previously known as pathological gambling) is classified under "substance-related and addictive disorders" in the *Diagnostic and Statistical Manual of Mental Disorders*, Fifth Edition, Text Revision (DSM-5-TR), thus emphasizing the validity of the concept of behavioral addiction. In the eleventh edition of the International Classification of Diseases (ICD-11), the "disorders due to substance use or addictive behaviors" category includes gambling disorder, gaming disorder, other specified disorders due to addictive behaviors, and disorders due to addictive behaviors unspecified (Petry et al., 2018). "Compulsive sexual behavior disorder" is also included in ICD-11. The conceptualization of this disorder is debated, with discussions on whether it is a behavioral addiction or related to impulse control, emotional dysregulation, or a sexual disorder (Lew-Starowicz & Coleman, 2022). The conceptualization of behavioral addictions in psychiatric classificatory systems is summarized in Table 12.1.

In this chapter, we explore the current literature on the connections between various behavioral addictions and neurodegenerative disorders (NDDs), examining potential mechanisms and treatment approaches. The discussion includes specific behavioral addictions, such as gambling disorder, inappropriate sexual behaviors (ISBs) (including sexual addiction), food addiction, and problematic internet use, highlighting how they intersect with NDDs. We also address strategies for assessment and management, outlining both pharmacological and non-pharmacological treatment options for effectively addressing behavioral addictions within this patient population.

BEHAVIORAL ADDICTIONS IN NEURODEGENERATIVE DISORDERS

Epidemiological studies show that the prevalence of behavioral addictions, such as gaming and gambling disorders, is increasing in older adults (Petry, 2002; Wang et al., 2024). The vulnerabilities associated with aging make older adults prone to the emergence or persistence of behavioral addictions. Various related risk factors identified in later life are loneliness, isolation, lack of autonomy, multi-morbidity, and structural changes in the brain, including dementia and other NDDs (Stewart et al., 2023). Additionally, medications for NDDs can lead to conditions that mimic behavioral addictions. The emergence of such addictions in older populations is associated with severe

DOI: 10.1201/9781003478485-12

TABLE 12.1
Behavioral Addictions as Described in Psychiatric Classificatory Systems

Gambling Disorder	
ICD-11	A pattern of persistent or recurrent gambling behavior, which may be online or offline, manifested by: • Impaired control over gambling (e.g., onset, frequency, intensity, duration, termination, and context). • Gambling increasingly prioritized to the extent that it takes precedence over other life interests and daily activities. • Continuation or escalation of gambling despite the occurrence of negative consequences. The behavior pattern is of sufficient severity to result in significant impairment in critical areas of functioning. The pattern may be episodic or recurrent. These features are evident over at least 12 months for diagnosis. The required duration may be shortened if all diagnostic requirements are met and symptoms are severe.
DSM-5-TR	Persistent and problematic gambling behavior patterns lead to significant distress or impairment. For diagnosis, an individual must meet at least four of these criteria within 12 months: • Preoccupation with gambling. • Needing to gamble with increasing amounts of money. • Unsuccessfully attempting to stop or control gambling. • Restlessness or irritability when attempting to stop gambling. • Gambling as an escape from problems or negative mood. • Chasing losses. • Lying to conceal the extent of gambling involvement. • Jeopardizing or losing essential opportunities. • Relying on others for financial bailouts. • Committing illegal acts. The severity can be mild, moderate, or severe. The pattern can be episodic or persistent. The DSM-5-TR includes the category "gambling disorder due to the direct physiological consequence of dopaminergic medications used to treat PD."
Sexual Addiction	
ICD-11	The related diagnosis is "Compulsive sexual behavior disorder," classified under impulse control disorders. This disorder is characterized by a persistent pattern of failure to control intense, repetitive sexual impulses or urges resulting in repetitive sexual behavior, manifested in one or more of the following: • Engagement in repetitive sexual behavior becoming a central focus of the individual's life to the point of neglecting health and personal care or other interests, activities, and responsibilities. • The individual making numerous unsuccessful efforts to control or significantly reduce repetitive sexual behavior. • The individual continuing to engage in repetitive sexual behavior despite adverse consequences. • The individual continuing to engage in repetitive sexual behavior even when they derive little or no satisfaction from it. The pattern of failure to control intense, repetitive sexual impulses or urges and resulting repetitive sexual behavior is manifested over an extended period (e.g., 6 months or more). Another mental disorder, medical condition, or the effects of a substance or medication do not better account for the symptoms. The ICD-11 excludes compulsive sexual behavior disorder symptoms caused by dopamine agonists. It also excludes symptoms caused by dementia, like failure to control sexual impulses, urges, or behaviors that are part of a more general pattern of disinhibition of impulse control due to neurocognitive impairment. Symptoms are associated with marked distress or significant impairment in critical areas of functioning. If functioning is maintained, it is only through significant additional effort.
DSM-5-TR	There is no related diagnosis in the DSM-5-TR.

consequences, including social, financial, and familial problems, as well as psychological problems, such as major depressive disorder and suicide (Subramaniam et al., 2015).

NDDs refer to a group of diseases characterized by progressive dysfunction in the brain that results in the loss of activities of daily living (ADLs). These include Alzheimer's disease (AD), frontotemporal dementia (FTD), dementia with Lewy bodies, Parkinson's disease (PD), Huntington's disease (HD), amyotrophic lateral sclerosis (ALS), motor neuron disease, progressive supranuclear palsy, corticobasal degeneration, multiple system atrophy, and other rare conditions.

Symptoms of behavioral addictions can occur at various junctures in individuals with NDDs—as a part of dementia itself, as a manifestation of behavioral and psychological symptoms of dementia, or as a side effect of medications. The behavioral addictions that are studied extensively in NDDs include gambling disorder, sexual addiction, internet-related addictions, food addiction, and compulsive shopping (Sixto-Costoya et al., 2021). The chapter will focus on gambling disorder and ISBs, including sexual addiction.

Gambling Disorder

Gambling refers to the wagering of money or something of value on chance activities with random uncertain outcomes. Gambling disorder is characterized by a preoccupation with gambling, progressively increasing the amount of money gambled to achieve the same excitement, repeated gambling to get even with the losses (chasing the losses), feeling irritable or restless when trying to cut down on gambling, and lying to conceal the gambling behavior (American Psychiatric Association, 2022)

Neurobiologically, dysfunctions in cognitive flexibility, inhibition response, planning, and decision-making, as well as cortical modifications in the frontal region, have all been linked with gambling disorder among older adults (Fontaine et al., 2023). Age-related cognitive decline is also associated with poor decision-making in gambling, as evidenced by poor performance in the Iowa Gambling Task (Fein et al., 2007). Hence, individuals with cognitive decline who gamble can accumulate significant losses before their family members discover their gambling behavior. Several case reports document disordered gambling in individuals with NDDs, as detailed below.

Frontotemporal Dementia

FTD is known to be associated with new-onset pathological gambling (Piguet & Hodges, 2013). Gambling disorder could also mask the underlying FTD and can be the first clinical presentation in some cases (Manes et al., 2010; Kloeters et al., 2013; Piguet & Hodges, 2013; Ozel-Kizil et al., 2013). The behavioral variant of FTD (bvFTD) is more commonly linked to problematic gambling than other dementias. It includes neuropsychiatric symptoms like apathy, disinhibited behavior, lack of empathy, compulsive behaviors, hyperorality, and executive function deficits (Younes & Miller, 2020). Based on available case reports, the clinical findings of gambling disorder in FTD can be summarized as follows:

- Disordered gambling may be the first symptom of bvFTD, especially in the early stages of the condition when the executive function is intact (Lo Coco & Nacci, 2004).
- Excessive indulgence in gambling can occur along with hoarding and abnormal sexual behavior (Nakaaki et al., 2007).
- Gambling can start at a lesser frequency and progress to a significant level, to the extent of selling properties and possessions, and can be accompanied by stealing (Manes et al., 2010).
- Although gambling is often seen with bvFTD, one case report revealed pathological gambling in a patient with semantic variant primary progressive aphasia that progressed over 2 years. The patient had normal executive functioning, and except for semantic deficits, the rest of their neuropsychological evaluation was regular (Battista et al., 2019).

- One case report discussed a patient with ALS and FTD. Initial symptoms included compulsive hand washing and behavioral changes, which later progressed to excessive gambling. The patient eventually exhibited motor symptoms suggestive of ALS (Ozel-Kizil et al., 2013).

Parkinson's Disease

PD is a neurodegenerative condition that impacts the dopaminergic neurons in the substantia nigra. It is characterized by resting tremors, bradykinesia, rigidity, and postural disturbances. PD is commonly associated with gambling disorder, which has a 12-month prevalence rate of 9.3% among those with severe PD (Crockford et al., 2008). This association is more prevalent among patients who are treated with dopaminergic therapy (Wong & Steiger, 2007). Indeed, gambling can worsen—with increased frequency and amount of money gambled—after the addition of dopamine agonists such as levodopa, carbidopa, pramipexole, and ropinirole (Leplow et al., 2023; Nadeau et al., 2017). Dopaminergic medication-induced gambling is hypothesized to result from drug-induced overstimulation of the relatively preserved mesocorticolimbic pathway in PD. Impulse dysregulation is thought to underlie problematic gambling in these situations (O'Sullivan & Lees, 2007). Discontinuation or dose reduction is known to be associated with an abrupt decrease in the urge to gamble.

Gamblers with PD have been found to exhibit hyperactivity in the basal ganglia, orbitofrontal cortex, hippocampus, amygdala, and insula (Cilia et al., 2008). Since gamblers are often secretive and family members may not suspect that gambling is related to PD or medications, the issue is typically not brought to attention until severe consequences arise.

Huntington's Disease

HD is a genetic trinucleotide repeat disease characterized by neurodegeneration in the striatum, cerebral cortex, thalamus, and cerebellum. Patients develop progressive motor dysfunction in the form of chorea, progressive cognitive decline, and psychiatric manifestations. Several case reports have documented the occurrence of gambling in HD (De Marchi et al., 1998; Sarafudheen et al., 2022). The psychological processes affected by HD are disinhibition, difficulties in emotional recognition, cognitive inflexibility, and poor judgment, resulting in impaired decision-making (Stout et al., 2001). Usually, these deficits become prominent in the intermediate stage of the disease. However, gambling behavior can start as early as young adulthood, as revealed by some case reports (De Marchi et al., 1998). Patients with HD and gambling disorder perform poorly on the Iowa Gambling and Cambridge Gambling Tasks, with risky decision-making in uncertain situations, decreased sensitivity to punishment, decreased ability to consider long-term rewards, and difficulty inhibiting impulsive responses to immediate rewards (Kalkhoven et al., 2014). Sarafudheen et al. (2022) reported a case of HD involving a long-standing gambling disorder, in which the patient accumulated significant debts of £200,000. Despite this vulnerability, certain factors, such as motor symptoms associated with reduced mobility and the social environment, may prevent individuals from developing pathological gambling (Kalkhoven et al., 2014).

Sexual Addiction and Inappropriate Sexual Behaviors

Sexual Addiction

Sexual addiction is an umbrella term that includes excessive non-paraphilic sexual behaviors such as excessive masturbation, excessive consumption of internet pornography, frequent demands for sex, preoccupation with sexual thoughts, excessive indulgence in sexual activity with consenting adults, telephonic sex, cyber-sex, sex with commercial sex workers, and frequently visiting strip clubs. Other related terms are hypersexual disorder and compulsive sexual behavior disorder. There is no consensus definition available for sexual addiction, and the term is often used interchangeably with hypersexuality and compulsive sexual behavior (Karila et al., 2014).

Sexual addiction is characterized by intense sexual urges/impulses, increased time spent in sexual activity, repetitive engagement in sexual behaviors for dysphoric mood states or stressful life events, difficulty controlling these behaviors, continuing the behaviors despite harm to self or others, and impairment in social, occupational, family, and personal functioning (Sahithiya & Kashyap, 2022). In psychiatric classificatory systems, the entity related to sexual addiction is compulsive sexual behavior disorder in the ICD-11, as discussed in Table 12.1. The DSM-5-TR does not include any diagnosis related to sexual addiction.

The Sexual Dependency Inventory 4.0 reports 10 possible items for diagnosing sexual addiction:

1. Repetitive failure to resist impulses to engage in specific sexual behaviors.
2. Frequent engagement in those behaviors to a greater extent or over a more extended period than intended.
3. Long-standing desire or unsuccessful efforts to stop, reduce, or control those behaviors.
4. Excessive time spent obtaining sex, being sexual, or recovering from sexual experiences.
5. Becoming obsessed with preparing for sexual activities.
6. Frequent engagement in sexual behavior at times when one is expected to be fulfilling occupational, academic, domestic, or social obligations.
7. Continuation of sexual behavior despite knowing it has caused or exacerbated persistent or recurrent social, financial, psychological, or physical problems.
8. The need to increase the intensity, rate, number, or risk of sexual behaviors to reach the desired effect.
9. Giving up social, occupational, or recreational activities secondary to sexual behavior.
10. Feeling upset, anxious, restless, or irritable if unable to engage in sexual behavior (IITAP, 2020).

NDDs affect sexual functioning, libido, expression of sexual desires, and sexual activity. Alternatively, ISBs, including excessive sexual urges, hypersexuality, and compulsive sexual behavior, can be seen in these disorders (Black et al., 2005).

Dysfunction of specific brain areas is known to be associated with hypersexuality in NDDs. These areas include the bilateral hippocampus and temporal region (Baird et al., 2007). The neuropsychological factors that affect sexuality in NDDs include processes like reward processing, decision-making, and empathy.

Sexual Disinhibition and Inappropriate Sexual Behaviors

Sexual disinhibition (SD) has been described in various ways in the literature, using terms such as ISBs, hypersexuality, disinhibited behavior, and aberrant behavior. When discussing terminology, it is essential to highlight the following:

1. The term "satyriasis," which refers to hypersexuality in men, is derived from *Satyr*, a figure in ancient Greek mythology known for its lecherous nature and depicted as half-human, half-goat. This concept may evoke images of figures like Casanova or Don Juan, who were thought to have psychodynamic issues stemming from unfulfilled love. In contrast, hypersexuality caused by organic brain dysfunction lacks such emotional conflicts, presenting instead as an excessive sexual drive that can likely be managed with medications aimed at directly suppressing sexual urges.
2. Hypersexuality is characterized by an excessive focus on sexual fantasies, urges, or behaviors, with the individual spending significant time planning or engaging in sexual activities. This is often coupled with repetitive, unsuccessful efforts to reduce or control these behaviors. Unlike other forms of SD, hypersexuality is distinct due to its compulsive nature and the inability to manage the behavior.

3. Inappropriate sexual advances can be either nonviolent or violent, reflecting unwanted sexual actions or expressed intentions toward another individual.
4. Lewd or aberrant sexual behavior refers to broadly inappropriate sexual actions that either represent a change from the individual's previous conduct or appear in contexts that are not appropriate.
5. Inappropriate sexual remarks can appear as excessive flirting, inappropriately sharing sexual fantasies, or recounting past sexual encounters. These comments can range from explicitly vulgar statements to more subtle suggestions.
6. Socially disruptive SD is defined not by specific behaviors but by its negative impact on the social dynamics between the patient and their caregiver or others around them. It can also include a preoccupation with sexual thoughts, sexual delusions, such as infidelity, or attempts to seduce others.

In one observational study, sexual behaviors were also categorized into three distinct types (Zeiss et al., 1996):

1. *Appropriate sexual behaviors*: These included actions such as sitting close to someone with physical contact (arms or legs touching), kissing, or gently stroking another person's face, hands, or arms.
2. *Ambiguous sexual behaviors*: This category involved more uncertain actions like being undressed outside of a bedroom or bathroom, rubbing against another person, or publicly touching oneself on the breasts or genitals.
3. *Inappropriate sexual behaviors*: These were defined as making explicit sexual remarks, touching someone other than a partner on the breasts or genitals, touching a partner inappropriately in public, or exposing one's breasts or genitals in public.

The inconsistency in terminology has created challenges in accurately identifying and comparing SD across studies. A systematic review by Chapman and Spitznagel (2015) identified 20 instruments for measurement of sexual disinhibition in FTD with five common content domains: hypersexuality, aberrant sexual behavior, inappropriate sexual advances, inappropriate sexual comments, and socially disruptive sexual behavior. As achieving a unified understanding and definition of SD, particularly in patients with dementia, is essential, Sachdeva et al. (2014) proposed a structured definition that includes four critical criteria:

- The presence of dementia.
- The behavior is recognized as sexual by others.
- Its inappropriate nature.
- The behavior must have developed after the onset of cognitive decline.

Statistics in Neurodegenerative Disorders

In individuals with NDDs, the dynamic evolution of paraphilias has been noted in the literature. This progression was observed alongside cognitive decline, deterioration in emotional regulation, and disruptions in social behavior, motivation, and emotional transmission (Nordvig et al., 2019).

The prevalence of hypersexuality in dementia ranges from 2% to 30% (Torrisi et al., 2017). It may be regarded as a neuropsychiatric symptom of the NDD or part of disinhibited behavior. In specific NDDs, up to 25% of patients with AD are estimated to have ISBs that include excessive sexual urges (Black et al., 2005). In bvFTD, hypersexuality occurs in 8% to 18% of cases and is associated with compulsive masturbation (Mendez & Shapira, 2013). However, it is important to keep in mind that the most prominent changes in bvFTD are hyposexual in nature, including decreased frequency of sexual relations, affection, and response to advances by partners (Ahmed et al., 2015). In individuals with PD, hypersexuality has an estimated prevalence of approximately 3.5% and is related to dopamine dysregulation syndrome, as well as higher doses of dopaminergic agents (Codling et al., 2015). It can manifest as compulsive pornography, excessive masturbation, increased libido, and extramarital relationships (Codling et al., 2015; Merims & Giladi, 2008). Hypersexuality in PD is managed by the reduction or discontinuation of dopaminergic agonists or switching to alternative agents. Deep brain stimulation may also be helpful (Nakum & Cavanna, 2016). In HD, hypersexuality is related to medial striatum involvement and manifests as increased sexual drive, paraphilia, and frequent sexual advances (Reininghaus & Lackner, 2015).

The presence of sexual addiction in patients with NDDs raises concerns over ethical and legal issues like consent, capacity to understand the consequences of the act, criminal responsibility, vulnerability to sexual abuse, protecting the sexual rights of the patient, and sexual aggression. It can also lead to significant distress and shame in caregivers.

Other Behavioral Addictions

Food Addiction

Food addiction is a condition characterized by excessive preoccupation with certain foods, binge eating, eating to reduce stress, and overeating. It is often accompanied by difficulty in controlling the urge to eat. Although patients with NDDs may not fulfill this definition, changes in eating behaviors similar to a food addiction can be seen. Patients with FTD, especially bvFTD, report an increase in appetite, preference for sweets and carbohydrate-rich foods, binge eating, and wanting to eat particular foods repeatedly. Hyperorality and dietary changes are found in up to 60% of patients with bvFTD (Ikeda et al., 2002). These symptoms predominantly occur during the earlier phase of FTD; thus, their presence should raise clinical suspicion.

Problematic Internet Use

Problematic internet use (PIU), also called internet addiction, refers to excessive engagement in internet-related activities, with repeated unsuccessful attempts to reduce its use, loss of control over internet use, and restlessness while attempting to cut down use (Aboujaoude, 2010). In older adults, PIU was noted in individuals with dysphoric mood using the internet to cope with negative emotions. However, older adults typically engage in less addictive online activities than younger individuals (Chen et al., 2017; Rochat et al., 2021). To our knowledge, there is only one case report of internet pornography addiction in a patient with FTD who exhibited hypersexuality (Mendez & Shapira, 2011).

ASSESSMENT OF AND APPROACH TO BEHAVIORAL ADDICTIONS IN NEURODEGENERATIVE DISORDERS

Assessment

The assessment involves taking a psychiatric history and conducting a comprehensive geriatric assessment (CGA). The psychiatric history should cover the history of the current illness, past psychiatric history, medical history, medication history, family history, personal history, sexual history, and premorbid personality.

Comprehensive Geriatric Assessment

Physical Health

A thorough physical health evaluation is essential, specifically focusing on a neurological examination. The evaluation should include motor function, sensory deficits, reflexes, and signs of parkinsonism or Alzheimer's-related neurological impairments that may co-occur with behavioral addictions.

Mental Status Examination

The Mental Status Examination (MSE) plays a vital role in diagnosing behavioral addictions in individuals with NDDs. The evaluation typically follows a standard format, focusing on:

- *General appearance and behavior*: Observing any signs of psychomotor agitation or retardation, impulsivity, or repetitive behaviors that may be related to addiction.
- *Language and thought disorders*: Assessing coherence, logical flow of speech, and any delusional thinking, which could influence addictive behavior, especially in cognitive decline.
- *Mood*: Evaluating mood stability, emotional lability, or depressive symptoms, as mood disturbances often co-occur with behavioral addictions in NDDs.
- *Perception*: Checking for hallucinations or distorted perceptions, specifically relevant in PD, where perceptual disturbances may complicate behavioral addictions.
- *Cognitive functions*: Focusing on attention, concentration, memory, orientation, and executive functioning. Deficits in these areas are common in NDDs and can exacerbate impulse control problems, leading to addictive behaviors.
- *Judgment and insight*: Assessing the patient's ability to recognize their compulsive behavior and its consequences. Impaired judgment and poor insight are common, contributing to the persistence of addictive behavior.

In addition to the MSE, cognitive screening tools such as the Mini-Mental State Examination can provide a standardized measure of cognitive decline, highlighting areas like memory deficits, spatial disorientation, and language impairments contributing to behavioral issues.

Functional Status

Assessing the patient's functional capacity is vital, particularly their ability to perform ADLs. Observing the patient's behavior in a home or simulated environment can provide insight into how well they manage daily tasks. In the context of NDDs, compulsive behaviors may interfere with ADLs, such as neglecting personal care due to excessive gambling.

Social and Economic Status

Understanding the patient's support system is essential for managing behavioral addictions. This includes family involvement, available caregivers, and economic resources. Behavioral addictions often result in financial problems, and for patients with NDDs, the risk of financial exploitation or neglect increases, especially in cases of legal issues or limited social support.

Environmental Characteristics

Evaluating the patient's living environment is crucial to ensuring their safety. Behavioral addictions, particularly those tied to cognitive impairment, may put patients at risk of self-harm, financial instability, or unsafe living conditions. An unsafe environment can exacerbate behavioral issues such as compulsive hoarding or poor dietary habits.

Assessing Competencies

Assessing competencies is a routine part of a CGA and includes determining whether the patient can give informed consent for medical treatment, manage their financial affairs, create advance directives, and make life-support decisions. Patients with behavioral addictions may lack the capacity to manage their personal and financial affairs, particularly in advanced stages of cognitive decline.

Laboratory and Radiological Investigations

Blood tests and neuroimaging provide a comprehensive picture of the patient's medical condition. Neuroimaging can help identify structural changes in the brain, such as atrophy or lesions, that correlate with the patient's behavioral symptoms.

Rating Scales

Several potential rating scales can be used, including the Neuropsychiatric Interview (NPI), Iowa Gambling Task, South Oaks Gambling Screen (SOGS), the 11-item Impulse Control Disorders (ICDs), Behavioral Pathology in Alzheimer's Disease Rating Scale (BEHAVE-AD), Dopamine Dysregulation Syndrome Patient and Caregiver Questionnaire (DDS-PC), Ryden Aggression Scale (RAS), and the Behavioral Syndromes Scale for Dementia. A more recent tool, the St. Andrew's Sexual Behavior Assessment (SASBA) (Knight et al., 2008), was developed to assess ISBs in patients with progressive neurological conditions and acquired brain injuries.

TREATMENT INTERVENTIONS

Depending on the type of behavioral addiction and NDD, therapeutic options include both non-pharmacological and pharmacological treatments. Non-pharmacological approaches encompass cognitive and emotion-oriented interventions, sensory stimulation techniques, behavior management strategies, and psychosocial activities. Pharmacological treatment options may include antidepressants, antipsychotics, antiepileptics, and, in some cases, hormonal therapies.

Treatment of Gambling Disorder

Pharmacological Treatment

Several studies explored the role of paroxetine (a selective serotonin reuptake inhibitor) (Grant et al., 2003) and naltrexone (an opioid antagonist) (Kim et al., 2001) in gambling disorder, among other pharmacological agents. Research has found variations in how different age groups respond to treatment, and a recent literature review summarized the available evidence, highlighting naltrexone and nalmefene as potentially promising but also emphasizing the current limitations, including the limited diversity in participant demographics such as older adults and those with NDDS (Mestre-Bach & Potenza, 2024).

Non-Pharmacological Treatment

Sharpe (1998) proposed that cognitive-behavioral therapy for problematic behaviors should follow a structured series of steps. These include stabilizing the issue, developing alternative behaviors for enjoyment and relaxation, enhancing awareness, incorporating relaxation techniques, improving problem-solving skills, and applying exposure therapy. This approach would then move on to addressing cognitive distortions and preventing relapse. One case report by Manes et al. (2010) discussed the use of occupational therapy and cognitive rehabilitation, alongside carbamazepine, to target gambling in a 69-year-old woman with FTD. Wynn et al. (2014) and de Lisle et al. (2012) suggested that mindfulness-based interventions and online treatment methods are vital for advancing research and therapeutic innovations in problematic gambling.

Treatment of Inappropriate Sexual Behaviors

Pharmacological Treatment

Case reports and series have shown successful treatment of ISBs using a variety of medications, such as antidepressants, antipsychotics, anticonvulsants, cholinesterase inhibitors, hormonal agents, and beta-blockers. While these drugs can be effective, each comes with potential side effects, and no single class has emerged as the first-choice treatment. When one type of medication proves ineffective, studies have explored switching to other medication classes. The quality of evidence remains limited to case reports and series, with no randomized controlled trials available. Additionally, research is challenged by the lack of a universally accepted definition of ISBs and limited objective methods to measure them (De Giorgi & Series, 2016). The findings are summarized in Table 12.2.

TABLE 12.2
Pharmacological Treatment Options for ISBs in NDDs

Pharmacological Treatment	Conditions Treated	Outcomes
Antidepressant: Paroxetine (20 mg/day) (Stewart & Shin, 1997)	Dementia, alcoholic dementia, ISBs	Within one week, staff at the living facility reported that the hypersexuality had improved dramatically after failing to respond to a variety of drugs (haloperidol, lithium, thioridazine, lorazepam, and nortriptyline). After 3 months, the improvement was maintained, with no substantial side effects.
Antidepressant: Citalopram (20 mg/day) (Chen, 2010)	Dementia, ISBs	Agitation decreased, and the patient returned to his hobby of woodwork and other usual activities. After 3 months, the improvement was maintained, with no substantial side effects.
Antidepressant: Mirtazapine (15–30 mg/day) (Bardell et al., 2011)	Vascular dementia, ISBs	No adverse effects were noted.
Antidepressant: Clomipramine (150 mg/day) (Leo & Kim, 1995)	Dementia, ISBs	After 4 weeks, the patient reported a significant decrease in his urge to expose himself. There was no report from collateral informants of any exhibitionism occurring during passes off grounds or after discharge.
Antidepressant: Trazodone (100–500 mg/day) (Simpson & Foster, 1986)	Severe organic brain syndromes, disturbed behaviors	The improvement in behavior might have occurred secondary to the sedating effect of trazodone.
Anxiolytic: Buspirone (5 mg three times daily) (Tiller et al., 1988)	Vascular dementia, ISBs	Improvement in ISB started within 5 days of initiating treatment. After 6 weeks of treatment, no relapse was observed within 1 month of stopping buspirone.
Antipsychotic: Haloperidol (Rosenthal et al., 2003)	AD, ISBs	A significant reduction in sexual disinhibition and hypersexuality was noted.
Antipsychotic: Quetiapine (25 mg/day) (MacKnight & Rojas-Fernandez, 2000)	Vascular dementia, ISBs	Symptoms disappeared within 2 days of starting quetiapine; the patient had failed to respond to cyproterone and paroxetine. The behavior did not recur at a 2-month follow-up.
Anticonvulsant: Carbamazepine (800 mg/day) (Poetter & Stewart, 2012)	FTD, ISBs	Carbamazepine was titrated to 800 mg/d, attaining a serum drug concentration of 6.0 mg/L. The sexual behavior completely resolved and never recurred for the subsequent 6 months.
Hormone: Medroxyprogesterone (MPA) (100–300 mg/week for 11 weeks) Leuprolide (7.5 mg/month for 8 weeks) (Bardell et al., 2011)	Vascular dementia, ISBs	Both drugs were found to be effective.
Hormone: Medroxyprogesterone (300 mg/week for 1 year) (Levitsky & Owens, 1999)	AD, ISBs	After failing thioridazine or chlorpromazine, the patient participated in a therapeutic trial of MPA. Undesirable sexual behaviors in these patients were eliminated completely within 10 to 14 days of starting MPA.

(*Continued*)

TABLE 12.2 (CONTINUED)

Pharmacological Treatment	Conditions Treated	Outcomes
Cholinesterase inhibitors: Rivastigmine, Donepezil (Bardell et al., 2011)	AD, ISBs	Rivastigmine reduced ISBs, but donepezil might have caused them. Side effects included gastrointestinal issues and sleep disturbances.
Beta-blockers: Propranolol, Pindolol (Bardell et al., 2011)	Aggressive and hypersexual behaviors	The agents were effective in reducing aggression but had limited efficacy for sexual behaviors. Side effects included coldness, fatigue, and sleep disturbances.
Finasteride (Na et al., 2009)	Vascular dementia, ISBs	Treatment (8 weeks) was effective for six patients. The remaining five required further interventions, such as propranolol or quetiapine.

Non-Pharmacological Treatment

Given the potential for adverse effects with pharmacological treatments in older adults, non-pharmacological interventions should be considered in treating ISBs. Common approaches include redirecting inappropriate behaviors, educating patients and caregivers, and counseling. In NDDs, however, these strategies can be challenging due to the impairment of new learning, making behavioral modification more difficult.

Non-pharmacological strategies involve using same-sex caregivers to minimize triggers, dressing patients in clothing that fastens in the back to reduce access to inappropriate touching, and creating structured environments that limit unmonitored interactions. While these approaches are helpful, success rates may vary, particularly in individuals with significant cognitive decline, emphasizing the necessity for individualized interventions based on the patient's needs (Wallace & Safer, 2009).

Psychoeducation for family, caregivers, and staff has shown positive outcomes in managing ISBs in individuals with NDDs. Educating these groups about sexuality and aging leads to more permissive and understanding attitudes, ultimately benefiting individuals with ISBs. Specifically, sex education programs that balance the importance of regular sexual expression while addressing inappropriate behaviors have been found to enhance the quality of life of people with NDDs by promoting healthier interactions and reducing misunderstandings about sexual needs. A brief, focused sexual education program can significantly improve attitudes toward sexuality in older adults (White & Catania, 1983). The process involves identifying whether increased libido, impulsivity, disinhibition, or social conduct issues drive the behavior. The therapy is structured to address both the patient and their environment, focusing also on the education and guidance of caregivers and healthcare staff. Techniques such as setting clear boundaries for acceptable behaviors, reducing sexual stimuli, and encouraging open discussions about sexual behaviors can help in managing and modifying the behavior (Derouesné, 2009).

Treatment of Food Addiction

The management of food addition involves keeping the patient engaged with playful activities to avoid loneliness and boredom, which often accompany excessive eating. Additionally, it is advised to divide the food portion into two and give the second portion only if the patient asks, along with increasing the amount of vegetables and salads in the diet (Adams et al., 2019).

CONCLUSION

This chapter summarizes the available evidence on behavioral addictions, particularly gambling disorder and ISBs, in NDDs. Research on behavioral addictions in this vulnerable population is continuously expanding. However, studying these phenomena in NDDs is complex due to various factors, including cognitive deficits that interfere with identification and treatment. Further studies should focus on methods for the early identification of behavioral addictions and the development of effective treatment strategies for this clinical group.

REFERENCES

Aboujaoude, E. (2010). Problematic internet use: An overview. *World Psychiatry: Official Journal of the World Psychiatric Association (WPA), 9*(2), 85–90. doi:https://doi.org/10.1002/j.2051-5545.2010.tb00278.x

Adams, R. C., Sedgmond, J., Maizey, L., Chambers, C. D., & Lawrence, N. S. (2019). Food addiction: Implications for the diagnosis and treatment of overeating. *Nutrients, 11*(9), 2086. https://doi.org/10.3390/nu11092086

Ahmed, R. M., Kaizik, C., Irish, M., Mioshi, E., Dermody, N., Kiernan, M. C., Piguet, O., & Hodges, J. R. (2015). Characterizing sexual behavior in frontotemporal dementia. *Journal of Alzheimer's Disease, 46*(3), 677–686.

American Psychiatric Association. (2022). *Diagnostic and Statistical Manual of Mental Disorders* (5th ed., text rev.). https://doi.org/10.1176/appi.books.9780890425787

Baird, A. D., Wilson, S. J., Bladin, P. F., Saling, M. M., & Reutens, D. C. (2007). Neurological control of human sexual behaviour: Insights from lesion studies. *Journal of Neurology, Neurosurgery, and Psychiatry, 78*(10), 1042–1049. doi:https://doi.org/10.1136/jnnp.2006.107193

Bardell, A., Lau, T., & Fedoroff, J. P. (2011). Inappropriate sexual behavior in a geriatric population. *International Psychogeriatrics, 23*(7), 1182–1188.

Battista, P., Capozzo, R., Rizzo, G., Zecca, C., Anastasia, A., De Blasi, R., & Logroscino, G. (2019). Early pathological gambling in co-occurrence with semantic variant primary progressive aphasia: A case report. *Clinical Interventions in Aging, 14*, 727–733. https://doi.org/10.2147/CIA.S197484

Black B., Muralee, S., & Tampi, R. R. (2005). Inappropriate sexual behaviors in dementia. *Journal of Geriatric Psychiatry and Neurology, 18*(3), 155–162. https://doi.org/10.1177/0891988705277541

Chapman, K. R., & Spitznagel, M. B. (2015). Measurement of sexual disinhibition in dementia: A systematic review. *International Journal of Geriatric Psychiatry, 34*(12), 1747–1757.

Chen, S. T. (2010, October 1). Treatment of a patient with dementia and inappropriate sexual behaviors with citalopram. *Alzheimer Disease & Associated Disorders, 24*(4), 402–403.

Chen, S. K., Lo, M. T., & Lin, S. S. J. (2017). Impulsivity as a precedent factor for problematic internet use: How can we be sure? *International Journal of Psychology 52*, 389. https://doi.org/10.1002/ijop.12231

Cilia, R., Siri, C., Marotta, G., Isaias, I. U., De Gaspari, D., Canesi, M., Pezzoli, G., & Antonini, A. (2008). Functional abnormalities underlying pathological gambling in Parkinson disease. *Archives of Neurology, 65*(12), 1604–1611. https://doi.org/10.1001/archneur.65.12.1604

Codling, D., Shaw, P., & David, A. S. (2015). Hypersexuality in Parkinson's disease: Systematic review and report of 7 new cases. *Movement Disorders Clinical Practice, 2*(2), 116–126. https://doi.org/10.1002/mdc3.12155

Crockford, D., Quickfall, J., Currie, S., Furtado, S., Suchowersky, O., & El-Guebaly, N. (2008). Prevalence of problem and pathological gambling in Parkinson's disease. *Journal of Gambling Studies, 24*(4), 411–422. https://doi.org/10.1007/s10899-008-9099-3

De Giorgi, R., & Series, H. (2016). Treatment of inappropriate sexual behavior in dementia. *Current Treatment Options in Neurology, 18*, 1–5.

de Lisle, S. M., Dowling, N. A., & Allen, J. S. (2012). Mindfulness and problem gambling: A review of the literature. *Journal of Gambling Studies, 28*, 719–739.

De Marchi, N., Morris, M., Mennella, R., La Pia, S., & Nestadt, G. (1998). Association of obsessive-compulsive disorder and pathological gambling with Huntington's disease in an Italian pedigree: Possible association with Huntington's disease mutation. A*cta psychiatrica Scandinavica, 97*(1), 62–65.

Derouesné, C. (2009, June 1). Comportements dits d'hypersexualité et démences [The so-called hypersexual behaviors in dementia]. *Psychologie & neuropsychiatrie du vieillissement, 7*(2), 101–108.

Reininghaus, E., & Lackner, N. (2015). Relationship satisfaction and sexuality in Huntington's disease. In F. B. David & B. Vodušek, *Handbook of clinical neurology* (1 ed., vol. 130, pp. 325–334). Elsevier. https://doi.org/10.1016/B978-0-444-63247-0.00018-3

Fein, G., McGillivray, S., & Finn, P. (2007). Older adults make less advantageous decisions than younger adults: Cognitive and psychological correlates. *Journal of the International Neuropsychological Society: JINS, 13*(3), 480–489. https://doi.org/10.1017/S135561770707052X

Fontaine, M., Lemercier, C., Bonnaire, C., Giroux, I., Py, J., Varescon, I., & Le Floch, V. (2023). Gambling and aging: An overview of a risky behavior. *Behavioral Sciences (Basel, Switzerland), 13*(6), 437. https://doi.org/10.3390/bs13060437

Grant, J. E., Kim, S. W., Potenza, M. N., Blanco, C., Ibanez, A., Stevens, L., Hektner, J. M., & Zaninelli, R. (2003). Paroxetine treatment of pathological gambling: A multi-centre randomized controlled trial. *International Clinical Psychopharmacology, 18*(4), 243–249.

Griffiths, M. (1996). Behavioural addiction: An issue for everybody? *Journal of Workplace Learning, 8*(3), 19–25.

IITAP. (2020). Sexual dependency inventory, revised 4.0. International Institute of Trauma and Addiction Professionals (IITAP). https://cdn.ymaws.com/iitap.com/resource/collection/B3DF8B36-47AB-4480-91DB-FB88B2EAA040/SDI_Therapist_Training_Manual.pdf

Ikeda, M., Brown, J., Holland, A. J., Fukuhara, R., & Hodges, J. R. (2002). Changes in appetite, food preference, and eating habits in frontotemporal dementia and Alzheimer's disease. *Journal of Neurology, Neurosurgery, and Psychiatry, 73*(4), 371–376. https://doi.org/10.1136/jnnp.73.4.371

Kalkhoven, C., Sennef, C., Peeters, A., & Van Den Bos, R. (2014). Risk-taking and pathological gambling behavior in Huntington's disease. *Frontiers in Behavioral Neuroscience, 8*, 103.

Karila, L., Wery, A., Weinstein, A., Cottencin, O., Petit, A., Reynaud, M., & Billieux, J. (2014). Sexual addiction or hypersexual disorder: Different terms for the same problem? A review of the literature. *Current Pharmaceutical Design, 20*(25), 4012–4020.

Kim, S. W., Grant, J. E., Adson, D. E., & Shin, Y. C. (2001). Double-blind naltrexone and placebo comparison study in the treatment of pathological gambling. *Biological Psychiatry, 49*(11), 914–921.

Kloeters, S., Bertoux, M., O'Callaghan, C., Hodges, J. R., & Hornberger, M. (2013). Money for nothing—Atrophy correlates of gambling decision making in behavioural variant frontotemporal dementia and Alzheimer's disease. *NeuroImage: Clinical, 2*, 263–272. https://doi.org/10.1016/j.nicl.2013.01.011

Knight, C., Alderman, N., Johnson, C., Green, S., Birkett-Swan, L., & Yorstan, G. (2008). The St Andrew's Sexual Behaviour Assessment (SASBA): Development of a standardised recording instrument for the measurement and assessment of challenging sexual behaviour in people with progressive and acquired neurological impairment. *Neuropsychological Rehabilitation, 18*(2), 129–159.

Leo, R. J., & Kim, K. Y. (1995). Clomipramine treatment of paraphilias in elderly demented patients. *Journal of Geriatric Psychiatry and Neurology, 8*(2), 123–124.

Leplow, B., Renftle, D., Thomas, M., Michaelis, K., Solbrig, S., Maetzler, W., Berg, D., & Liepelt-Scarfone, I. (2023). Characteristics of behavioural addiction in Parkinson's disease patients with self-reported impulse control disorder and controls matched for levodopa equivalent dose: A matched case-control study. *Journal of Neural Transmission, 130*(2), 125–133.

Levitsky, A. M., & Owens, N. J. (1999). Pharmacologic treatment of hypersexuality and paraphilias in nursing home residents. *Journal of the American Geriatrics Society, 47*(2), 231–234.

Lew-Starowicz, M. & Coleman, E. (2022). Mental and sexual health perspectives of the International Classification of Diseases (ICD-11) Compulsive sexual behavior disorder. *Journal of Behavioral Addictions, 11*(2), 226–229. https://doi.org/10.1556/2006.2022.00032

Lo Coco, D., & Nacci, P. (2004). Frontotemporal dementia presenting with pathological gambling. *Journal of Neuropsychiatry and Clinical Neurosciences, 16*(1), 117–118. https://doi.org/10.1176/jnp.16.1.117

MacKnight, C., & Rojas-Fernandez, C. (2000). Quetiapine for sexually inappropriate behavior in dementia. *Journal of the American Geriatrics Society, 48*(6), 707

Manes, F., Torralva, T., Roca, M., Gleichgerrcht, E., Bekinschtein, T. A., & Hodges, J. R. (2010). Frontotemporal dementia presenting as pathological gambling. *Neurology, 6*(6), 347–352.

Mendez, M. F., & Shapira, J. S. (2011). Internet pornography and frontotemporal dementia. *Journal of Neuropsychiatry and Clinical Neurosciences, 23*(2), E3–E3.

Mendez, M. F., & Shapira, J. S. (2013). Hypersexual behavior in frontotemporal dementia: A comparison with early-onset Alzheimer's disease. *Archives of Sexual Behavior, 42*(3), 501–509.

Merims, D., & Giladi, N. (2008). Dopamine dysregulation syndrome, addiction and behavioral changes in Parkinson's disease. *Parkinsonism & Related Disorders, 14*(4), 273–280.

Mestre-Bach, G., & Potenza, M. N. (2024). Pharmacological management of gambling disorder: An update of the literature. *Expert Review of Neurotherapeutics, 24*(4), 391–407.

Na, H., Lee, J., Park, S., Ko, S., Kim, S., & Cho, S. (2009). Inappropriate sexual behaviors in patients with vascular dementia: Possible response to finasteride. *Journal of the American Geriatrics Society, 57*(11), 2161.

Nakum, S., & Cavanna, A. E. (2016). The prevalence and clinical characteristics of hypersexuality in patients with Parkinson's disease following dopaminergic therapy: A systematic literature review. *Parkinsonism & Related Disorders, 25*, 10–16.

Nadeau, D., Giroux, I., Simard, M., Jacques, C., & Dupré, N. (2018). Gambling habits of people with Parkinson's disease. *Journal of Gambling Issues,* (37), 149–171. http://dx.doi.org/10.4309/jgi.2018.37.7

Nakaaki, S., Murata, Y., Sato, J., Shinagawa, Y., Hongo, J., Tatsumi, H., Mimura, M., & Furukawa, T. A. (2007). Impairment of decision-making cognition in a case of frontotemporal lobar degeneration (FTLD) presenting with pathologic gambling and hoarding as the initial symptoms. *Cognitive and Behavioral Neurology: Official Journal of the Society for Behavioral and Cognitive Neurology, 20*(2), 121–125. https://doi.org/10.1097/WNN.0b013e31804c6ff7

Nordvig, A. S., Goldberg, D. J., Huey, E. D., & Miller, B. L. (2019). The cognitive aspects of sexual intimacy in dementia patients: A neurophysiological review. *Neurocase, 25*(1–2), 66–74. https://doi.org/10.1080/13554794.2019.1603311

O'Sullivan, S. S., & Lees, A. J. (2007). Pathological gambling in Parkinson's disease. *Lancet Neurology, 6*(5), 384–386. https://doi.org/10.1016/S1474-4422(07)70089-3

Ozel-Kizil, E., Sakarya, A., Arica, B., & Haran, S. (2013). A case of frontotemporal dementia with amyotrophic lateral sclerosis presenting with pathological gambling. *Journal of Clinical Neurology (Seoul, Korea), 9*(2), 133–137. https://doi.org/10.3988/jcn.2013.9.2.133

Petry, N. M., Zajac, K., & Ginley, M. K. (2018). Behavioral addictions as mental disorders: To be or not to be? *Annual Review of Clinical Psychology, 14*, 399–423. https://doi.org/10.1146/annurev-clinpsy-032816-045120

Piguet, O. &Hodges, J. R. (2013). Behavioural-variant frontotemporal dementia: An update. *Dementia & Neuropsychologia, 7*(1), 10–18. https://doi.org/10.1590/S1980-57642013DN70100003

Poetter, C. E., & Stewart, J. T. (2012). Treatment of indiscriminate, inappropriate sexual behavior in frontotemporal dementia with carbamazepine. *Journal of Clinical Psychopharmacology, 32*(1), 137–138.

Robbins, T. & Clark, L. (2015). Behavioral addictions. *Current Opinion in Neurobiology, 30*, 66–72. https://doi.org/10.1016/j.conb.2014.09.005

Rochat, L., Wilkosc-Debczynska, M., Zajac-Lamparska, L., Rothen, S., Andryszak, P., Gaspoz, J., Colombo, L., Khazaal, Y., & Achab, S. (2021). Internet use and problematic use in seniors: A comparative study in Switzerland and Poland. *Frontiers in Psychiatry, 12,* 609190. https://doi.org/10.3389/fpsyt.2021.609190

Rosenthal, M., Berkman, P., Shapira, A., Israel, G., & Abramovitz, J. (2003). Urethral masturbation and sexual disinhibition in dementia: A case report. *Israel Journal of Psychiatry, 40*(1), 67.

Sachdeva, J., Harbishettar, V., Barraclough, M., McDonald, K., & Leroi, I. (2014, January 1). Clinical profile of compulsive sexual behaviour and paraphilia in Parkinson's disease. *Journal of Parkinson's Disease, 4*(4), 665–670.

Sahithiya, B. R., & Kashyap, R. S. (2022). Sexual addiction disorder—A review with recent updates. *Journal of Psychosexual Health, 4*(2), 95–101. https://doi.org/10.1177/26318318221081080

Sarafudheen, S., Shoka, A., & Kathirgamachelvam, J. (2022). Pathological gambling and impulsivity related to Huntington's disease. *Progress in Neurology and Psychiatry, 26*(3), 12–15.

Sharpe, L. (1998, January 1). Cognitive-behavioural treatment of problem gambling. In *International handbook of cognitive and behavioral treatments for psychological disorders* (pp. 393–416). Elsevier. https://doi.org/10.1016/b978-008043433-9/50015-8.

Simpson, D. M., & Foster, D. (1986). Improvement in organically disturbed behavior with trazodone treatment. *Journal of Clinical Psychiatry, 47*(4), 191–193.

Sixto-Costoya, A., Castelló-Cogollos, L., Aleixandre-Benavent, R., & Valderrama-Zurián, J. C. (2021). Global scientific production regarding behavioral addictions: An analysis of the literature from 1995 to 2019. *Addictive Behaviors Reports, 14*, 100371. https://doi.org/10.1016/j.abrep.2021.100371

Stewart, J. T., & Shin, K. J. (1997, October 1). Paroxetine treatment of sexual disinhibition in dementia. *American Journal of Psychiatry, 154*(10), 1474.

Stewart, S. A., Copeland, A. L., & Cherry, K. E. (2023). Risk factors for substance use across the lifespan. *Journal of Genetic Psychology, 184*(2), 145–162. https://doi.org/10.1080/00221325.2022.2130025

Stout, J. C., Rodawalt, W. C., & Siemers, E. R. (2001). Risky decision making in Huntington's disease. *Journal of the International Neuropsychological Society, 7*(1), 92–101.

Subramaniam, M., Wang, P., Soh, P., Vaingankar, J. A., Chong, S. A., Browning, C. J., & Thomas, S. A. (2015). Prevalence and determinants of gambling disorder among older adults: A systematic review. *Addictive Behaviors, 41*, 199–209. https://doi.org/10.1016/j.addbeh.2014.10.007

Tiller, J. W., Dakis, J. A., & Shaw, J. M. (1988). Short-term buspirone treatment in disinhibition with dementia. *Lancet, 332*(8609), 510.

Torrisi, M., Cacciola, A., Marra, A., De Luca, R., Bramanti, P., & Calabrò, R. S. (2017). Inappropriate behaviors and hypersexuality in individuals with dementia: An overview of a neglected issue. *Geriatrics & Gerontology International, 17*(6), 865–874.

Wallace, M., & Safer, M. (2009). Hypersexuality among cognitively impaired older adults. *Geriatric Nursing, 30*(4), 230–237.

Wang, D., Liu, X., Chen, K., Gu, C., Zhao, H., Zhang, Y., & Luo, Y. (2024). Risks and protection: a qualitative study on the factors for internet addiction among elderly residents in Southwest China communities. *BMC Public Health, 24*(1), 531.

White, C. B., & Catania, J. A. (1983). Psychoeducational intervention for sexuality with the aged, family members of the aged, and people who work with the aged. *The International Journal of Aging and Human Development, 15*(2), 121–138.

Wong, S. H. & Steiger, M. J. (2007). Pathological gambling in Parkinson's disease. *BMJ (Clinical Research Ed.), 334*(7598), 810–811. https://doi.org/10.1136/bmj.39176.363958.80

Wynn, J., Hudyma, A., Hauptman, E., Nicole Houston, T., & Michael Faragher, J. (2014). Treatment of problem gambling: Development, status, and future. *Drugs and Alcohol Today, 14*(1), 42–50.

Younes, K. & Miller, B. L. (2020). Neuropsychiatric aspects of frontotemporal dementia. *Psychiatric Clinics of North America, 43*(2), 345–360. https://doi.org/10.1016/j.psc.2020.02.005

Zeiss, A. M., Davies, H. D., & Tinklenberg, J. R. (1996). An observational study of sexual behavior in demented male patients. *Journals of Gerontology Series A: Biological Sciences and Medical Sciences, 51*(6), M325–M329.

13 Addressing Addiction in Individuals with Neurodegenerative Disorders

Approach, Screening, and Management

Karine Kahil, Lea Daou, Hala Itani, Rawan Koleilat, and Farid Talih

INTRODUCTION

At present, over 57 million individuals globally are living with dementia, with predictions indicating that this figure will exceed 152 million by 2050 (International, 2018). The prevalence of substance use disorders (SUDs) in older adults is a growing concern, with studies indicating an upward trend in this demographic (Lin et al., 2023). This prevalence varies across different neurodegenerative disorders (NDDs), with certain conditions exhibiting higher susceptibility. The use of sedatives/hypnotics, stimulants (cocaine and amphetamines), opioids, and cannabis is prevalent among individuals with Parkinson's disease (PD) and is linked to a more severe course of illness and loss of functioning (Kaur et al., 2021). Behavioral addictions, including gambling disorder, have also been reported in older adults, with increasing prevalence over the years (Guillou Landreat et al., 2019). Individuals with NDDs may be at increased risk of developing SUDs due to factors such as pain, mood disturbances, anxiety, and cognitive decline. These factors can contribute to self-medication behaviors and vulnerability to addictive behaviors. Notably, while SUDs may coexist with NDDs, there is conflicting evidence as to whether SUDs are implicated in the onset or progression of these conditions in younger individuals. Some research shows that individuals who consumed mild or moderate amounts of alcohol exhibited a 17% and 21% reduction in the risk of all-cause dementia, respectively. In comparison, individuals who consistently consumed alcohol heavily had an 8% increased risk of developing all-cause dementia (Jeon et al., 2023). A recent study from 2023 reported that alcohol, cocaine, and methamphetamine use deleteriously affects the brain and may trigger NDDs, such as Alzheimer's disease (AD), PD, and amyotrophic lateral sclerosis (Guo et al., 2023). Therefore, chronic substance misuse may be implicated in accelerating neurodegenerative processes.

SUD in older adults is frequently overlooked, yet it represents a serious public health issue with a multitude of challenges complicated by ageism, stigma, and inadequate healthcare infrastructure (Rao & Arora, 2015). As individuals age, they become increasingly susceptible to societal biases and stereotypes, leading to failure to screen for SUDs in older adults (Rao & Arora, 2015). Furthermore, the lack of adequate training among healthcare workers in addressing SUD in this demographic, perpetuates barriers to effective treatment and support (Rao et al., 2019). Triggering factors for substance use in older adults stem from challenging circumstances such as chronic pain, grief, and loneliness, underscoring the importance of holistic and compassionate care approaches (NIDA, 2020). It is crucial to validate and communicate the diagnosis of SUD as a medical condition rather

 DOI: 10.1201/9781003478485-13

than a moral defect, emphasizing the neurobiological underpinnings and complexities involved (Heilig et al., 2021). Thus, a model shift towards person-centered language and care models is essential to foster dignity, empowerment, and inclusivity in addressing SUDs among older adults.

In this chapter, we will explore SUDs in older adults with NDDs, addressing the importance of screening and the different challenges in screening, as well as providing an overview of the management and potential treatment options.

IMPORTANCE OF SCREENING FOR SUBSTANCE USE DISORDERS IN OLDER ADULTS

Screening for SUDs in older adults is crucial due to the bidirectional relationship between substance use and NDDs. Substance use increases the risk of neurodegeneration, and on the other hand, certain conditions affecting older adults put them at a higher risk for substance use.

Impact of Substances on Neurodegeneration

Substance misuse can have a detrimental impact on the aging brain, as advancing age increases individuals' vulnerability to the adverse cognitive effects of substances. Multiple substances, including alcohol, are known to cause or worsen neurodegeneration through multiple mechanisms, including neuroinflammation and toxicity. Chronic alcohol consumption can generate a persistent neuroinflammatory response, which itself can induce neurodegeneration (Guo et al., 2023; Kaur et al., 2021). Common NDDs associated with alcohol use disorder include AD and PD (Kamal et al., 2020). Moreover, excessive alcohol use can result in higher oxidative stress via augmented production of reactive oxygen species, which in turn can contribute to a process called ferroptosis; this is another mechanism by which substances can affect the aging brain (Guo et al., 2023; Kamal et al., 2020). Ferroptosis is defined as regulatory cell necrosis via iron ion catalysis and lipid peroxidation, and there is growing evidence that ferroptosis is highly linked to the emergence and progression of NDDs. Ferroptosis is associated not only with alcohol use but with other substance use, including cocaine and methamphetamine (Guo et al., 2023).

Impact of Substance Use on Disease Outcome and Prognosis

Substance use not only increases the risk of NDDs but also affects disease outcomes and is associated with a poorer prognosis in those already diagnosed. A retrospective study was conducted in 2021 using the National Inpatient Sample (NIS) database in the United States to explore SUDs in older adults diagnosed with PD (Kaur et al., 2021). The results of this study showed an increase in the severity of illness in adults with PD with a comorbid SUD, including opioid, barbiturate, alcohol, and stimulant use disorders, but not those with comorbid cannabis use disorder. Specifically, significant loss of body functioning was increased by 1.6 times in patients with PD and alcohol use disorder and by almost fourfold in patients with PD and opioid use disorder. Substance use was also associated with prolonged care, as the length of hospital stay was increased in PD patients with comorbid SUDs, more so in barbiturate use disorder (Kaur et al., 2021). On the other hand, treatment of comorbid SUDs might have a favorable effect on disease outcomes in patients with NDDs. A study exploring the impact of an alcohol intervention service on cognition in older adults revealed a significant improvement in cognitive functioning following treatment of alcohol use (Seddon et al., 2021).

Neurodegenerative Disorders as Risk Factors for Substance Misuse

On the other hand, certain NDDs put patients at increased risk of substance misuse. For example, patients with frontotemporal dementia suffer from behavioral disinhibition, which can lead to substance misuse, particularly in the behavioral variant subtype (Kalapatapu & Sullivan, 2010). Moreover, there were several cases reported in the literature about frontotemporal dementia primarily manifesting as a behavioral addiction, such as gambling disorder. Similarly to frontotemporal dementia, pathological gambling was found to be associated with frontal lobe impairment, specifically orbitofrontal cortex dysfunction (Cipriani et al., 2016; Tondo et al., 2017).

Elder Abuse and Caregiver Substance Misuse

Screening for elder abuse, particularly in the context of substance misuse by caregivers, is also a critical public health issue. Research indicates that caregiver stress has been linked to a higher risk of elder abuse (Dong et al., 2014). A caregiver may use alcohol or benzodiazepines to try to control an older adult's challenging behavior (Rockville, 2020c). For instance, a caregiver may provide the older adult with alcohol to help them sleep or to calm down when angry or upset. This form of over-medication can lead to physical dependence, dizziness, sedation, confusion, and recurrent falls (Virnes et al., 2022).

Routine screening for abuse should be integrated into healthcare visits, mainly when there are signs of physical or behavioral changes in the older adult that could be attributed to substance misuse. Caregivers may not reveal that they are stressed or frustrated; hence, asking them directly about burnout is essential. Moreover, discussing the phenomenon that some individuals might over-medicate older adult patients to control their behaviors may help destigmatize the topic. It is crucial to educate caregivers about the risks of over-medication discussed above and to explain that it may worsen behavioral disturbances (Rockville, 2020c).

Misuse of Prescription Medications

It is also important to note that other conditions common in older adults, including insomnia and chronic pain, can make them particularly vulnerable to the misuse of prescription medications. At least one in four older adults in the United States is prescribed a psychoactive drug with the potential for misuse. Among these prescription drugs are analgesics, including opioids, sedatives, and hypnotics, such as benzodiazepines and gabapentinoids (Ates Bulut & Isik, 2022).

In summary, screening for SUDs in older adults, particularly those with NDDs, is essential to improve patients' outcomes and prognosis.

CHALLENGES IN SCREENING FOR SUBSTANCE USE DISORDERS IN OLDER ADULTS

Screening for SUDs in older adults might be hindered by multiple challenges, some of which are detailed below.

Role of Primary Care Physicians and Geriatricians

It is important to note that the core of the screening process lies in the hands of primary care physicians and geriatricians before reaching psychiatrists. In fact, older adults are most commonly followed by family medicine physicians, neurologists, or geriatricians, and these have a crucial role in bridging the gap between screening and referral to treatment (Rockville, 2020b). Based on the Treatment Improvement Protocol (TIP) consensus panel, it is recommended to carry out universal screening in all older adults during health visits, at least yearly, for all substance use, including alcohol, tobacco, prescription drugs, and illicit substances (Rockville, 2020b). More frequent screening might be warranted in certain situations raising concern for substance misuse, such as repeated requests for prescription drugs, frequent falls, and medical conditions that might be associated with alcohol or drug use, such as hypertension, insomnia, and sexually transmitted diseases.

Furthermore, even following a negative screen, occasional rescreening is recommended by the TIP consensus panel (Rockville, 2020b).

Screening Tools and Criteria

Screening tools used in psychiatry are based on well-defined criteria listed in internationally used references, such as the *Diagnostic and Statistical Manual of Mental Disorders*, Fifth Edition, Text Revision (DSM-5-TR) and the International Classification of Diseases, 11th Revision (ICD-11).

The DSM-5-TR defines SUD as a problematic pattern of use of a mind-altering substance leading to clinically significant impairment, as demonstrated by at least 2 out of 11 criteria within a 12-month period. The criteria fall broadly into four categories: impaired control, social problems, risky use, and physical dependence (American Psychiatric Association, 2022).

Similarly, the ICD-11 requires a pattern of recurrent episodic or continuous use of the substance with evidence of impaired regulation of substance use as manifested by 2 or more of the following 4 criteria (World Health Organization, 2021):

1. Impaired control over substance use.
2. Increasing precedence of the substance over different aspects of life, including the maintenance of health, everyday activities, and responsibilities.
3. Physiological features indicative of neuroadaptation to the substance, including tolerance, withdrawal symptoms, or repeated use to prevent or alleviate withdrawal symptoms.
4. The features of dependence are evident over at least 1 year, but the diagnosis may be made if use is continuous (daily or almost daily) for at least 3 months.

Limitations of DSM-5-TR Criteria in the Geriatric Population

The DSM-5-TR has been widely criticized for its SUD criteria, particularly in the geriatric population. Older adults who misuse substances may not meet full DSM-5-TR criteria for a specific SUD but may be dependent on various substances or engaging in risky substance use (Rockville, 2020a).

The DSM-5-TR cites recurrent use of the substance, resulting in an inability to fulfill significant role obligations at home, work, or college. Often, older adults with cognitive deficits are not employed and thus are not necessarily required to perform specific duties (Ghantous et al., 2022). As a result, the above-mentioned functionality criteria may not be best suited to evaluate SUDs in older adults with NDDs. Similarly, this population might not be initially involved in important social, occupational, or recreational activities that would be given up in the event of SUDs in younger populations (Ghantous et al., 2022).

Older adults also have a lower tolerance for substances like alcohol than younger adults, given the sensitivity of their receptors to substances. Thus, they may experience harmful effects, as well as develop tolerance and withdrawal symptoms with lower consumption of substances (Ghantous et al., 2022; Rockville, 2020b). Case in point, the National Institute on Alcohol Abuse and Alcoholism recommends no more than 7 standard drinks weekly for men aged 65 years or older versus 14 drinks weekly for younger men (NIAAA). Based on the criteria above, the DSM-5-TR might not be optimal for the diagnosis of SUDs in the geriatric population, particularly for patients with NDDs. Table 13.1 highlights the different challenges of the DSM-5-TR criteria for SUD in older adults with NDDs, with potential suggestions to aid in the screening.

Memory Impairment and Its Impact on Screening

Another challenge in screening for SUDs in adults with NDDs is memory problems, as cognitive impairment can impact adequate self-reporting and answering the questions of any screening tool. In addition, insight into the disorder, judgment, and capacity to identify hazardous situations

TABLE 13.1
DSM-5-TR Criteria for SUD, Challenges Faced by Older Adults with NDDs, and Respective Suggestions

DSM-5-TR Criteria for SUD	Challenges for Older Adults with NDDs	Suggestions
The substance is often taken in larger amounts or over a longer period than was intended.	Self-monitoring in older adults with NDDs is limited.	Rely on caregivers' reports, when applicable.
A persistent desire or unsuccessful efforts to cut down or control use.	Self-awareness in older adults with NDDs might be limited.	Rely on caregivers' reports, when applicable, and modify criteria for "observed" persistent desire.
Much time is spent on activities necessary to obtain, use, or recover from the substance's effects.	Older adults may take longer to recover from a given substance's effects. Further, they may take even more time to obtain and use substances due to cognitive limitations.	Rely on caregivers' reports, when applicable, and modify criteria "observed" a great deal of time.
Craving or a strong desire or urge to use the substance.	Older adults with NDDs may not reliably recognize or report cravings.	Rely on observed behaviors associated with cravings, e.g., restlessness and irritability.
Recurrent use results in a failure to fulfill significant role obligations at work, school, or home.	Role obligations may differ in older adults with NDDs relative to the general adult population.	Focus on role obligations pertinent for older adults, e.g., caring for themselves or a pet.
Continued use despite persistent or recurrent social or interpersonal problems caused or exacerbated by its effects.	Older adults with NDDs may not reliably recognize or attribute problems to substance use.	Rely on caregivers' reports, e.g., relationship changes from baseline functioning.
Important social, occupational, or recreational activities are given up or reduced because of substance use.	Older adults with NDDs typically engage in fewer activities regardless of substance use, making this criterion challenging to establish.	Consider eliminating this criterion in older adults with NDDs.
Recurrent use in situations in which it is physically hazardous.	Older adults with NDDs may not understand or report physically hazardous situations.	Rely on caregivers' reports, when applicable, e.g., the risk of falling after consuming a substance.
Continued use despite knowledge of having a persistent or recurrent physical or psychological problem that is likely to have been caused or exacerbated by the substance.	Self-awareness in older adults with NDDs is limited.	–
Tolerance	Because of the increased sensitivity to substances as they age, older adults might seem to have lowered rather than increased tolerance.	–
Withdrawal	Withdrawal symptoms can manifest in more "subtle and protracted" ways.	–

Sources: American Psychiatric Association, 2022; Barry et al., 2002; Han & Moore, 2018; Kuerbis et al., 2014.

where they have been consuming might be affected to different extents (Ghantous et al., 2022). Interviewing a caregiver can help overcome this challenge, but a reliable caregiver is not always available.

Alternative Screening Approaches

In memory impairment, the clinician might rely on reported risk factors and signs rather than on specific questionnaires based on criteria to screen for SUDs. For example, the clinician can ask a family member about some risk factors and signs of opioid use disorder, which the family member might be able to answer even if they do not live with the patient and do not have daily contact with them (Duggirala et al., 2022). Risk factors of opioid use disorder include long-term opioid therapy for chronic non-cancer pain, history of opioid misuse or opioid use disorder, current or past history of other substance misuse and/or SUD, family history of SUD and/or psychiatric disorders, history of adverse childhood experiences, and history of traumatic life events (Duggirala et al., 2022).

Signs of opioid use disorder that can be retrieved from family members and might facilitate screening for opioid use disorder include recurrent visits to the emergency department for opioid-related adverse effects, especially respiratory or central nervous system depression, family members noticing frequent drowsiness or slurred speech, use of prescription opioids for motives other than pain management, for example, to manage behavioral symptoms of NDDs, and signs of opioid withdrawal (Duggirala et al., 2022).

In general, it is judicious to suspect and screen for SUDs in older patients with dementia presenting repetitively for falls, breathing problems, sleeping problems, cognitive changes, seizures, or changes in blood pressure (Rockville, 2020c).

Along the same lines, objective questions and measures might be valuable to the clinician in detecting SUDs in older adults with NDDs. For example, a random urine drug screen is recommended in high-risk patients as a tool to supplement clinical data (Duggirala et al., 2022). Other objective screening tools that are more specific for alcohol use include blood testing for blood alcohol levels, as well as breathalyzers. The presence of macrocytosis and elevated gamma-glutamyl transferase are also important objective indicators of chronic alcohol consumption (Agrawal et al., 2022).

Challenges with Subjective Self-Report Tools

Concerning subjective self-report screening tools, some screening tools are validated in the geriatric population, but none are validated in older adults with NDDs. The latter are often excluded from validation studies, and most published work is registry-based without any mention of screening tools (Kalapatapu & Sullivan, 2010; Kaur et al., 2021). Among the scales or questionnaires used in the geriatric population, the Alcohol Use Disorders Identification Test (AUDIT) and the "Cut Down, Annoyed, Guilty, Eye-Opener" Questionnaire (CAGE) are the most commonly used instruments. The Michigan Alcohol Screening Test and its geriatric version, the Alcohol-Related Problems Survey, the Severity of Dependence Scale, and the Alcohol, Smoking, and Substance Involvement Screening Test (ASSIST) are also used (Sarkar et al., 2021).

In older adults with cognitive difficulties, questionnaires with dichotomous (yes/no) answers may be more appropriate and easier to administer than surveys with a multiple-point Likert-type scale where the patient must express a certain degree of agreement or disagreement with a statement. For example, the Drug Abuse Screening Test (DAST-10) is a good option to be used by older adults with NDDs, since each question requires only a yes/no response, and the tool can be completed in a few minutes (Skinner, 1982). However, this scale does not ask about tobacco or alcohol but rather about controlled medications and/or illicit drugs. The various classes of drugs include cannabis (e.g., marijuana, hash), solvents, tranquilizers (e.g., Valium), barbiturates, cocaine, stimulants (e.g., speed), hallucinogens (e.g., LSD), or narcotics (e.g., heroin) (Skinner, 1982). Adaptation and validation studies for the DAST-10 could be carried out in patients with NDDs to screen for SUDs.

In NDDs, answers to the scales may need to be double-checked with the caregiver since interviewing the patient alone might not be sufficient to exclude an ongoing SUD (Ghantous et al., 2022).

Importance of Screening for Behavioral Addictions

Finally, as previously mentioned, screening for behavioral addictions such as gambling disorder is also crucial and often overlooked. For this reason, screening tools should be refined to better assess these disorders in older adults, specifically those with NDDs (Guillou Landreat et al., 2019).

OVERVIEW OF MANAGEMENT FOR SUBSTANCE USE DISORDERS IN OLDER ADULTS WITH NEURODEGENERATIVE DISORDERS

Limited Access to Specialized Treatment Programs

Older adults with SUDs, including those without NDDs, have limited access to treatment options compared to their younger counterparts. Only around 23% of SUD treatment programs in the United States offer a program or group specifically tailored to older adults (Rockville, 2020d). The available information in the literature suggests that specialized treatment and care programs for older adults with SUDs are rare in Europe (*Drugs & Addiction, 2010*). To our knowledge, there is no information about specialized treatment programs for older adults with SUDs in low- and middle-income countries. One reason for this is the common assumption that older adults have a limited capacity to engage in treatment and change, which holds for those with cognitive impairment (Ghantous et al., 2022; Rao et al., 2019).

Impact of Cognitive Impairment on Treatment Outcomes

In SUDs, treatment completion has been consistently associated with a favorable outcome. However, unfortunately, individuals with cognitive impairment may be less likely than others to complete treatment, hence not benefiting fully (Brorson et al., 2013). Moreover, as older adults with NDDs often experience difficulties in planning, memory, problem-solving, and self-regulation, this affects their daily functioning, including the capacity to understand and apply concepts and skills delivered in substance use treatment (Collings et al., 2022; Fernández-Serrano et al., 2010). Therefore, cognitive impairment in itself is a major limiting factor in the treatment of SUDs in individuals with NDDs.

Physical Limitations and Their Effects on Treatment

Moreover, older adults with SUDs and NDDs may be physically limited. Often, older adults have sensory impairments, medical illnesses (e.g., kidney failure, liver failure), and impaired mobility—all of which may significantly impact treatment response, including but not limited to therapy, physician visits, and medication options (e.g., acamprosate and naltrexone) (Jensen et al., 2012; Mohd Rosnu et al., 2022). Furthermore, older adults with SUDs and NDDs are sometimes limited financially or by transportation means (Choi et al., 2014; Hadley Strout et al., 2016; Osborn et al., 2017).

Role of Caregivers in Managing Substance Use

As a result of the aforementioned limitations, a considerable portion of the care of individuals with NDDs and SUDs will fall upon their caregivers. Care initially starts with psychoeducation, as caregivers may not see alcohol and drug use, e.g., cannabis, as a problem. They may have normalized the habit of substance use and rationalized that it may be harmless or even pleasurable and beneficial (Briggs et al., 2011; Rao & Arora, 2015).

Older Adults' Sensitivity to Substance Effects

The reality is that older adults are much more vulnerable than younger or middle-aged individuals to the adverse physical effects of medications, illicit drugs, and alcohol. Their central nervous system is often more sensitive to the effects of substances, and it takes longer for their organs (e.g., liver and kidneys) to eliminate these substances. Moreover, as older adults have less lean body mass and total body water, they can become intoxicated even with small amounts of substances (Barry & Blow, 2016; Rockville, 2020c). Drug–drug interactions are also more likely in older adults and have been associated with risks of falls, seizures, internal hemorrhage, and fatal overdoses (Holton et al., 2017; Moore et al., 2007; Zanjani et al., 2016). As a matter of fact, a national cross-sectional study conducted in 2014 including 1,210 older adult Malaysian subjects with dementia found that substance misuse is among the factors significantly increasing the risk of falls in this population (Eshkoor et al., 2014). Furthermore, this population can experience cardiovascular and cerebrovascular accidents, sexually transmitted infections, and premature death while using heroin, cocaine, and methamphetamines (Ghantous et al., 2022; Rivers et al., 2004; Yarnell, 2015).

Psychoeducation and Caregiver Support

Psychoeducation will serve to raise awareness and shed light on the risks of substance use to both older adults and caregivers and to communicate the goal of abstinence. On the one hand, medical and psychological sequelae would be limited upon abstinence. However, on the other hand, caregivers might have to deal with the burden of withdrawal symptoms such as cravings, agitation, and irritability—which increases their risk of burnout (Sattar et al., 2007). To better support caregivers of older adults with NDDs and substance use, one can provide them with resources for psychosocial and educational programs, as well as caregiver support groups (Brodaty & Donkin, 2009; Gilhooly et al., 2016; Martín-Carrasco et al., 2014).

Pharmacological Treatment Options

There is a general paucity of evidence regarding the use of pharmacological treatments for SUDs in older adults, and to our knowledge, no research has been explicitly conducted among those with NDDs (Agrawal et al., 2022).

Treatment of SUDs depends on the substance misused. For the treatment of cocaine or stimulant use disorders, there are no currently approved pharmacologic treatments. A recent in-depth analysis exploring the effectiveness of various drugs like antidepressants, antiepileptics such as topiramate, dopamine agonists, and antipsychotics did not show any clear superiority over placebo (Ghantous et al., 2022).

As for alcohol use disorder, there seems to be conflicting evidence regarding the efficacy of naltrexone in older adults. The use of disulfiram is not recommended due to the increased risk of delirium and cardiovascular adverse effects in older adults. On the other hand, despite the lack of clinical trials exploring its efficacy in older adults, acamprosate use is considered safe, but additional caution is crucial in patients with impaired renal function (Agrawal et al., 2022). Based on the Canadian guidelines on alcohol use disorder in older adults, it is recommended to use either naltrexone for alcohol reduction and abstinence or acamprosate for abstinence (Butt et al., 2020).

For opioid use disorders, both methadone and buprenorphine can be used. On the one hand, methadone has been shown to decrease cravings, improve mood, and decrease the risk of death in older adults with opioid use disorder (Kakko et al., 2019; Rosen et al., 2011; Salmond & Allread, 2019). On the other hand, buprenorphine decreases symptoms of dysphoria as well as the risk of death (Kakko et al., 2019; Salmond & Allread, 2019). In Singapore, one study revealed that methadone maintenance treatment was cost-effective in treating older adults (mean age 74.8 years) who use opium (Guo et al., 2010). While the usual dose of methadone for the treatment of heroin misuse ranges between 60 and 120 mg, only a mean dose of 9 mg was used in this older group for safety reasons. Psychiatrists collaborated closely with caregivers/family members, providing monthly prescriptions of methadone to be administered by caregivers (Guo et al., 2010). Dose adjustments may be required for older adults taking other medications that might interact with methadone (e.g., antivirals, antidepressants, and antiepileptics) or for older adults with severe liver or kidney disease (Volpe et al., 2018). Due to cumulative effects, caution is taken when respiratory-depressant medications such as benzodiazepines are co-prescribed with either methadone or buprenorphine (Dürsteler-MacFarland et al., 2011). Adverse reactions associated with methadone use include prolonged corrected QT interval, drug–drug interactions via the cytochrome P450 (CYP)-3A4 enzyme, and toxicity due to the long elimination half-life, which older adults are particularly liable to. Buprenorphine seems to be a more reasonable option in older adults, as it does not require dose adjustment in renal failure, does not prolong the QTc interval, and may be notably safer in those with cardiac and respiratory diseases (Gazelka et al., 2020). Recently, long-acting injectable formulations for naltrexone and buprenorphine have become available. These formulations may be attractive options for older adults with NDDs, mainly due to potentially enhanced compliance.

It is important to keep in mind that inpatient detoxification might be required in some instances, particularly in patients with safety concerns, multiple substance use, medical complications (such as diabetes mellitus complications), or failure of outpatient treatment. For alcohol or benzodiazepine detoxification, short-acting benzodiazepines such as lorazepam may be preferred in older adults; lower dosing and slower taper should also be considered. Similarly, low-dose clonidine can be used in acute opioid detoxification, in addition to supportive treatment, with close monitoring for potential adverse effects, including hypotension and confusion, as older adults might be more susceptible to developing these (Agrawal et al., 2022).

Treatment of Medical Sequelae and Comorbidities

An additional important aspect of the treatment of older adults with SUDs and NDDs is the management of the aforementioned associated sequelae—including but not limited to urinary tract infections, hemorrhages, and cardiovascular accidents.

In addition, it is noteworthy to mention that both substance use and dementia are risk factors for developing delirium across different clinical populations. Delirium is a common, serious, costly, under-recognized, and often fatal condition for older patients. Dementia and cognitive impairment, as well as a history of alcohol misuse, are all risk factors for delirium consistently identified at admission in both medical and non-cardiac surgery populations (Inouye et al., 2014).

Non-Pharmacological Treatment Options

Apart from pharmacologic treatments, several non-pharmacologic interventions have demonstrated some benefits. Cognitive remediation, a behavioral intervention aiming to restore cognitive abilities and improve functional outcomes, can potentially improve cognition and treatment outcomes for those with SUDs, as well as behavioral interventions, including motivational interviewing (Nardo et al., 2022). Furthermore, simple environmental adaptations in residential treatments, e.g., simplified written material, role play, accessible designs, and person-centered principles, have been positively correlated with treatment retention in older adults with SUDs (Collings et al., 2022).

CONCLUSION

In conclusion, the intersection of SUDs and NDDs in older adults presents a multifaceted challenge with profound implications for public health and individual well-being. With a global increase in both dementia cases and substance misuse among older populations, there is a pressing need to address the bidirectional relationship between these conditions. Ultimately, addressing SUDs in older adults with NDDs requires a paradigm shift in healthcare delivery characterized by person-centered care models that prioritize dignity, empowerment, and inclusivity. By taking a comprehensive approach to understanding substance use in the context of aging and neurodegeneration, we can work towards better outcomes, improved quality of life, and healthier aging for everyone facing these complex issues.

REFERENCES

Agrawal, A., Touma Sawaya, R., Ojeahere, M., Padilla, V., & El Hayek, S. (2022). Neurocognitive and substance use disorders in older adults: Challenges and evidence. *Adv Dual Diagn*, *15*(2), 79–99. https://doi.org/10.1108/ADD-01-2022-0004

American Psychiatric Association (2022). *Diagnostic and statistical manual of mental disorders* (5th edition, Text Revision ed.).

Ates Bulut, E., & Isik, A. T. (2022). Abuse/misuse of prescription medications in older adults. *Clin Geriatr Med*, *38*(1), 85–97. https://doi.org/10.1016/j.cger.2021.07.004

Barry, K. L., & Blow, F. C. (2016). Drinking over the lifespan: Focus on older adults. *Alcohol Res*, *38*(1), 115–120.

Barry, K. L., Blow, F. C., & Oslin, D. W. (2002). Substance abuse in older adults: Review and recommendations for education and practice in medical settings. *Subst Abus*, *23*(3 Suppl), 105–131. https://doi.org/10.1080/08897070209511510

Briggs, W., Magnus, V., Lassiter, P., Patterson, A., & Smith, L. (2011). Substance use, misuse, and abuse among older adults: Implications for clinical mental health counselors. *J Ment Health Couns*, *33*(2), 112–127. https://doi.org/10.17744/mehc.33.2.y107266w86215440

Brodaty, H., & Donkin, M. (2009). Family caregivers of people with dementia. *Dialogues Clin Neurosci*, *11*(2), 217–228. https://doi.org/10.31887/DCNS.2009.11.2/hbrodaty

Brorson, H. H., Ajo Arnevik, E., Rand-Hendriksen, K., & Duckert, F. (2013). Drop-out from addiction treatment: A systematic review of risk factors. *Clin Psychol Rev*, *33*(8), 1010–1024. https://doi.org/10.1016/j.cpr.2013.07.007

Butt, P. R., White-Campbell, M., Canham, S., Johnston, A. D., Indome, E. O., Purcell, B., Tung, J., & Van Bussel, L. (2020). Canadian guidelines on alcohol use disorder among older adults. *Can Geriatr J*, *23*(1), 143–148. https://doi.org/10.5770/cgj.23.425

Choi, N. G., DiNitto, D. M., & Marti, C. N. (2014). Treatment use, perceived need, and barriers to seeking treatment for substance abuse and mental health problems among older adults compared to younger adults. *Drug Alcohol Depend*, *145*, 113–120. https://doi.org/10.1016/j.drugalcdep.2014.10.004

Cipriani, G., Cammisuli, D. M., Danti, S., & Di Fiorino, M. (2016). Disordered gambling and dementia. *Eur Geriatr Med*, *7*(5), 474–478. https://doi.org/https://doi.org/10.1016/j.eurger.2016.02.009

Collings, S., Allan, J., & Munro, A. (2022). Improving treatment for people with cognitive impairment and substance misuse issues: Lessons from an inclusive residential treatment program pilot in Australia. *Disabil Health J*, *15*(2s), 101295. https://doi.org/10.1016/j.dhjo.2022.101295

Dong, X., Chen, R., & Simon, M. A. (2014). Elder abuse and dementia: A review of the research and health policy. *Health Aff (Millwood)*, *33*(4), 642–649. https://doi.org/10.1377/hlthaff.2013.1261

Drugs, E. M. C. f., & Addiction, D. (2010). *Treatment and care for older drug users—EMCDDA Selected Issues*. Publications Office. https://doi.org/doi/10.2810/39905

Duggirala, R., Khushalani, S., Palmer, T., Brandt, N., & Desai, A. (2022). Screening for and management of opioid use disorder in older adults in primary care. *Clin Geriatr Med*, *38*(1), 23–38. https://doi.org/10.1016/j.cger.2021.07.001

Dürsteler-MacFarland, K. M., Vogel, M., Wiesbeck, G. A., & Petitjean, S. A. (2011). There is no age limit for methadone: A retrospective cohort study. *Subst Abuse Treat Prev Policy*, *6*, 9. https://doi.org/10.1186/1747-597x-6-9

Eshkoor, S. A., Hamid, T. A., Nudin, S. S., & Mun, C. Y. (2014). Does substance abuse contribute to further risk of falls in dementia? *Neuropsychol Dev Cogn B Aging Neuropsychol Cogn*, *21*(3), 317–324. https://doi.org/10.1080/13825585.2013.819067

Fernández-Serrano, M. J., Pérez-García, M., Schmidt Río-Valle, J., & Verdejo-García, A. (2010). Neuropsychological consequences of alcohol and drug abuse on different components of executive functions. *J Psychopharmacol*, *24*(9), 1317–1332. https://doi.org/10.1177/0269881109349841

Gazelka, H. M., Leal, J. C., Lapid, M. I., & Rummans, T. A. (2020). Opioids in older adults: Indications, prescribing, complications, and alternative therapies for primary care. *Mayo Clin Proc*, *95*(4), 793–800. https://doi.org/10.1016/j.mayocp.2020.02.002

Ghantous, Z., Ahmad, V., & Khoury, R. (2022). Illicit drug use in older adults: An invisible epidemic? *Clin Geriatr Med*, *38*(1), 39–53. https://doi.org/10.1016/j.cger.2021.07.002

Gilhooly, K. J., Gilhooly, M. L., Sullivan, M. P., McIntyre, A., Wilson, L., Harding, E., Woodbridge, R., & Crutch, S. (2016). A meta-review of stress, coping and interventions in dementia and dementia caregiving. *BMC Geriatr*, *16*, 106. https://doi.org/10.1186/s12877-016-0280-8

Guillou Landreat, M., Cholet, J., Grall Bronnec, M., Lalande, S., & Le Reste, J. Y. (2019). Determinants of gambling disorders in elderly people—A systematic review. *Front Psychiatry*, *10*, 837. https://doi.org/10.3389/fpsyt.2019.00837

Guo, C., Chen, L., & Wang, Y. (2023). Substance abuse and neurodegenerative diseases: Focus on ferroptosis. *Arch Toxicol*, *97*(6), 1519–1528. https://doi.org/10.1007/s00204-023-03505-4

Guo, S., Winslow, M., Manning, V., & Thane, K. K. (2010). Monthly take-home methadone maintenance regime for elderly opium-dependent users in Singapore. *Ann Acad Med Singap*, *39*(6), 429–434.

Hadley Strout, E., Fox, L., Castro, A., Haroun, P., Leavitt, B., Ross, C., Sayan, M., Delaney, T., Platzer, A., Hutchins, J., & Carney, J. K. (2016). Access to transportation for Chittenden County, Vermont older adults. *Aging Clin Exp Res*, *28*(4), 769–774. https://doi.org/10.1007/s40520-015-0476-3

Han, B. H., & Moore, A. A. (2018). Prevention and screening of unhealthy substance use by older adults. *Clin Geriatr Med*, *34*(1), 117–129. https://doi.org/10.1016/j.cger.2017.08.005

Heilig, M., MacKillop, J., Martinez, D., Rehm, J., Leggio, L., & Vanderschuren, L. (2021). Addiction as a brain disease revised: Why it still matters, and the need for consilience. *Neuropsychopharmacology*, *46*(10), 1715–1723. https://doi.org/10.1038/s41386-020-00950-y

Holton, A. E., Gallagher, P., Fahey, T., & Cousins, G. (2017). Concurrent use of alcohol interactive medications and alcohol in older adults: A systematic review of prevalence and associated adverse outcomes. *BMC Geriatr*, *17*(1), 148. https://doi.org/10.1186/s12877-017-0532-2

Inouye, S. K., Westendorp, R. G., & Saczynski, J. S. (2014). Delirium in elderly people. *Lancet*, *383*(9920), 911–922. https://doi.org/10.1016/s0140-6736(13)60688-1

International, A. s. D. (2018). *World Alzheimer Report 2018, The state of the art of dementia research: New frontiers*. A. s. D. I. (ADI).

Jensen, C. J., Lukow Ii, H. R., & Heck, A. L. (2012). Identifying barriers to care for older adults with substance use disorders and cognitive impairments. *Alcoholism Treatment Quarterly*, *30*(2), 211–223. https://doi.org/10.1080/07347324.2012.663302

Jeon, K. H., Han, K., Jeong, S. M., Park, J., Yoo, J. E., Yoo, J., Lee, J., Kim, S., & Shin, D. W. (2023). Changes in alcohol consumption and risk of dementia in a nationwide cohort in South Korea. *JAMA Netw Open*, *6*(2), e2254771. https://doi.org/10.1001/jamanetworkopen.2022.54771

Kakko, J., Alho, H., Baldacchino, A., Molina, R., Nava, F. A., & Shaya, G. (2019). Craving in opioid use disorder: From neurobiology to clinical practice. *Front Psychiatry*, *10*, 592. https://doi.org/10.3389/fpsyt.2019.00592

Kalapatapu, R. K., & Sullivan, M. A. (2010). Prescription use disorders in older adults. *Am J Addict*, *19*(6), 515–522. https://doi.org/10.1111/j.1521-0391.2010.00080.x

Kamal, H., Tan, G. C., Ibrahim, S. F., Shaikh, M. F., Mohamed, I. N., Mohamed, R. M. P., Hamid, A. A., Ugusman, A., & Kumar, J. (2020). Alcohol use disorder, neurodegeneration, Alzheimer's and Parkinson's disease: Interplay between oxidative stress, neuroimmune response and excitotoxicity. *Front Cell Neurosci*, *14*, 282. https://doi.org/10.3389/fncel.2020.00282

Kaur, J., Sandhu, R. K., Kubra, K. T., Canenguez Benitez, J. S., Onyeaka, H. K., Akter, S., & Amuk Williams, O. C. (2021). Substance use disorders in patients with Parkinson's disease and adverse hospitalization outcomes: A national inpatient study. *Cureus*, *13*(6), e16033. https://doi.org/10.7759/cureus.16033

Kuerbis, A., Sacco, P., Blazer, D. G., & Moore, A. A. (2014). Substance abuse among older adults. *Clin Geriatr Med*, *30*(3), 629–654. https://doi.org/10.1016/j.cger.2014.04.008

Lin, J., Arnovitz, M., Kotbi, N., & Francois, D. (2023). Substance use disorders in the geriatric population: A review and synthesis of the literature of a growing problem in a growing population. *Curr Treat Options Psychiatry*, 1–20. https://doi.org/10.1007/s40501-023-00291-9

Martín-Carrasco, M., Ballesteros-Rodríguez, J., Domínguez-Panchón, A. I., Muñoz-Hermoso, P., & González-Fraile, E. (2014). Interventions for caregivers of patients with dementia. *Actas Esp Psiquiatr*, *42*(6), 300–314.

Mohd Rosnu, N. S., Singh, D. K. A., Mat Ludin, A. F., Ishak, W. S., Abd Rahman, M. H., & Shahar, S. (2022). Enablers and barriers of accessing health care services among older adults in South-East Asia: A scoping review. *Int J Environ Res Public Health*, *19*(12). https://doi.org/10.3390/ijerph19127351

Moore, A. A., Whiteman, E. J., & Ward, K. T. (2007). Risks of combined alcohol/medication use in older adults. *Am J Geriatr Pharmacother*, *5*(1), 64–74. https://doi.org/10.1016/j.amjopharm.2007.03.006

Nardo, T., Batchelor, J., Berry, J., Francis, H., Jafar, D., & Borchard, T. (2022). Cognitive remediation as an adjunct treatment for substance use disorders: A systematic review. *Neuropsychol Rev*, *32*(1), 161–191. https://doi.org/10.1007/s11065-021-09506-3

NIAAA. Older Adults and Drinking National Institute on Alcohol Abuse and Alcoholism (NIAAA). Retrieved August from https://www.niaaa.nih.gov/alcohols-effects-health/alcohol-topics/older-adults

NIDA. (2020). *Substance use in older adults drug facts*. https://nida.nih.gov/publications/drugfacts/substance-use-in-older-adults-drugfacts

Osborn, R., Doty, M. M., Moulds, D., Sarnak, D. O., & Shah, A. (2017). Older Americans were sicker and faced more financial barriers to health care than counterparts in other countries. *Health Aff (Millwood)*, *36*(12), 2123–2132. https://doi.org/10.1377/hlthaff.2017.1048

Rao, R., & Arora, A. (2015). *Substance misuse in older people: An information guide*. Royal College of Psychiatrists. Cross-Faculty Report OA/AP/01.

Rao, R., Crome, I., Crome, P., & Iliffe, S. (2019). Substance misuse in later life: Challenges for primary care: A review of policy and evidence. *Prim Health Care Res Dev*, *20*, e117. https://doi.org/10.1017/s1463423618000440

Rivers, E., Shirazi, E., Aurora, T., Mullen, M., Gunnerson, K., Sheridan, B., Eichhorn, L., & Tomlanovich, M. (2004). Cocaine use in elder patients presenting to an inner-city emergency department. *Acad Emerg Med*, *11*(8), 874–877. https://doi.org/10.1111/j.1553-2712.2004.tb00771.x

Rockville. (2020a). Treating substance use disorder in older adults: Updated 2020. In *Treatment Improvement Protocol (TIP) Series, No. 26*. https://www.ncbi.nlm.nih.gov/books/NBK571033/

Rockville. (2020b). Treating substance use disorder in older adults: Updated 2020. In *Treatment Improvement Protocol (TIP) Series, No. 26*. https://www.ncbi.nlm.nih.gov/books/NBK571036/

Rockville. (2020c). Treating substance use disorder in older adults: Updated 2020. In *Treatment Improvement Protocol (TIP) Series, No. 26*. https://www.ncbi.nlm.nih.gov/books/NBK571039/

Rockville. (2020d). Treating substance use disorder in older adults: Updated 2020. In *Treatment Improvement Protocol (TIP) Series, No. 26*. https://www.ncbi.nlm.nih.gov/books/NBK571034/

Rosen, D., Hunsaker, A., Albert, S. M., Cornelius, J. R., & Reynolds, C. F., 3rd. (2011). Characteristics and consequences of heroin use among older adults in the United States: A review of the literature, treatment implications, and recommendations for further research. *Addict Behav*, *36*(4), 279–285. https://doi.org/10.1016/j.addbeh.2010.12.012

Salmond, S., & Allread, V. (2019). A population health approach to America's opioid epidemic. *Orthop Nurs*, *38*(2), 95–108. https://doi.org/10.1097/nor.0000000000000521

Sarkar, S., Sood, E., Bhad, R., & Mishra, A. K. (2021). Validated scales for substance use disorders in the geriatric population: A scoping review. *J Geriatr Ment Health*, *8*, 70. https://doi.org/10.4103/jgmh.jgmh_38_21

Sattar, S. P., Padala, P. R., McArthur-Miller, D., Roccaforte, W. H., Wengel, S. P., & Burke, W. J. (2007). Impact of problem alcohol use on patient behavior and caregiver burden in a geriatric assessment clinic. *J Geriatr Psychiatry Neurol*, *20*(2), 120–127. https://doi.org/10.1177/0891988706297091

Seddon, J., Wadd, S., Elliott, L., & Madoc-Jones, I. (2021). Cognitive impairment and treatment outcomes amongst people attending an alcohol intervention service for those aged 50+. *Adv Dual Diagn*, *14*(2), 58–69. https://doi.org/10.1108/ADD-02-2021-0003

Skinner, H. (1982). *Guide for using the drug abuse screening test (DAST)*. Centre for Addiction and Mental Health. https://socwel.ku.edu/sites/socwel/files/documents/Research%20Projects/Family%20First/Survey%20Measures/DAST%20Guide%202019%20-%20Remediated.pdf

Tondo, G., De Marchi, F., Terazzi, E., Sacchetti, M., & Cantello, R. (2017). Frontotemporal dementia presenting as gambling disorder: When a psychiatric condition is the clue to a neurodegenerative disease. *Cogn Behav Neurol*, *30*(2), 62–67. https://doi.org/10.1097/wnn.0000000000000122

Virnes, R. E., Tiihonen, M., Karttunen, N., van Poelgeest, E. P., van der Velde, N., & Hartikainen, S. (2022). Opioids and falls risk in older adults: A narrative review. *Drugs Aging*, *39*(3), 199–207. https://doi.org/10.1007/s40266-022-00929-y

Volpe, D. A., Xu, Y., Sahajwalla, C. G., Younis, I. R., & Patel, V. (2018). Methadone metabolism and drug-drug interactions: In vitro and in vivo literature review. *J Pharm Sci*, *107*(12), 2983–2991. https://doi.org/10.1016/j.xphs.2018.08.025

World Health Organization. (2021). *International Classification of Diseases*, 11th Revision (ICD-11).

Yarnell, S. C. (2015). Cocaine abuse in later life: A case series and review of the literature. *Prim Care Companion CNS Disord*, *17*(2). https://doi.org/10.4088/PCC.14r01727

Zanjani, F., Smith, R., Slavova, S., Charnigo, R., Schoenberg, N., Martin, C., & Clayton, R. (2016). Concurrent alcohol and medication poisoning hospital admissions among older rural and urban residents. *Am J Drug Alcohol Abuse*, *42*(4), 422–430. https://doi.org/10.3109/00952990.2016.1154966

14 Psychedelic-Inspired Approaches for Treating Neurodegenerative Disorders

Mahdieh Saeidi, Victor Pereira-Sanchez, Maryam Noroozian, Mohammadreza Shalbafan, Mona Salehi, and Sasidhar Gunturu

INTRODUCTION

Neurodegenerative disorders (NDDs) encompass a range of conditions involving the progressive degeneration of neurons, synapses, and glial cells, leading to cognitive, motor, and behavioral impairments. These disorders pose significant challenges for individuals, families, and society, requiring innovative therapeutic approaches (Nichols et al., 2019). Key examples of NDDs include:

- *Alzheimer's disease (AD)* is characterized by amyloid-beta plaques and tau tangles, leading to cognitive decline and memory loss (Ashrafian et al., 2021).
- *Frontotemporal dementia (FTD)* is marked by behavioral, personality, and language changes due to the degeneration of the frontal and temporal lobes (Bang et al., 2015).
- *Parkinson's disease (PD)* is caused by the loss of dopaminergic neurons, resulting in motor symptoms like tremors and bradykinesia (Wakabayashi et al., 2006).
- *Amyotrophic lateral sclerosis (ALS)* is a fatal condition characterized by the progressive loss of motor neurons, leading to muscle weakness, speech difficulties, and respiratory failure (Feldman et al., 2022).

Prevalence, Economic Burden, and Treatment Challenges

The prevalence of NDDs is rising due to the aging population. Deaths from dementia in the European Union increased by nearly 60% from 2011 to 2017, and the number of people with AD in the United States is expected to double by 2050 (Ciurea et al., 2023). Regions like North Africa/Middle East and Latin America show the highest dementia prevalence, linked to socioeconomic factors and healthcare access. The economic burden of NDDs is significant, with AD and PD costing over $216 and $14.4 billion annually in the US, respectively, expected to double by 2040 (Zahra et al., 2020).

Treatment for NDDs focuses on symptom management, as no cure exists. Non-pharmacological options like cognitive training, physical exercise, and the MIND diet can improve cognitive function and slow progression. Pharmacological treatments, such as cholinesterase inhibitors and memantine, provide symptom relief but do not stop the disease's advancement (Walczak-Nowicka and Herbet, 2021; H. Chen et al., 2024; Bernini et al., 2024).

Given the limitations of current treatments in halting the progression of NDDs, there is a growing interest in exploring alternative approaches. This chapter delves into the historical perspective, classifications, and mechanisms of action of psychedelics, examining their potential role in NDDs, safety, and tolerability, and the ethical and practical considerations involved.

DOI: 10.1201/9781003478485-14

HISTORICAL PERSPECTIVE ON PSYCHEDELICS

In 1957, Dr. Humphry Osmond coined the term "psychedelic" to describe mind-expanding drugs, initially known only to a small group of researchers. Since then, these substances have gained wider recognition, with media often highlighting their risks while popular culture has embraced them (Aaronson, Osmond, et al., 1970). Psychedelics have been used for millennia in indigenous rituals for healing and spiritual purposes. Pre-Columbian Mesoamerican cultures, including the Olmec, Maya, and Aztecs, used substances like peyote, ololiuhqui, *Bufo* toad toxins, and sacred mushrooms in religious ceremonies. Archaeological evidence traces peyote use to over 5,000 years ago. Today, Mesoamerican shamans continue these traditions (Carod-Artal, 2015).

Modern research into psychedelics began with Albert Hofmann's discovery of lysergic acid diethylamide (LSD) in 1943, sparking further studies in the 1950s and 1960s. Aldous Huxley's 1954 book *The Doors of Perception* popularized mescaline's effects, leading to research on psychedelics' impact on consciousness (Huxley, 1968; Hofmann, 1970; Guttmann, 1936). Huxley framed psychedelics as tools for exploring human consciousness (Huxley, 1968).

The "psychedelic renaissance" began after a pause in research during the 1970s, mainly due to the 1970 Controlled Substances Act in the United States, which classified LSD, psilocybin, and other psychedelics as Schedule I substances with no accepted medical use and high potential for abuse (Hall, 2022). This was part of the broader "War on Drugs" under President Nixon, which stifled research for decades. Interest in psychedelics resurfaced in the late 20th century and accelerated in the early 21st century (Nutt, 2019). A pivotal moment came in 2000 when Roland Griffiths' team at Johns Hopkins University found that psilocybin could induce lasting, meaningful mystical experiences and improve behavior and attitudes (Griffiths et al., 2006).

CLASSIFICATION AND MECHANISMS OF ACTION OF PSYCHEDELICS

Psychedelics are diverse substances that significantly affect perception, mood, and cognition. They can be categorized into classic psychedelics, dissociative agents, atypical hallucinogens, and entactogens based on their chemical structures and receptor interactions.

- Classic psychedelics include LSD, psilocybin (from "magic mushrooms"), mescaline (from the peyote cactus), and N,N-dimethyltryptamine (DMT). These substances mainly interact with serotonin receptors, especially 5-HT_{2A}, inducing vivid sensory experiences and altered states of consciousness.
- Dissociative agents, such as ketamine and phencyclidine (PCP), act as N-methyl-D-aspartate (NMDA) receptor antagonists, causing a sense of detachment from reality.
- Atypical hallucinogens like salvinorin-target kappa opioid receptors and entactogens such as 3,4-methylenedioxymethamphetamine (MDMA or ecstasy) promote emotional openness rather than hallucinations (Roth and Gumpper, 2023).

Psilocybin

Psilocybin, found in *Psilocybe* mushrooms, is metabolized into psilocin, which activates serotonin receptors, particularly 5-HT_{2A}. Traditionally used in spiritual rituals, it has recently gained attention for its potential to treat psychiatric disorders by enhancing neuroplasticity and reducing neuroinflammation. In humans, psilocybin induces hallucinations, altered time perception, and mood changes (Saeger and Olson, 2022; Tylˇs et al., 2014).

Psilocybin promotes neuroplasticity and neurogenesis and reduces neuroinflammation through several mechanisms:

- *Serotonin receptors*: It targets 5-HT_{2A} receptors, enhancing brain flexibility in the prefrontal cortex and hippocampus (Desouza et al., 2021).

- *Brain-derived neurotrophic factor (BDNF)*: It increases brain BDNF, supporting neuron survival and growth (Almeida et al., 2019).
- *Synaptogenesis*: It promotes new neuron connections and dendritic growth, improving communication between brain cells (Cameron et al., 2021).
- *Immediate early genes*: It activates genes like c-Fos and Arc, which are crucial for learning and memory (Shao et al., 2021).
- *Glutamate system*: It increases glutamate release, stimulating other receptors and boosting BDNF production (González-Maeso et al., 2007).

Lysergic Acid Diethylamide

LSD is a semi-synthetic psychedelic derived from ergot alkaloids, synthesized by Albert Hofmann in 1938, with its psychedelic properties discovered in 1943. It primarily acts as a partial agonist at 5-HT_{2A} receptors, abundant in the cerebral cortex and crucial in regulating perception, mood, and cognition. This interaction enhances glutamate release, especially in the prefrontal cortex, altering brain network connectivity and promoting neural plasticity. Additionally, LSD influences dopamine pathways by interacting with D1 and D2 receptors, further contributing to its psychoactive effects (Hofmann, 1970; Pahnke et al., 1970; Holze et al., 2021).

The combination of serotonin receptor agonism and dopamine interaction produces LSD's characteristic hallucinogenic effects, altered consciousness, and changes in sensory perception. Its ability to modulate both serotonin and dopaminergic systems results in diverse neuropsychological effects, ranging from altered cognition and perception to enhanced introspection and emotional experiences, explaining its therapeutic potential (González-Maeso et al., 2007; Holze et al., 2021).

N,N-Dimethyltryptamine

DMT, the active component in ayahuasca, produces short-lived effects when smoked. However, when consumed as part of ayahuasca, its effects are prolonged, often leading to mystical insights. DMT is nonaddictive and alters perception and cognition by binding to serotonin receptors, particularly the 5-HT_{2A} receptor (Barker, 2022).

DMT interacts with multiple receptors:

- *5-HT_{2A} receptors*: DMT binds with high affinity to 5-HT_{2A} receptors, significantly influencing its hallucinogenic effects. It also affects other 5-HT subtypes, including 5-HT_{1A}, 5-HT_{1B}, and 5-HT_{2C}, indicating its broad serotonergic activity.
- *Sigma-1 receptors*: As an agonist of sigma-1 receptors, DMT helps regulate stress responses and calcium signaling between the endoplasmic reticulum (ER) and mitochondria, potentially offering neuroprotective benefits, especially under hypoxic conditions.
- *Trace amine-associated receptors (TAARs)*: DMT is an agonist for TAAR1, activating adenylyl cyclase and cAMP pathways, although the full role of this mechanism is not entirely understood.

This interaction with multiple receptors highlights DMT's complex mechanism, contributing to its profound effects on perception, consciousness, and neuroplasticity (Barker, 2022).

NEUROBIOLOGICAL EFFECTS OF PSYCHEDELICS AND ROLE IN NEURODEGENERATION RESEARCH

As current treatments for NDDs remain limited, there is growing interest in exploring alternative approaches, such as psychedelics, for their management. Traditionally used in religious practices, psychedelics are now being investigated for their ability to alter cognition, perception, and emotions, as well as for their neuroplastic and neuroprotective effects. In particular, microdosing—using sub-hallucinogenic doses with minimal side effects—has gained attention as a promising, effective, and safe strategy (DiVito and Leger, 2020). Current clinical trials are underway to evaluate these compounds in various NDDs (ICTRP Search Portal, n.d.) (Table 14.1).

TABLE 14.1
Overview of Clinical Trials Investigating Psychedelics in NDDs

Title	Intervention	Conditions	Brief Summary
"Pilot Study of Serotonin 2A Receptor (5-HT_{2A}) Agonist Psilocybin for Depression in Patients with Mild Cognitive Impairment or Early Alzheimer's Disease" • *ID*: NCT04123314 • *Sponsor*: Johns Hopkins University	Psilocybin *Two sessions*: • *Week 4*: 15 mg/70 kg dose • *Week 6*: 15 or 25 mg/70 kg dose	Depression, AD, mild cognitive impairment (MCI)	• This open-label pilot study (8-week treatment) is investigating the safety and effectiveness of psilocybin in treating depression in individuals with MCI or early AD. • The study is exploring whether psilocybin can improve participants' quality of life. • Follow-up assessments are being done up to 6 months post-treatment.
"The Efficacy of Psilocybin Therapy for Depression in Parkinson's Disease" • *ID*: NCT06455293 • *Sponsor*: University of California, San Francisco	Psilocybin *Two sessions*: 1–25 mg	PD, depression	• A randomized controlled trial assessing the efficacy of psilocybin therapy for depression in individuals aged 40–80 with early to moderate-stage PD. • Depression, motor and non-motor symptoms, quality of life, and biomarkers are being evaluated through clinical assessments, neuroimaging, brain stimulation, and blood samples. • Follow-up is continuing for 3 months after the second session, focusing on efficacy, safety, and tolerability.
"Psilocybin Therapy for Depression and Anxiety in Parkinson's Disease: A Pilot Study" • *ID*: NCT04932434 • *Sponsor*: University of California, San Francisco	Psilocybin *Two sessions*: • *First session*: Low-to-moderate dose of 10 mg • *Second session (2 weeks later)*: Higher dose of 25 mg if the first session is well-tolerated	PD, depression, anxiety	• An open-label, single-arm pilot study is investigating the safety, tolerability, and feasibility of psilocybin therapy for depression and anxiety in individuals with early-stage PD. • Participants aged 40–75 with a diagnosis of PD and depressive or anxiety disorders are being enrolled. • Follow-up assessments are being done for 3 months to evaluate safety and efficacy. • Exploratory endpoints include changes in depressive and anxiety symptoms, as well as functional measures.

(*Continued*)

TABLE 14.1 (CONTINUED)

Title	Intervention	Conditions	Brief Summary
"Examining Ketamine Effects on Depression, Neuroplasticity, and Inflammation in Veterans with Parkinson's Disease" • *ID*: NCT06231563 • *Sponsor*: VA Office of Research and Development	Ketamine Intravenous (IV) infusion at 0.5 mg/kg Remimazolam IV infusion at 0.25 mg/kg	PD	• This double-masked, placebo-controlled, randomized trial will compare the effects of IV ketamine versus remimazolam in 80 veterans with PD and depression. • The trial aims to assess ketamine's safety, tolerability, and rapid antidepressant effects within 24 hours. • It is also investigating whether ketamine's antidepressant effects are linked to improved neuroplasticity and reduced inflammation. • Clinical assessments, noninvasive brain stimulation, and blood-based cytokine measurements are being used to evaluate depression, neuroplasticity, and systemic inflammation.
"Does Psilocybin Change Synaptic Density in Amnestic Mild Cognitive Impairment?" • *ID*: NCT06041152 • *Sponsor*: Centre for Addiction and Mental Health/ Toronto, Ontario, Canada	Psilocybin • *Psilocybin group*: Two 25 mg doses of psilocybin, separated by 1 week • *Placebo group*: Two placebo doses, also separated by 1 week	Amnestic MCI (aMCI)	• This trial examines psilocybin's effect on synaptic vesicular density (SVD) using the PET radiotracer 18F-SynVesT-1 in participants with aMCI and healthy controls. The hypothesis is that psilocybin will increase SVD and improve cognition compared to placebo. • Baseline and follow-up 18F-SynVesT-1 PET scans for tauopathy and clinical and neuropsychological testing will be done at 1, 4, and 12 weeks post-treatment.

"Phase I Open-Label Study of Low-Dose Ketamine Infusion Treatment in Levodopa-Induced Dyskinesia in Parkinson's Disease"
• *ID*: NCT06021756
• *Sponsor*: University of Arizona
	Ketamine:
	IV infusion at a rate ranging from 0.05 mg/kg/h to 0.30 mg/kg/h
		PD, levodopa-induced dyskinesia	• This trial evaluates the tolerability and efficacy of low-dose ketamine infusions for treating levodopa-induced dyskinesias, both immediately and during the post-infusion period (weeks 2–6), using patient diaries and the UDysRS.
			• It aims to:
			  • Assess ketamine's impact on Parkinson's symptoms and levodopa side effects, including "off," "on without dyskinesia," and "troublesome dyskinesia" periods.
			  • Evaluate effects on pain, quality of life, and general symptoms using the Unified Parkinson's Disease Rating Scale.
			• Positive outcomes could pave the way for new treatments for dyskinesia and other Parkinson's symptoms, including depression and pain.

Psychedelics enhance dendritic spine and neurite formation by modulating the mechanistic target of rapamycin (mTOR) and tropomyosin receptor kinase B (TrkB) pathways, which are crucial for BDNF production and neuroplasticity in the prefrontal cortex (Inserra et al., 2021; Saeger and Olson, 2022). Jones et al. first demonstrated that psychedelics affect structural plasticity, finding that 2,5-dimethoxy-4-iodoamphetamine (DOI) temporarily enlarged dendritic spines in pyramidal neurons—for 60 minutes—through the activation of 5-HT_{2A} and 5-HT_{2C} receptors (Jones et al., 2009; Mi et al., 2017). Yoshida et al. (2011) later showed that a 24-hour exposure to DOI increased dendritic spine density in rat cortical cultures. In 2024, a study revealed that a single dose of DOI induced structural changes in the brains of mice and improved cognitive flexibility, particularly in adapting to new tasks. This suggests that DOI-treated mice learned from previously overlooked cues, enhancing their cognitive adaptability (Sabanović et al., 2024). These findings imply that psychedelics can enhance learning and cognitive adaptation, potentially aiding treatment for conditions like depression and anxiety characterized by perseverative behaviors.

Alzheimer's Disease, Related Dementia, and Psychedelics

A comprehensive 2023 review of studies, including preclinical and limited observational research, highlighted the promising effects of psychedelics like psilocybin, LSD, and DMT on behavior, cognition, learning, and memory in the context of AD and related dementias. While no human clinical trials have been conducted, animal models suggest that classic psychedelics promote synaptic, structural, and behavioral plasticity through mechanisms such as serotonergic modulation, GABAergic activity, and upregulation of neurotrophic factors (Winkelman et al., 2023). One study using molecular tools demonstrated that psychedelics promote neuroplasticity by increasing dendritic spine density and activating intracellular 5-HT_{2A} receptors (Vargas et al., 2023). The serotonergic system plays a crucial role in AD and dementia, particularly in learning and memory. Significant reductions in serotonin receptors in brain regions like the neocortex, hippocampus, and amygdala are associated with cognitive decline and neurodegeneration in patients with AD (Garcia-Romeu et al., 2021). Psychedelics are thought to modulate brain activity primarily through their impact on serotonin 5-HT_{2A} receptors (Roth and Gumpper, 2023).

Regarding psilocybin, a 2024 review of multiple animal studies demonstrated that it increases adult hippocampal neurogenesis, enhances dendritic plasticity, and improves fear extinction in mice, suggesting psilocybin's potential to elevate BDNF levels and promote neuroplasticity while reducing depressive-like behaviors (Haniff et al., 2024). A 2013 animal study revealed that low-dose psilocybin enhanced hippocampal neurogenesis and facilitated the extinction of conditioned fear responses, while higher doses decreased neuron formation, highlighting the importance of its dosage (Catlow et al., 2013). A 2023 hypothesis suggested psilocybin may enhance synaptogenesis in patients with aMCI, potentially serving as an early intervention for AD by increasing synaptic density in the hippocampus and prefrontal cortex via 5-HT_{2A} receptor activation (Song et al., 2023). A longitudinal, open-label study in Denmark found that a single dose of psilocybin led to lasting changes in brain connectivity, specifically reducing resting-state functional connectivity within the executive control network (ECN) 1 week after administration. Although these effects diminished by 3 months, they were linked to sustained improvements in mindfulness and well-being, suggesting that psilocybin may enhance cognitive flexibility and emotional processing by modulating ECN connectivity (McCulloch et al., 2022).

Chronic neuroinflammation is another critical factor in AD that contributes to neuronal dysfunction and death. Psychedelics like psilocybin have also shown anti-inflammatory properties by reducing pro-inflammatory cytokines such as tumor necrosis factor-alpha (TNF-α), interleukin-(IL-) 6, and IL-1β through serotonin receptor activity, offering potential neuroprotection (Saeger & Olson, 2022). A 2023 study found no significant changes in inflammatory biomarkers after a single psilocybin dose in healthy participants (Burmester et al., 2023). However, a 2021 in vitro study demonstrated that psilocybin-containing mushroom extracts inhibited pro-inflammatory mediators like TNF-α and COX-2, supporting its anti-inflammatory potential (Nkadimeng et al., 2021).

Alternatively, DMT has shown promise in restoring cellular homeostasis by improving cross-talk between the ER and mitochondria. A 2024 study using 48 3×TG-AD transgenic mice, an AD model, found that chronic DMT administration (2 mg/kg for 3 weeks) significantly alleviated cognitive impairment, reduced amyloid-beta (Aβ) accumulation, and restored sigma-1 receptor function, thereby enhancing ER–mitochondria contact, regulating calcium homeostasis, and improving mitochondrial function (Cheng et al., 2024).

A 2021 review highlighted that both ketamine and classical psychedelics enhance neuroplasticity in pyramidal neurons, particularly in the prefrontal cortex. By interacting with glutamate and serotonin pathways, ketamine promotes glutamate release, activates AMPA receptors, and engages BDNF and mTOR signaling, fostering synaptogenesis. These mechanisms help "reset" neural circuits, counteracting synaptic deficits and neuronal atrophy seen in depression, offering sustained therapeutic benefits with minimal dosing (Inserra et al., 2021). Ketamine has also demonstrated neuroprotective properties by enhancing autophagy and safeguarding neurons against apoptosis, thereby improving motor and cognitive functions in PD models (Fan et al., 2017).

Amyotrophic Lateral Sclerosis

Psychedelics have emerged as promising therapeutic agents for ALS due to their potential anti-inflammatory, neuroprotective, and antioxidant properties. Psychedelics interact with 5-HT_{1A}, 5-HT_{2A}, sigma-1, and TAAR1 receptors, triggering several mechanisms that could be beneficial in ALS management (Gold et al., 2023; Ono et al., 2014):

- *Protection against oxidative stress*: Psychedelics may activate antioxidant pathways by stimulating 5-HT_{1A}, 5-HT_{2A}, and sigma-1 receptors, initiating the Nrf2 signaling pathway and leading to the upregulation of crucial antioxidant genes and proteins, such as heme oxygenase-1 (HO-1), NAD(P)H quinone dehydrogenase 1 (NQO1), catalase, superoxide dismutase 1 (SOD1), and superoxide dismutase 2 (SOD2). By enhancing the expression of these antioxidant proteins, psychedelics may provide a defense against oxidative stress, which is a significant factor in ALS progression.
- *Protection against ER stress*: ER stress can disrupt protein folding and mitochondrial function, contributing to neuronal damage in ALS. The activation of the sigma-1 receptor by psychedelics may help protect against ER stress by regulating the expression of proteins such as C/EBP homologous protein (CHOP), activating transcription factor 4 (ATF4), and activating transcription factor 6 (ATF6). This regulation helps balance pro-apoptotic (BAX) and anti-apoptotic (BCL2) factors, promoting cell survival and potentially reducing neurodegeneration.
- *Maintenance of blood–brain barrier (BBB) integrity and anti-inflammatory effects*: In ALS and other NDDs, inflammation and BBB disruption contribute to disease progression. Psychedelics may exert anti-inflammatory effects by stimulating 5-HT receptors, sigma-1, and TAAR1. These actions reduce pro-inflammatory cytokines (e.g., IL-1, IL-6, TNF-α) while increasing anti-inflammatory factors such as BDNF, glial cell line-derived neurotrophic factor (GDNF), and IL-10. Psychedelics may also enhance the expression of tight junction proteins (e.g., zonula occludens-1, occludin, and claudin), which helps maintain BBB integrity and prevents further neuroinflammation.
- *Neurotrophic support and reduced toxic protein aggregation*: By activating 5HT_{2A}, 5-HT_{2B}, sigma-1, and TAAR1 receptors, psychedelics may support neurotrophy and protect neurons from reactive oxygen species, neuroinflammation, and toxic protein aggregation. This process involves the downregulation of indoleamine 2,3-dioxygenase (IDO) and nuclear factor of activated T cells (NFAT) while promoting the upregulation of

neurotrophic factors such as c-Fos and cAMP response element-binding protein (CREB) (Kozlowska et al., 2022).

Frontotemporal Dementia

Psychedelic-assisted therapy shows promise for treating FTD, mainly due to its potential to enhance cortical plasticity—a significant factor given FTD's characteristic cortical atrophy (Saeger & Olson, 2022). A key mechanism involves the modulation of BDNF. Research indicates that certain psychedelics, such as DOI, can increase BDNF transcription in the cortex, although this effect is inhibited in the hippocampus (Vaidya et al., 1997). The upregulation of BDNF in the cortex is thought to occur through increased glutamate release and activation of AMPA receptors. The latter are ion channels that mediate fast synaptic transmission in the central nervous system. When glutamate binds to these receptors, the channels open, allowing sodium ions to flow into the neuron. This process contributes to neuronal depolarization and is essential for synaptic plasticity, learning, and memory. AMPA receptors, therefore, facilitate rapid signaling between neurons, which is critical for various cognitive and neural processes (Zuccato & Cattaneo, 2009).

SAFETY AND TOLERABILITY

The safety and tolerability of psychedelics are critical considerations in their therapeutic application, especially in NDDs. A meta-analysis in 2024 examined six randomized, double-blind clinical trials involving 528 participants to assess the safety profile of psilocybin. The findings indicated that while therapeutic doses of psilocybin were associated with significant adverse effects, including headaches, nausea, anxiety, dizziness, and elevated blood pressure, these effects were generally tolerable and resolved within 48 hours (Yerubandi et al., 2024). DMT is also associated with common side effects, including nausea, vomiting, increased heart rate, and elevated blood pressure (White et al., 2024). LSD, although possessing therapeutic potential, can produce adverse effects, particularly in uncontrolled environments. Acute reactions may include anxiety, paranoia, and psychosis-like symptoms. These are dose-dependent, typically transient, and resolve without sequelae (Holze et al., 2022). Long-term adverse effects from LSD are rare but can arise with repeated or high-dose use, manifesting as physical symptoms (e.g., increased pulse, elevated blood pressure, dilated pupils, and slight fever), psychological symptoms (e.g., intense emotional responses, including terror or depression, often referred to as "bad trips," which typically subside within 6 to 12 hours (McWilliams & Tuttle, 1973), and persistent effects. The latter involves a small percentage of users who may develop persistent psychotic disorders, particularly if they have pre-existing vulnerabilities, flashbacks, or hallucinogen-persisting perception disorder characterized by enduring perceptual changes triggered by stress or other substances (Lerner & Ran, 2015). However, in clinical settings with controlled dosages, LSD is generally safe, and proper screening and support can effectively mitigate risks (Holze et al., 2022).

ETHICAL AND PRACTICAL CONSIDERATIONS

The ethical administration of psychedelics for managing NDDs is essential to ensure that these treatments are safe, effective, and accessible. As interest in this area grows, several considerations emerge:

- *Informed consent*: Cognitive decline in dementia complicates informed consent, requiring patients to fully understand the risks and experimental nature of psychedelic treatments. With effects like "oceanic boundlessness" (a sense of unity with the universe) and ego dissolution, a thorough consent process is crucial for patient comprehension (Smith & Appelbaum, 2022).

- *Patient vulnerability*: Individuals with severe NDDs are particularly vulnerable to adverse effects, such as anxiety or paranoia, during psychedelic therapy. Rigorous screening, trained facilitators, and robust support systems are essential to minimize risks and protect patient's well-being (Barber & Dike, 2023; Bonnieux et al., 2023).
- *Desperation*: The lack of effective treatments can leave patients and families vulnerable to exaggerated claims about the benefits of psychedelics. Transparent communication and ethical guidance are vital to prevent exploitation (Peterson et al., 2023).
- *Equity and access*: NDDs frequently result in a range of physical and sensory disabilities, such as those seen in ALS, cerebral palsy, muscular dystrophy, spinal cord injury, and spinal muscular atrophy. Unfortunately, individuals with these disabilities are often excluded from clinical trials involving psychedelics, reinforcing structural ableism and limiting their access to potentially transformative treatments (Mintz et al., 2022; Schwartz & Unni, 2021). Equitable access to psychedelic treatments is essential, particularly for marginalized communities with limited healthcare. The unregulated nature of many psychedelic practices highlights the need for inclusive healthcare policies and standardized protocols (Belouin et al., 2022).
- *Clinical versus nonclinical use*: The rise of nonclinical psychedelic use, often without medical oversight, underscores the need for guidelines that extend ethical principles to unregulated settings to ensure patient safety. While psychedelics can promote personal growth, misuse and unrealistic expectations present psychological risks, making public education and harm reduction essential. Public-private partnerships are crucial for developing comprehensive ethical guidelines, ensuring that commercialization does not overshadow patient safety and ethical considerations (Barber & Dike, 2023; Smith & Appelbaum, 2022; Bonnieux et al., 2023; Belouin et al., 2022).
- *Self-medication trends*: Studies indicate a growing trend of adults, including older individuals, using psychedelics like psilocybin and LSD for self-medication, mainly through microdosing (Haijen et al., 2022). Older adults with multiple chronic conditions are especially vulnerable, particularly those with a history of substance use from the 1960s counterculture. Given these risks, healthcare providers must regularly inquire about self-medication practices and closely monitor for potential complications (Hoffman, 2010).
- *Caregiving dynamics*: Caregivers play a central role in psychedelic therapy. Their involvement in the treatment process and research is essential for achieving optimal outcomes (Peterson et al., 2023).
- *Drug–drug interactions*: As novel treatments, psychedelics' interactions with other medications remain unclear, especially for vulnerable populations with NDDs. A thorough evaluation of potential interactions is necessary to ensure safe treatment protocols.

CURRENT STATE OF RESEARCH

While animal studies have provided valuable insights into the potential effects of psychedelic compounds, translating these findings to human applications is a complex and nonlinear journey. The recent decision by the Food and Drug Administration Psychopharmacologic Drugs Advisory Committee to reject an MDMA-assisted therapy study for post-traumatic stress disorder in the US underscores the significant challenges psychedelics face in gaining acceptance as legitimate treatments, even when preclinical data appears encouraging. This decision highlights broader concerns about ensuring long-term efficacy, addressing biases like functional unblinding, and managing potential safety risks, especially cardiovascular and physiological effects. These challenges underscore the urgent need for more rigorous, comprehensive research to determine the true potential of psychedelics in clinical settings, identifying where they can be safely integrated into treatment protocols or where their use may be inappropriate (Richmond, 2024). Currently, no published human studies have conclusively proven the efficacy of psychedelics in treating ND. However, as outlined

TABLE 14.2
Psychedelic Drug Developments and Patents for NDDs

Psychedelic Patent	Company	Potential Treatment	Formulation	Status
AKO004	Core One Labs/ Akome Biotech (Core One Labs Inc., n.d.)	PD	DMT and a naturally occurring plant bioactive	The company has filed a provisional patent application for AKO004 with the US Patent and Trademark Office
BMND08	BioMind (Biomind Labs I Next-Gen Pharma, n.d.)	Depression and anxiety in patients with AD	Metoxi-N,N-dimethyltryptamine (5-MeO-DMT) sublingual formulation	Phase II
BMND09	BioMind (Biomind Labs I Next-Gen Pharma, n.d.)	PD	N/A	Phase I
Ketamine IND	PharmaTher (PharmaTher Holdings Ltd., 2021)	PD	N/A	Phase II

in Table 14.1, several ongoing open-label studies are investigating the safety and effectiveness of different psychedelics in various NDDs. Table 14.2 summarizes psychedelic drug developments and patents for NDDs (McConnell, 2024).

The scientific community must approach this area with caution, conducting thorough research before confidently discussing the mechanisms of action, potential effects, and side effects of psychedelics, particularly in vulnerable populations. Through such comprehensive studies, we can clearly understand their safety, efficacy, and appropriate clinical applications.

CONCLUSION

While large-scale clinical trials are crucial to confirm early findings, current research lacks sufficient evidence to validate the therapeutic potential of psychedelics in NDDs. Robust in vivo and in vitro research is essential, particularly since NDDs often begin years before symptoms appear and progress at the cellular level. Before significant synaptic damage occurs, early detection and intervention are critical for preserving brain function. Future studies must consider individual risk factors and comorbidities to develop tailored treatments. While psychedelics affect general pathways, a deeper understanding of specific mechanisms is needed to optimize clinical applications. Future research should focus on safety, efficacy, optimal dosing, and the mechanisms behind psychedelics' effects. These therapies could revolutionize treatments for millions affected by NDDs. While some studies suggest psychedelic-induced cognitive and affective changes can last from one to 25 years, more data are needed on the duration and potential need for booster doses (Aday et al., 2020). These substances enhance neuroplasticity, reduce neuroinflammation, and protect neural circuits. However, as nonmedical use rises, clear regulations must distinguish therapeutic, medically supervised use from recreational use to ensure the public understands the risks and proper contexts for these substances. Collaboration among researchers, clinicians, and policymakers is vital to ensure ethical considerations, risk-benefit evaluations, and safe integration into healthcare systems. Continued research could pave the way for innovative treatments that significantly improve the quality of life for those with NDDs (Bonnieux et al., 2023).

ACKNOWLEDGMENT

We thank Dr Albert Garcia-Romeu from the Johns Hopkins Center for Psychedelic and Consciousness Research for his invaluable insights that greatly improved this chapter. For more information, visit https://hopkinspsychedelic.org/.

DISCLOSURE STATEMENT

The authors contributed to the development of this chapter in an entirely voluntary and uncompensated capacity. No payments, stipends, grants, or institutional resources were provided or received in connection with this work. The activities involved were solely academic and collaborative in nature, conducted outside the scope of any formal employment, training, or contractual obligations.

REFERENCES

Aaronson, B. S., Osmond, H., et al. (1970). *Psychedelics: The uses and implications of hallucinogenic drugs.* Doubleday Garden City, New York.

Aday, J. S., Bloesch, E. K., & Davoli, C. C. (2020). Can psychedelic drugs attenuate age-related changes in cognition and affect? *Journal of Cognitive Enhancement*, *4*(2), 219–227.

Almeida, R. N. d., Galva˜o, A. C. d. M., da Silva, F. S., Silva, E. A. d. S., Palhano-Fontes, F., Maia-de-Oliveira, J. P., de Arau´jo, L.-S. B., Loba˜o-Soares, B., & Galva˜oCoelho, N. L. (2019). Modulation of serum brain-derived neurotrophic factor by a single dose of Ayahuasca: Observation from a randomized controlled trial. *Frontiers in Psychology*, *10*, 1234.

Ashrafian, H., Zadeh, E. H., & Khan, R. H. (2021). Review on alzheimer's disease: Inhibition of amyloid beta and tau tangle formation. *International Journal of Biological Macromolecules*, *167*, 382–394.

Bang, J., Spina, S., & Miller, B. L. (2015). Frontotemporal dementia. *Lancet*, *386*(10004), 1672–1682.

Barber, G. S., & Dike, C. C. (2023). Ethical and practical considerations for the use of psychedelics in psychiatry. *Psychiatric Services*, *74*(8), 838–846.

Barker, S. A. (2022). Administration of N, N-dimethyltryptamine (DMT) in psychedelic therapeutics and research and the study of endogenous DMT. *Psychopharmacology*, *239*(6), 1749–1763.

Belouin, S. J., Averill, L. A., Henningfield, J. E., Xenakis, S. N., Donato, I., Grob, C. S., Berger, A., Magar, V., Danforth, A. L., & Anderson, B. T. (2022). Policy considerations that support equitable access to responsible, accountable, safe, and ethical uses of psychedelic medicines. *Neuropharmacology*, *219*, 109214.

Bernini, S., Riccietti, C., Farris, E. R., & Bottiroli, S. (2024). Neurodegenerative diseases and cognitive interventions: What is the future? In: *Prospects of sustainability: Yesterday, today and tomorrow.* D'Ambrosio, I. (Ed.). Virtus Interpress. https://doi.org/10.22495/psytt.

Biomind Labs | Next-Gen Pharma. (n.d.). Biomind Labs. https://www.biomindlabs.com/

Bonnieux, J. N., VanderZwaag, B., Premji, Z., Garcia-Romeu, A., & Garcia-Barrera, M. A. (2023). Psilocybin's effects on cognition and creativity: A scoping review. *Journal of Psychopharmacology*, *37*(7), 635–648.

Burmester, D. R., Madsen, M. K., Szabo, A., Aripaka, S. S., Stenbæk, D. S., Frokjaer, V. G., Elfving, B., Mikkelsen, J. D., Knudsen, G. M., & Fisher, P. M. (2023). Subacute effects of a single dose of psilocybin on biomarkers of inflammation in healthy humans: An open-label preliminary investigation. *Comprehensive Psychoneuroendocrinology*, *13*, 100163.

Cameron, L. P., Tombari, R. J., Lu, J., Pell, A. J., Hurley, Z. Q., Ehinger, Y., Vargas, M. V., McCarroll, M. N., Taylor, J. C., Myers-Turnbull, D., et al. (2021). A non-hallucinogenic psychedelic analogue with therapeutic potential. *Nature*, *589*(7842), 474–479.

Carod-Artal, F. J. (2015). Hallucinogenic drugs in Pre-Columbian Mesoamerican cultures. *Neurología (English Edition)*, *30*(1), 42–49.

Catlow, B. J., Song, S., Paredes, D. A., Kirstein, C. L., & Sanchez-Ramos, J. (2013). Effects of psilocybin on hippocampal neurogenesis and extinction of trace fear conditioning. *Experimental Brain Research*, *228*, 481–491.

Chen, H., Dunk, M. M., Wang, B., Zhao, M., Shen, J., Zong, G., Pan, Y., Tong, L., Xu, W., & Yuan, C. (2024). Associations of the Mediterranean-DASH intervention for neurodegenerative delay diet with brain structural markers and their changes. *Alzheimer's & Dementia*, *20*(2), 1190–1200.

Cheng, D., Lei, Z.-G., Chu, K., Lam, O. J. H., Chiang, C. Y., & Zhang, Z.-J. (2024). N, N-dimethyltryptamine, a natural hallucinogen, ameliorates alzheimer's disease by restoring neuronal sigma-1 receptor-mediated endoplasmic reticulum-mitochondria crosstalk. *Alzheimer's Research & Therapy*, *16*(1), 95.

Ciurea, V. A., Covache-Busuioc, R.-A., Mohan, A. G., Costin, H. P., & Voicu, V. (2023). Alzheimer's disease: 120 years of research and progress. *Journal of Medicine and Life*, *16*(2), 173.

Core One Labs Inc. (n.d.). https://core1labs.com/

Desouza, L. A., Benekareddy, M., Fanibunda, S. E., Mohammad, F., Janakiraman, B., Ghai, U., Gur, T., Blendy, J. A., & Vaidya, V. A. (2021). The hallucinogenic serotonin2A receptor agonist, 2, 5-dimethoxy-4-iodoamphetamine, promotes CAMP response element binding protein-dependent gene expression of specific plasticity-associated genes in the rodent neocortex. *Frontiers in Molecular Neuroscience*, *14*, 790213.

DiVito, A. J., & Leger, R. F. (2020). Psychedelics as an emerging novel intervention in the treatment of substance use disorder: A review. *Molecular Biology Reports*, *47*(12), 9791–9799.

Fan, J.-C., Song, J.-J., Wang, Y., Chen, Y., & Hong, D.-X. (2017). Neuron-protective effect of subanesthetic-dosage ketamine on mice of Parkinson's disease. *Asian Pacific Journal of Tropical Medicine*, *10*(10), 1007–1010.

Feldman, E. L., Goutman, S. A., Petri, S., Mazzini, L., Savelieff, M. G., Shaw, P. J., & Sobue, G. (2022). Amyotrophic lateral sclerosis. *Lancet*, *400*(10360), 1363–1380.

Garcia-Romeu, A., Darcy, S., Jackson, H., White, T., & Rosenberg, P. (2021). Psychedelics as novel therapeutics in Alzheimer's disease: Rationale and potential mechanisms. *Disruptive Psychopharmacology*, *56*, 287–317.

Gold, N. D., Mallard, A. J., Hermann, J. C., Zeifman, R. J., Pagni, B. A., Bogenschutz, M. P., & Ross, S. (2023). Exploring the potential utility of psychedelic therapy for patients with amyotrophic lateral sclerosis. *Journal of Palliative Medicine*, *26*(10), 1408–1418.

Gonza'lez-Maeso, J., Weisstaub, N. V., Zhou, M., Chan, P., Ivic, L., Ang, R., Lira, A., Bradley-Moore, M., Ge, Y., Zhou, Q., et al. (2007). Hallucinogens recruit specific cortical 5-HT_{2A} receptor-mediated signaling pathways to affect behavior. *Neuron*, *53*(3), 439–452.

Griffiths, R. R., Richards, W. A., McCann, U., & Jesse, R. (2006). Psilocybin can occasion mystical-type experiences having substantial and sustained personal meaning and spiritual significance. *Psychopharmacology*, *187*, 268–283.

Guttmann, E. (1936). Artificial psychoses produced by mescaline. *Journal of Mental Science, 82*(338), 203–221.

Haijen, E. C., Hurks, P. P., & Kuypers, K. P. (2022). Microdosing with psychedelics to self-medicate for adhd symptoms in adults: A prospective naturalistic study. *Neuroscience Applied*, *1*, 101012.

Hall, W. (2022). Why was early therapeutic research on psychedelic drugs abandoned? *Psychological Medicine*, *52*(1), 26–31.

Haniff, Z. R., Bocharova, M., Mantingh, T., Rucker, J. J., Velayudhan, L., Taylor, D. M., Young, A. H., Aarsland, D., Vernon, A. C., & Thuret, S. (2024). Psilocybin for dementia prevention? The potential role of psilocybin to alter mechanisms associated with major depression and neurodegenerative diseases. *Pharmacology & Therapeutics*, 108641.

Hoffman, J. (2010). The psychedelic 1960s, hippies in their 60s: Substance abuse in the elderly. *Consultant Pharmacist*, *25*(9), 570–576.

Hofmann, A. (1970). The discovery of LSD and subsequent investigations on naturally occurring hallucinogens. *Discoveries in Biological Psychiatry*, 91–106.

Holze, F., Avedisian, I., Varghese, N., Eckert, A., & Liechti, M. E. (2021). Role of the 5-HT_{2A} receptor in acute effects of lsd on empathy and circulating oxytocin. *Frontiers in Pharmacology*, *12*, 711255.

Holze, F., Caluori, T. V., Vizeli, P., & Liechti, M. E. (2022). Safety pharmacology of acute LSD administration in healthy subjects. *Psychopharmacology*, 1–13.

Huxley, A. (1968). *The doors of perception*. Windus London.

ICTRP search portal. (n.d.). https://www.who.int/clinical-trials-registry-platform/the-ictrp-search-portal

Inserra, A., De Gregorio, D., & Gobbi, G. (2021). Psychedelics in psychiatry: Neuroplastic, immunomodulatory, and neurotransmitter mechanisms. *Pharmacological Reviews*, *73*(1), 202–277.

Jones, K. A., Srivastava, D. P., Allen, J. A., Strachan, R. T., Roth, B. L., & Penzes, P. (2009). Rapid modulation of spine morphology by the 5-HT_{2A} serotonin receptor through Kalirin-7 signaling. *Proceedings of the National Academy of Sciences*, *106*(46), 19575–19580.

Kozlowska, U., Nichols, C., Wiatr, K., & Figiel, M. (2022). From psychiatry to neurology: Psychedelics as prospective therapeutics for neurodegenerative disorders. *Journal of Neurochemistry*, *162*(1), 89–108.

Lerner, A. G., & Ran, S. L. (2015). LSD-associated "Alice in Wonderland Syndrome" (AIWS): A hallucinogen persisting perception disorder (HPPD) case report. *Israel Journal of Psychiatry*, *52*(1), 67.

McConnell, P., & McConnell, P. (2024, January 13). List of companies creating new DMT therapies—Tripsitter. *Tripsitter*. https://tripsitter.com/dmt/companies/

McCulloch, D. E.-W., Madsen, M. K., Stenbaek, D. S., Kristiansen, S., Ozenne, B., Jensen, P. S., Knudsen, G. M., & Fisher, P. M. (2022). Lasting effects of a single psilocybin dose on resting-state functional connectivity in healthy individuals. *Journal of Psychopharmacology*, *36*(1), 74–84.

McWilliams, S. A., & Tuttle, R. J. (1973). Long-term psychological effects of LSD. *Psychological Bulletin*, *79*(6), 341.

Mi, Z., Si, T., Kapadia, K., Li, Q., & Muma, N. A. (2017). Receptor-stimulated transamidation induces activation of Rac1 and CDC42 and the regulation of dendritic spines. *Neuropharmacology*, *117*, 93–105.

Mintz, K. T., Gammer, B., Khan, A. J., Shaub, G., Levine, S., & Sisti, D. (2022). Physical disability and psychedelic therapies: An agenda for inclusive research and practice. *Frontiers in Psychiatry*, *13*, 914458.

Nichols, E., Szoeke, C. E., Vollset, S. E., Abbasi, N., Abd-Allah, F., Abdela, J., Aichour, M. T. E., Akinyemi, R. O., Alahdab, F., Asgedom, S. W., et al. (2019). Global, regional, and national burden of Alzheimer's disease and other dementias, 1990–2016: A systematic analysis for the global burden of disease study 2016. *Lancet Neurology*, *18*(1), 88–106.

Nkadimeng, S. M., Steinmann, C. M., & Eloff, J. N. (2021). Anti-inflammatory effects of four psilocybin-containing magic mushroom water extracts in vitro on 15-lipoxygenase activity and on lipopolysaccharide-induced cyclooxygenase-2 and inflammatory cytokines in human U937 macrophage cells. *Journal of Inflammation Research*, 3729–3738.

Nutt, D. (2019). Psychedelic drugs—A new era in psychiatry? *Dialogues in Clinical Neuroscience*, *21*(2), 139–147.

Ono, Y., Tanaka, H., Takata, M., Nagahara, Y., Noda, Y., Tsuruma, K., Shimazawa, M., Hozumi, I., & Hara, H. (2014). Sa4503, a sigma-1 receptor agonist, suppresses motor neuron damage in in vitro and in vivo amyotrophic lateral sclerosis models. *Neuroscience Letters*, *559*, 174–178.

Pahnke, W. N., Kurland, A. A., Unger, S., Savage, C., & Grof, S. (1970). The experimental use of psychedelic (LSD) psychotherapy. *JAMA*, *212*(11), 1856–1863.

PharmaTher Holdings Ltd. (2021, May 17). PharmaTher announces FDA approval of ketamine IND in the treatment of Parkinson's disease. *GlobeNewswire News Room*. https://www.globenewswire.com/news-release/2021/05/17/2230622/0/en/PharmaTher-Announces-FDA-Approval-of-Ketamine-IND-In-The-Treatment-of-Parkinson-s-Disease.html

Peterson, A., Largent, E. A., Lynch, H. F., Karlawish, J., & Sisti, D. (2023). Journeying to Ixtlan: Ethics of psychedelic medicine and research for Alzheimer's disease and related dementias. *AJOB Neuroscience*, *14*(2), 107–123.

Richmond, L. M. (2024). Psychedelics' use as medicine hits significant roadblock. *Psychiatric News*, *59*(07). https://doi.org/10.1176/appi.pn.2024.07.7.54

Roth, B. L., & Gumpper, R. H. (2023). Psychedelics as transformative therapeutics. *American Journal of Psychiatry*, *180*(5), 340–347.

Sabanović, M., Lazari, A., Blanco-Pozo, M., Tisca, C., Tachrount, M., Martins-Bach, Á. B., Lerch, J. P., Walton, M. E., & Bannerman, D. M. (2024). Lasting dynamic effects of the psychedelic 2,5-dimethoxy-4-iodoamphetamine ((±)-DOI) on cognitive flexibility. *Molecular Psychiatry*, 1–14.

Saeger, H. N., & Olson, D. E. (2022). Psychedelic-inspired approaches for treating neurodegenerative disorders. *Journal of Neurochemistry*, *162*(1), 109–127.

Schwartz, J. K., & Unni, E. (2021). Inclusion of people with disabilities in research to improve medication adherence: A systematic review. *Patient Preference and Adherence*, 1671–1677.

Shao, L.-X., Liao, C., Gregg, I., Davoudian, P. A., Savalia, N. K., Delagarza, K., & Kwan, A. C. (2021). Psilocybin induces rapid and persistent growth of dendritic spines in frontal cortex in vivo. *Neuron*, *109*(16), 2535–2544.

Smith, W. R., & Appelbaum, P. S. (2022). Novel ethical and policy issues in psychiatric uses of psychedelic substances. *Neuropharmacology*, *216*, 109165.

Song, J., Kambari, Y., Amaev, A., Ueno, F., Torres Carmona, E., De Luca, V., Pollock, B., Flint, A., Husain, M. I., Graff-Guerrero, A., & Gerretsen, P. (2023). Psilocybin to promote synaptogenesis in the brains of patients with mild cognitive impairment. *Medical Hypotheses*, *175*, 111068. https://doi.org/https://doi.org/10.1016/j. mehy.2023.111068

Tylš, F., Páleníček, T., & Horáček, J. (2014). Psilocybin—Summary of knowledge and new perspectives. *European Neuropsychopharmacology*, *24*(3), 342–356.

Vaidya, V. A., Marek, G. J., Aghajanian, G. K., & Duman, R. S. (1997). 5-HT_{2A} receptor-mediated regulation of brain-derived neurotrophic factor mRNA in the hippocampus and the neocortex. *Journal of Neuroscience*, *17*(8), 2785–2795.

Vargas, M. V., Dunlap, L. E., Dong, C., Carter, S. J., Tombari, R. J., Jami, S. A.,Cameron, L. P., Patel, S. D., Hennessey, J. J., Saeger, H. N., et al. (2023). Psychedelics promote neuroplasticity through the activation of intracellular 5-HT_{2A} receptors. *Science*, *379*(6633), 700–706.

Wakabayashi, K., Mori, F., & Takahashi, H. (2006). Progression patterns of neuronal loss and Lewy body pathology in the substantia nigra in Parkinson's disease. *Parkinsonism & Related Disorders*, *12*, S92–S98.

Walczak-Nowicka, L. J., & Herbet, M. (2021). Acetylcholinesterase inhibitors in the treatment of neurodegenerative diseases and the role of acetylcholinesterase in their pathogenesis. *International Journal of Molecular Sciences*, *22*(17), 9290.

White, E., Kennedy, T., Ruffell, S., Perkins, D., & Sarris, J. (2024). Ayahuasca and dimethyltryptamine adverse events and toxicity analysis: A systematic thematic review. *International Journal of Toxicology*, *43*(3), 327–339.

Winkelman, M. J., Szabo, A., & Frecska, E. (2023). The potential of psychedelics for the treatment of Alzheimer's disease and related dementias. *European Neuropsychopharmacology*, *76*, 3–16. https://doi .org/https://doi.org/10.1016/j.euroneuro. 2023.07.003

Yerubandi, A., Thomas, J. E., Bhuiya, N. M. A., Harrington, C., Zapata, L. V., & Caballero, J. (2024). Acute adverse effects of therapeutic doses of psilocybin: A systematic review and meta-analysis. *JAMA Network Open*, *7*(4), e245960–e245960.

Yoshida, H., Kanamaru, C., Ohtani, A., Li, F., Senzaki, K., & Shiga, T. (2011). Subtype specific roles of serotonin receptors in the spine formation of cortical neurons in vitro. *Neuroscience Research*, *71*(3), 311–314.

Zahra, W., Rai, S. N., Birla, H., Singh, S. S., Dilnashin, H., Rathore, A. S., & Singh, S. P. (2020). The global economic impact of neurodegenerative diseases: Opportunities and challenges. *Bioeconomy for Sustainable Development*, 333–345.

Zuccato, C., & Cattaneo, E. (2009). Brain-derived neurotrophic factor in neurodegenerative diseases. *Nature Reviews Neurology*, *5*(6), 311–322.

15 Environmental Factors and Challenges in Managing Addiction and Neurodegenerative Disorders

Ruta Karaliuniene, Margaret Ojeahere, Chonnakarn Jatchavala, Sachin Nagendrappa, Ramyadarshni Vadivel, and Bita Vahdani

INTRODUCTION

Neurodegenerative disorders (NDDs) and substance use disorders (SUDs) are major public health concerns, contributing significantly to morbidity and mortality worldwide. As the global population continues to age, the proportion of older adults is growing, leading to increased vulnerability to NDDs. While addiction and SUDs affect individuals of all ages, their impact is particularly pronounced in those over 65. The psychological effects of environmental influences, such as climate change, migration, and conflict, further complicate these health challenges. Migration, often driven by climate-related disasters and conflicts, adds to the psychological burden faced by vulnerable populations, exacerbating the risks for addiction and neurodegeneration. The intersection of these environmental factors with SUDs and NDDs requires multifaceted health interventions and comprehensive treatment approaches. This chapter examines the complex relationships between climate change, conflict, and migration and their effects on NDDs and SUDs. It also explores the impact of these stressors on general mental health and provides recommendations for mitigating comorbidities, with a focus on social determinants and access to culturally sensitive healthcare.

THE UNRAVELING RELATIONSHIP BETWEEN CLIMATE CHANGE-RELATED STRESSORS, NEURODEGENERATIVE DISORDERS, AND SUBSTANCE USE DISORDERS

Introduction

Climate change is the leading public health threat of the 21st century, associated with complex and deleterious health consequences (WHO, 2023). Accumulating changes in the climate have led to rapidly evolving ecosystems, which affect the occurrence, incubation periods, survival, distribution, and transmission of pathogens and vector-borne diseases (Pörtner et al., 2022). The changing ecosystem has also been associated with increasing atmospheric carbon dioxide, changes in precipitation, water balance, ocean chemistry, and an increasing frequency and magnitude of extreme events (Malhi et al., 2020).

In recent years, there has been a significant rise in average life expectancy, accompanied by a surge in chronic diseases typically associated with aging (Vasto & Caruso, 2004). Previous studies show that human vulnerability to NDDs increases with age, with a higher prevalence among older adults (Tong et al., 2017). Increasing heat stress and inflammation from climate change, which

DOI: 10.1201/9781003478485-15

depletes immature neuronal hormones and degrades neurotransmitters, may accelerate the onset of NDDs and promote the use of psychoactive substances in affected individuals and their caregivers (Erkkinen et al., 2018).

General Concepts on Climate Stressors

Climate change is a significant determinant of complex, interconnected factors that range from acute to chronic stressors (Cianconi et al., 2020). Acute stressors refer to sudden, intense events that can cause considerable mental and emotional distress, such as higher average summer temperatures, more frequent and severe heat waves, droughts, wildfires, storms, floods, and hurricanes (Walinski et al., 2023; Cianconi et al., 2020).

Chronic stressors, on the other hand, are long-term, persistent factors that negatively affect individuals and communities over extended periods. These include socioeconomic and health challenges such as water scarcity, food insecurity, famine, civil unrest, forced migration, war, changes in biodiversity, the spread of insect-borne diseases, prolonged allergy seasons, and the alteration or loss of landscapes (Vergunst et al., 2023). Extant evidence links climate-related stressors to mental health disorders (Albrecht et al., 2007).

Impact on Mental Health

For years, attention has been focused on the physical burden of climate change, but an increasing volume of scientific research now demonstrates its substantial effects on mental health (Ching et al., 2022; Ansah et al., 2024). The effects of climate change on mental health can occur directly with immediate impacts (e.g., heat waves), indirectly in the short term during extreme events (e.g., floods, tornadoes, hurricanes), and indirectly in the long term due to environmental changes (e.g., prolonged droughts, rising sea levels, deforestation, forced migration) (Helm et al., 2018). These effects can be categorized into three groups: direct effects on mental health, short-term indirect effects, and long-term effects through physical conditions (Karaliuniene et al., 2022).

Through these various mechanisms, climate change leads to measurable impacts on mental health (Walinski et al., 2023; Cianconi et al., 2020), resulting in conditions such as anxiety, depression, post-traumatic stress disorder (PTSD), aggression, SUDs, and neurological disorders. Notably, much of the research on climate change and mental health has focused on depression, PTSD, and anxiety (Palinkas & Wong, 2020), while conditions such as NDDs and SUDs remain underexplored. For instance, a cross-sectional study in Puerto Rico on the effects of extreme weather found that 54.1% of participants showed symptoms suggestive of major depression, 48.6% had generalized anxiety disorder, and 41.9% exhibited PTSD symptoms (Scaramutti et al., 2019). Similar findings were reported in a study among participants exposed to Hurricane Katrina in 2006, where about 49.1% experienced anxiety or mood disorders, and 30.3% had PTSD (Galea et al., 2007). The interplay between acute and chronic stressors of climate change and their tangible effects is further illustrated in Table 15.1.

Relationship between Acute and Chronic Stressors of Climate Change and Neurodegenerative Disorders

Climatic changes, including extreme weather events and ecosystem shifts, impact NDDs both directly and indirectly (Walinski et al., 2023). Although the underlying mechanisms remain unclear, evidence suggests that climate change may contribute to the rising incidence of NDDs through multiple pathways (Vergunst et al., 2023). It is hypothesized that environmental factors can influence the brain by crossing the blood–brain barrier, potentially disrupting central and peripheral nervous system functions and triggering psychological and neurological conditions (Ruszkiewicz et al., 2019).

TABLE 15.1
Summary of Acute and Chronic Stressors of Climate Change and Their Tangible Effects

Acute and Chronic Stressors	Tangible Effects
Extreme weather events: Hurricanes, floods, mudslides, wildfires, droughts, heat waves	*Direct effects on mental health*: Post-traumatic stress disorder and increased levels of stress, anxiety, and adjustment disorders after catastrophic events.
Short- and long-term economic impact	*Indirect effects on mental health*: Due to political turmoil, forced migration, marital discord, and property destruction.
Long-term consequences to health	*Effects on mental health through physical conditions*: Triple burden of malnutrition, dehydration, and vector-borne diseases.

Extreme weather conditions, such as altered precipitation patterns and temperature variability, have been shown to cause heat stress and inflammation, which may lead to hormonal depletion and neurotransmitter degradation in individuals with NDDs (Ayeni et al., 2022). Such harsh conditions can accelerate or worsen the progression of disorders like Parkinson's disease (PD) and Alzheimer's disease (AD). For example, prolonged exposure to high temperatures has been linked to worsened symptoms in these populations (Bongioanni et al., 2021). Furthermore, fine particulate matter from wildfire smoke can penetrate deeply into the lungs and enter the bloodstream, exacerbating NDDs through inflammation and oxidative stress (Heaney et al., 2022).

Environmental contaminants from highways and power plant emissions may also pose neurotoxic risks. Short- and long-term exposure to air pollution is associated with an increased risk of stroke, silent infarctions, and brain atrophy (Shetty et al., 2023). In addition, extreme events such as floods, hurricanes, wildfires, and heat waves can disrupt healthcare access, creating significant challenges for individuals at risk of or currently receiving treatment for NDDs. Displacement due to these acute stressors of climate change not only affects communities through increased temperatures and resource scarcity but also limits access to treatment options, disrupts livelihoods, and contributes to unemployment, including among caregivers for individuals with NDDs (Schraufnagel et al., 2019; Amiri et al., 2021).

Relationship between Acute and Chronic Stressors of Climate Change and Substance Use Disorders

SUDs, including the harmful use of tobacco, alcohol, recreational drugs (such as cannabis, cocaine, heroin, and lysergic acid diethylamide [LSD]), and certain prescription medications (e.g., opioids, benzodiazepines, and barbiturates), account for approximately 3.3% of the global disease burden and are linked to 16.8 million deaths annually (McLellan, 2017). Epidemiological studies reveal a strong correlation between exposure to climate hazards—such as hurricanes and wildfires—and increased substance misuse (Hobden et al., 2023). A widely accepted psychological model suggests that individuals may turn to substance use as a self-medicating strategy to cope with climate-induced negative emotions (Khantzian, 1997).

While the psychological model emphasizes self-medication for managing climate-related stress, social epidemiology points to the impact of social context and early exposure to substance use, and the neurobiological approach highlights reward pathway sensitization, which fuels addiction through cravings and compulsive behaviors (Vergunst et al., 2023). Indeed, a complex interplay of persistent and interrelated stressors across the lifespan increases susceptibility to substance use through multiple distinct pathways (Walinski et al., 2023). Unchecked and accelerating climate change stressors could further amplify this susceptibility (Cianconi et al., 2020). For instance, rising global temperatures may lead to increased outdoor activity, which, combined with dehydration from heat, can result in heightened use of psychoactive substances and, consequently, increased hospitalizations, accidents, and, in severe cases, death (Amaro et al., 2021). Climate-related grief, another factor contributing to SUDs, is associated with the emotional impact of climate change, leading individuals to use psychoactive substances to cope with losses related to life, property, or identity tied to their environment (Cunsolo & Ellis, 2018; Comtesse et al., 2021).

The connection between food shortage, NDDs, and SUDs remains underexplored. However, it is suggested that climate change-induced food insecurity and malnutrition in individuals with SUDs may contribute to insulin resistance, cognitive decline, and disruptions in the gut microbiota-brain axis, thus exacerbating NDDs (Liang et al., 2022).

COVID-19, Addiction, and Neurodegenerative Disorders

COVID-19 exemplifies the impact of ecosystem changes and climate-borne diseases on public health (Sadeghmousavi & Rezaei, 2020). Although the relationship between COVID-19 and NDDs is not fully understood, inflammatory responses appear to be a common mechanism contributing to tissue damage across several age-related diseases. Numerous biological studies suggest that various viruses may play a role in classical NDDs; however, the exact influence of viruses on these disorders remains uncertain and continues to be an active area of research, particularly in the context of environmental factors (Garjani et al., 2021). Additionally, individuals with SUDs faced a heightened risk of severe COVID-19 outcomes. The pandemic saw a surge in addictive behaviors, including both new cases and relapses, as well as an increase in behavioral addictions. People with SUDs also encountered significant barriers to healthcare access, resulting in increased morbidity and overdose risk (Dubey et al., 2020).

DEALING WITH THE DUAL CHALLENGES OF ARMED CONFLICT: NEURODEGENERATION AND ADDICTION

Traumatic Brain Injury in Conflict Zones

Traumatic brain injury (TBI) is a significant health concern in conflict zones (Okie, 2005). Exposure to explosive blasts and concussions from improvised explosive devices significantly impacts military personnel, with estimates suggesting that 15.2% to 22.8% of returning service members experience TBI, affecting up to 320,000 troops (Omalu et al., 2011). Around 80% of TBIs are classified as mild, also referred to as concussions, while the remaining range from moderate to severe and can be penetrating (O'Neil et al., 2013).

TBI in war zones is a recognized risk factor for NDDs. Although awareness of the health impacts of armed conflict is increasing, there is still a limited understanding of the specific needs of individuals with neurodegeneration in these settings (Hou et al., 2019). Research shows that secondary injury mechanisms following TBI—such as neuronal cell loss, axonal injury, and synaptic dysfunction—can lead to cognitive impairment (Walker & Tesco., 2013). Clinically, a meta-analysis found that TBI raises the risk of AD by approximately 1.5 times (Fleminger et al., 2003). Likewise, the risk of PD is higher after a single TBI, regardless of its severity, as shown in a study of 320,000 individuals (Gardner et al., 2018). Additionally, a single TBI has been modestly linked to an increased risk of amyotrophic lateral sclerosis (ALS) (Watanabe & Watanabe, 2017). Research also indicates that 10%–15% of individuals experience prolonged post-concussive syndrome (PCS), with subtle impairments in executive functions (Walker & Tesco, 2013).

SUDs compound these challenges by exacerbating cognitive impairments such as memory, learning, executive function, visuospatial skills, processing speed, and motor coordination and balance. This further complicates the cognitive issues faced by individuals with TBI and other NDDs (Pagulyan et al., 2016). Lastly, as in climate crisis scenarios, individuals with TBI and NDDs in conflict zones encounter considerable barriers to diagnosis, treatment, and ongoing support (National Academies of Sciences, Engineering, and Medicine, 2022; Jaradat et al., 2024).

Psychological Distress in Conflict Zones

The prevalence of mental health conditions is rising in conflict zones, yet effective treatments and prevention measures remain lacking, even in relatively stable regions (Hou et al., 2019). The prolonged and intense violence in these areas severely impacts mental well-being, contributing to the emergence of trauma-related conditions, such as depression, PTSD, and addiction (Lamaisaard et al., 2021). Throughout history, substances have been used in wars and conflict zones to debilitate adversaries, energize soldiers for combat, and alleviate post-war psychological distress (Bergen-Cico, 2015). In this context, evidence indicates that substance misuse—particularly of psychostimulants and alcohol—not only exacerbates mental health issues but is also linked to NDDs through specific molecular mechanisms (Guo et al., 2023).

The Current Mental Health Situation in Conflict Zones

Populations in conflict zones, regardless of civilian or military status, face an increased risk of trauma-related disorders due to factors such as exposure to violence, poverty, and daily stressors (Internal Displacement Monitoring Centre, 2015). While SUDs are a significant concern among military veterans globally, with an estimated prevalence of 11% in the United States, research indicates that civilians exposed to conflict are even more vulnerable to increased substance misuse, including alcohol and tobacco (Lo et al., 2015; Teeters et al., 2017). The mental health challenges resulting from trauma and violence can disrupt individuals' coping mechanisms, often leading to self-medication through alcohol and illicit drug use (Jain et al., 2022).

A significant challenge for healthcare systems in these areas is prioritizing mental health needs. The predominant model in conflict-affected regions is the "humanitarian emergency relief" approach, which prioritizes immediate health needs such as food, water, sanitation, shelter, and disease control (Huang et al., 2015). However, only a few models have been developed to address non-communicable diseases, including mental health and substance misuse, despite their growing significance in conflict-affected or post-conflict areas. As a result, healthcare workers face significant challenges in addressing mental health issues, particularly among vulnerable groups such as young people, women, older adults, and those with pre-existing mental illnesses. Moreover, government and non-governmental agencies struggle to provide adequate health services, including brief SUD interventions, in conflict zones (Mattoo et al., 2018). Weak policy enforcement and governance also create opportunities for the marketing and availability of legal substances such as tobacco and alcohol, as observed in Montenegro (Deliso, 2017).

Frameworks for Addressing Neurodegenerative and Substance Use Disorders in Conflict Zones

Research on SUDs in war-torn areas indicates that individuals often use multiple substances, leading to more severe cognitive impairment and a heightened risk for TBI (Jatchavala & Vittayanont, 2017). Therefore, in such settings, the framework for Screening, Brief Intervention, and Referral to Treatment (SBIRT) must consider polysubstance use (Sarkar et al., 2020). Tools like the Alcohol Use Disorder Identification Test (AUDIT) and technology-based interventions should also be considered (Mattoo et al., 2018). In this regard, telemedicine for addiction has proven effective in the management of several vulnerable populations, including LGBTQIA+ individuals, pregnant women, children, and adolescents (Avery & Khan, 2023).

However, no established framework currently exists for managing neuropsychiatric disorders, including NDDs and SUDs, in conflict zones. The "multi-pillar collaborative approach" model (Figure 15.1) was developed to address this gap based on the World Health Organization's Mental Health Gap Action Program (mhGAP) *Intervention Guide* (Jain et al., 2022).

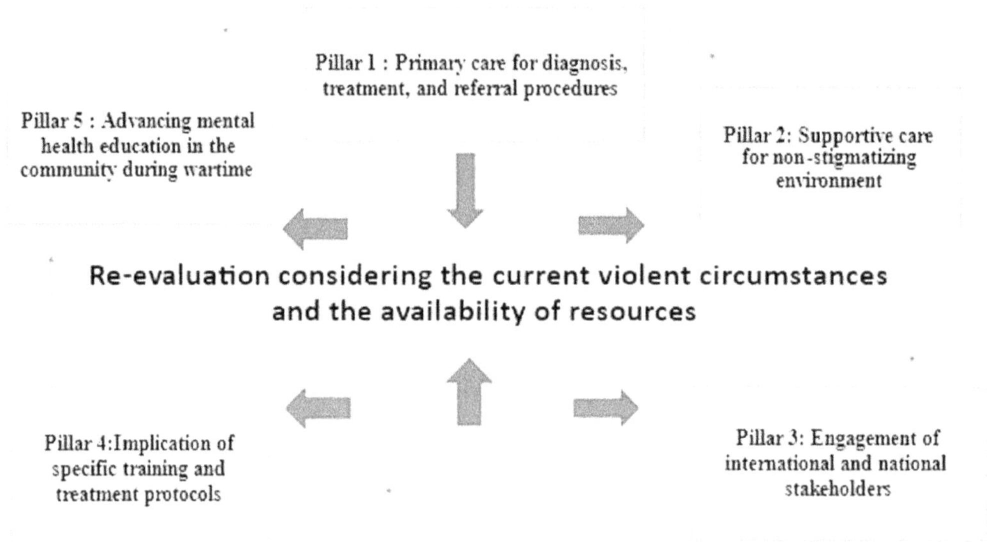

FIGURE 15.1 Five-pillar collaborative model for managing substance use and neuropsychiatric disorders in war-affected regions (based on the mhGAP *Intervention Guide*).

In this model, Pillar 1 focuses on involving all primary care physicians in the diagnostic process during wartime while integrating other healthcare providers for screening, brief intervention, and referrals within the SBIRT framework, considering available resources (Thoele et al., 2021). In conflict-affected areas, severe cases of SUDs and other neuropsychiatric conditions, such as TBI and NDDs, should be prioritized for referral.

Pillar 2 emphasizes creating a non-stigmatizing care environment, ideally rooted in social cohesion and cultural sensitivity established before the conflict. However, violent situations may temporarily alter certain norms and cultures, necessitating a reevaluation before fostering a non-stigmatizing environment in war zones, along with long-term planning post-conflict (Sibley et al., 2024).

To activate Pillar 3, national and international stakeholders, such as the WHO, International Red Cross, and Doctors Without Borders, can assist local healthcare providers in training for Pillar 4 and developing treatment guidelines specific to the region, encompassing the principles of Pillars 1 and 2. Campaigns for Pillars 2, 3, and 4 should focus on de-stigmatization and promoting sustainable mental health education in the region through Pillar 5 (Committee on the Science of Changing Behavioral Health Social Norms, 2016). Technological interventions should be integrated across all pillars, especially Pillars 2, 4, and 5, utilizing social media and web-based learning (Avery & Khan, 2023).

Implementing the "multi-pillar collaborative approach" model within the "humanitarian emergency relief model" can help prioritize life-saving measures, especially for vulnerable groups such as women, children, and older adults (Huang et al., 2015; Jain et al., 2022).

IMPACT OF MIGRATION ON SUBSTANCE USE AND NEURODEGENERATIVE DISORDERS

Introduction

Numerous studies have highlighted the profound impact of migration on mental health, emphasizing the heightened vulnerability of migrant communities (Close et al., 2016; Virupaksha et al., 2014). Indeed, factors such as acculturative stress and social isolation contribute to an increased risk of mental health disorders, including depression, anxiety, and PTSD (Kirmayer et al., 2011). Forced displacement introduces additional stressors that span pre-migration, migration, and post-migration phases. Pre-migration factors such as cultural backgrounds, trauma, economic hardship, and pre-existing mental health conditions compound the challenges faced by migrants. During migration, stressors like violence, harsh transportation conditions, temporary camp placements before assimilation, and family separation intensify the burden. Post-migration, the pressures of cultural assimilation, discrimination, stigma, and loneliness further affect mental well-being (Keyes et al., 2011).

Relationship between Migration, Substance Misuse, and Dementia

The prevalence of substance use among migrant populations is a significant concern, with studies revealing notable disparities in substance use patterns and access to addiction treatment services (Horyniak et al., 2016; López Jaramillo et al., 2023). Migration-related stressors, along with social marginalization and economic hardships, contribute to increased substance use, resulting in a higher prevalence of SUDs within these communities (Kyaw et al., 2023). A study conducted in Finland highlighted the complex relationship between migration-related stressors and substance use, underscoring the need for tailored public health interventions that address the unique experiences of different migrant groups (Salama et al., 2019).

On the other hand, recent meta-analyses and systematic reviews showed a significant association between migration and dementia, with migrant populations exhibiting a higher risk compared to non-migrant groups (Selten et al., 2021; Kovaleva et al., 2021). The mechanisms underlying the link between migration and NDDs are complex and involve a combination of genetic, environmental, and social determinants (World Migration Report, 2020).

Recommendations for Policymakers and Healthcare Providers

Limitations of Current Research

As previously mentioned, limited literature explores the relationship between NDDs, SUDs, and environmental factors. The intersection of climate change, neurodegeneration, and addiction is an emerging area of research that highlights how environmental changes can influence both mental and physical health. Addressing these interconnected issues requires a holistic approach, including policies to reduce emissions, safeguard the rights of displaced communities, improve mental health support, and expand treatment services. Understanding these links is essential for developing strategies to mitigate the impacts of environmental challenges on health and well-being.

Much of the current research on substance use, neurodegeneration, and their sequelae primarily focuses on forced migration, displacement due to wars and disasters, and refugees in high-income countries. However, these studies often lack representativeness, particularly in recognizing the impact of migration in low-income countries. Several studies also lack appropriate control groups, limiting the ability to compare substance use patterns between disadvantaged populations. Additionally, while some studies examine risk factors for substance use through multivariable analysis, they focus on individual-level characteristics while neglecting structural factors, such as acculturation, which are particularly relevant in migration. Finally, the scarcity of intervention studies significantly limits the evidence base for addressing SUDs and NDDs across various environmental challenges (Piebre et al. 2016).

Funding Future Research

Exploring how migration—especially forced migration due to conflict, displacement, or environmental disasters—affects the onset, progression, and management of NDDs is critical. The cumulative stress of migration, combined with disrupted access to healthcare, may exacerbate addiction, particularly among vulnerable groups. Longitudinal studies tracking these effects are essential for understanding how migration interacts with SUDs and NDDs, thereby facilitating informed policy and treatment strategies.

Furthermore, it is essential to investigate migration patterns from low- and middle-income countries (LMICs) to high-income countries, as well as migration patterns within LMICs, to better explore their impact on mental health. Cultural factors, such as the normalization of alcohol consumption in Western populations and culturally accepted practices (e.g., khat chewing in some African countries or drinking kawa in the Pacific Islands) that may be perceived differently in countries of settlement, need to be studied. These dynamics significantly affect the assimilation process and post-migration settlement, which, in turn, influences substance use behaviors and the mental health of migrant populations.

Multi-country studies incorporating standardized recruitment, data collection, and analysis across diverse settings would provide a robust approach to producing meaningful data at both site-specific and cross-contextual levels (Horyniak et al., 2016). These studies could also enhance our understanding of how various environmental challenges, such as climate change and the devastation caused by war, impact SUDs and NDDs in different settings.

In addition to investigating risk factors, future research should focus on identifying protective factors at both individual and community levels. Understanding these mechanisms will inform the development of culturally appropriate and effective interventions. Examining the impact of these interventions through longitudinal studies, including randomized controlled trials (RCTs), will be essential to determine their efficacy in mitigating the effects of SUDs and NDDs across diverse populations.

Addressing Social Determinants

This chapter has highlighted various social determinants that influence SUDs and NDDs, including poverty, employment status, education, housing, and access to healthcare. A public health approach is essential when addressing substance use in populations affected by climate change and armed conflicts, as these factors often exacerbate both addiction and neurodegeneration. In this regard, ensuring equitable access to healthcare is paramount to facilitating timely assessment and intervention for these conditions.

Services must be tailored to the diverse needs of affected populations. This includes offering gender-responsive, culturally sensitive, trauma-informed, and inclusive treatments that cater to the unique challenges faced by individuals in these vulnerable communities.

Overcoming Discrimination and Stigmatization

A significant barrier to accessing care is the stigma and shame often associated with seeking help, particularly for those affected by conflict or climate change. Migrants and refugees from these regions report feeling isolated and discriminated against in their new communities, which may further compound their substance use or exacerbate neurodegenerative conditions. Social reintegration services play a critical role in facilitating the integration of migrants, particularly those with SUDs, into host communities (Priebe et al., 2016). These services typically address factors such as housing, education, and employment, which can help lower barriers to treatment and support long-term recovery (Lemmens et al., 2017; van Selm et al., 2023).

The development of integrated health services that collaborate with housing and social support systems is also essential for effectively addressing the health needs of migrants, including those with SUDs and NDDs (Roberts et al., 2015). Additionally, immigration laws and policies that impact healthcare access, such as the eligibility for opioid substitution treatment, should be carefully reviewed to ensure that non-citizens are not excluded from receiving necessary care.

Alternatively, translation services are vital in enhancing the healthcare experience for foreign-language speakers, contributing to better retention and adherence to treatment programs (Guerrero et al., 2012). Cultural mediators can provide crucial insights into the values, assumptions, and socio-cultural conventions that may lead to misunderstandings in healthcare settings. They help clarify culture-specific expressions and concepts, ensuring that treatment approaches align with the host country's legal framework and cultural expectations.

Investing in these services will enable healthcare providers to adapt treatment protocols to better address the variations in drug laws and societal norms surrounding drug use while also addressing the unique needs of those living with NDDs.

Applying Early Intervention

Substance use and consumption patterns can change over time due to various environmental factors. Therefore, providing adequate support during transitional periods is crucial for facilitating early intervention strategies, particularly in vulnerable populations. It is essential to establish screening and needs assessment procedures for individuals seeking international protection and migrants, particularly upon arrival (Lemmens et al., 2017). These assessments should include measures to evaluate mental health conditions, potential SUDs, NDDs, as well as other conditions that may be exacerbated by the migration process.

Establishing contact through home visits and fostering community connections can be critical components of an effective early intervention strategy. Research suggests that brief interventions offered within facilities housing migrants can contribute to prevention strategies and play a vital role in community-based or multi-component approaches (Greene et al., 2019). In this context, brief interventions aim to prevent the progression from recreational substance use to high-risk use and from high-risk use to SUDs. These interventions may also help mitigate the onset or progression of NDDs by addressing environmental stressors early on.

Workforce Training and Empowerment

Healthcare frameworks must be strengthened at all levels of care, from grassroots key workers and primary healthcare providers to specialist services. Cultural competency training empowers healthcare professionals to be culturally sensitive, enabling them to reflect on their cultural backgrounds and assumptions and understand how these may influence their clinical interactions with populations affected by environmental challenges. This awareness is crucial for developing treatment plans that respect patients' diverse beliefs and expectations, both at the individual and organizational levels (Guerrero et al., 2015). Healthcare organizations must also consider structural changes to effectively accommodate their workforce's training (Priebe et al., 2016).

CONCLUSION

Addressing the challenges faced by individuals with neurodegeneration and addiction requires a multifaceted approach that integrates medical, psychological, social, and environmental interventions. Areas affected by climate change and armed conflicts often experience complex socioeconomic, religious, cultural, and political issues. Therefore, a tailored model for managing addiction may be more appropriate than a universal framework, as the latter may not be suited to the specific context. However, it is crucial for both policymakers and healthcare providers to evaluate such models thoroughly using standardized methodologies.

In affected regions, weakened legal enforcement and governance create increased opportunities for access to both legal and illegal substances, especially for individuals diagnosed with neuropsychiatric illnesses, including NDDs and SUDs. The aging global population presents an additional challenge for healthcare systems, including humanitarian relief efforts, which may need to be redesigned to address the needs of older adult migrants. By investing in research, policy development, and healthcare infrastructure, policymakers and medical providers can enhance care for affected individuals, improve support systems, and foster community resilience.

REFERENCES

Albrecht, G., Sartore, G. M., Connor, L., Higginbotham, N., Freeman, S., Kelly, B., et al. (2007). Solastalgia: The distress caused by environmental change. *Australas Psychiatry,* 15, 595–598. https://doi.org/10.1080/10398560701701288.

Amaro, H., Sanchez, M., Bautista, T., & Cox, R. (2021). Social vulnerabilities for substance use: Stressors, socially toxic environments, and discrimination and racism. *Neuropharmacology*, 188, 108518. https://doi.org/10.1016/j.neuropharm.2021.108518.

Amiri, M., Peinkhofer, C., Othman, M. H., De Vecchi, T., Nersesjan, V., & Kondziella, D. (2021). Global warming and neurological practice: Systematic review. *PeerJ*, 9, e11941. https://doi.org/10.7717/peerj.11941.

Ansah, E. W., Amoadu, M., Obeng, P., et al. (2024). Health systems response to climate change adaptation: A scoping review of global evidence. *BMC Public Health,* 24. https://doi.org/10.1186/s12889-024-19459-w.

Avery, J. D., & Khan, M. (Eds.). (2023). *Technology-assisted interventions for substance use disorders.* Cham: Springer.

Ayeni, E. A., Aldossary, A. M., Ayejoto, D. A., et al. (2022). Neurodegenerative diseases: Implications of environmental and climatic influences on neurotransmitters and neuronal hormones activities. *Int J Environ Res Public Health,* 19(19), 12495. https://doi.org/10.3390/ijerph191912495.

Bergen-Cico, D. K. (2015). *War and drugs: The role of military conflict in the development of substance abuse.* Routledge. https://doi.org/10.4324/9781315631226.

Bongioanni, P., Del Carratore, R., Corbianco, S., Diana, A., Cavallini, G., Masciandaro, S. M., Dini, M., & Buizza, R. (2021). Climate change and neurodegenerative diseases. *Environmental Research*, 201, 111511. doi:10.1016/j.envres.2021.111511.

Cianconi, P., Betrò, S., & Janiri, L. (2020). The impact of climate change on mental health: A systematic descriptive review. *Frontiers in Psychiatry*, 11, 74. https://doi.org/10.3389/fpsyt.2020.00074.

Close, C., Kouvonen, A., Bosqui, T. et al. (2016). The mental health and well-being of first generation migrants: A systematic-narrative review of reviews. *Global Health,* 12, 47. https://doi.org/10.1186/s12992-016-0187-3.

Committee on the Science of Changing Behavioral Health Social Norms. (2016). *Ending discrimination against people with mental and substance use disorders: The evidence for stigma change.* National Academies Press.

Comtesse, H., Ertl, V., Hengst, S. M. C., Rosner, R., & Smid, G. E. (2021). Ecological grief as a response to environmental change: A mental health risk or functional response? *International Journal of Environmental Research and Public Health*, 18(2), 734. https://doi.org/10.3390/ijerph18020734.

Cunsolo, A., & Ellis, N. R. (2018). Ecological grief as a mental health response to climate change-related loss. *Nature Climate Change,* 8, 275–281. https://doi.org/10.1038/s41558-018-0092-2.

Deliso, C. (2017, August 4). *Letter from Montenegro: Organized crime's state of play.* The American Interest. https://www.the-american-interest.com/2015/10/27/letter-from-montenegro-organized-crimes-state-of-play/.

Dubey, M. J., Ghosh, R., Chatterjee, S., Biswas, P., Chatterjee, S., & Dubey, S. (2020). COVID-19 and addiction. *Diabetes & Metabolic Syndrome: Clinical Research & Reviews*, 14(5), 817–823.

Erkkinen, M. G., Kim, M. O., & Geschwind, M. D. (2018). Clinical neurology and epidemiology of the major neurodegenerative diseases. *Cold Spring Harbor Perspectives in Biology*, 10(4), a033118. https://doi.org/10.1101/cshperspect.a033118.

Fleminger, S., Oliver, D. L., Lovestone, S., Rabe-Hesketh, S., & Giora, A. (2003). Head injury as a risk factor for Alzheimer's disease: The evidence 10 years on; a partial replication. *Journal of Neurology, Neurosurgery & Psychiatry*, 74(7), 857–862. https://doi.org/10.1136/jnnp.74.7.857.

Galea, S., Brewin, C. R., Gruber, M., Jones, R. T., King, D. W., King, L. A., McNally, R. J., Ursano, R. J., Petukhova, M., & Kessler, R. C. (2007). Exposure to hurricane-related stressors and mental illness after Hurricane Katrina. *Archives of General Psychiatry*, 64(12), 1427–1434. https://doi.org/10.1001/archpsyc.64.12.1427.

Gardner, R. C., Byers, A. L., Barnes, D. E., Li, Y., Boscardin, J., & Yaffe, K. (2018). Mild TBI and risk of Parkinson disease: A chronic effects of neurotrauma consortium study. *Neurology*, 90(20), e1771–e1779. https://doi.org/10.1212/WNL.0000000000005522.

Garjani, A., Middleton, R. M., Hunter, R., et al. (2021). COVID-19 is associated with new symptoms of multiple sclerosis that are prevented by disease-modifying therapies. *Multiple Sclerosis and Related Disorders*, 52, 102939. https://doi.org/10.1016/j.msard.2021.102939.

Greene, M. C., Ventevogel, P., & Kane, J. C. (2019). Substance use services for refugees. *Bulletin of the World Health Organization*, 97(4), 246–246A.

Guerrero, E. G., et al. (2012). Do cultural and linguistic competence matter in Latinos completion of mandated substance abuse treatment? *Substance Abuse Treatment, Prevention, and Policy,* 7(1), 34.

Guerrero, E. G., et al. (2015). Paths to improving engagement among racial and ethnic minorities in addiction health services. *Substance Abuse Treatment, Prevention, and Policy,* 10(1), 40.

Guo, C., Chen, L., & Wang, Y. (2023). Substance abuse and neurodegenerative diseases: Focus on ferroptosis. *Archives of Toxicology*, 97(6), 1519–1528. https://doi.org/10.1007/s00204-023-03505-4.

Heaney, A., Stowell, J. D., Liu, J. C., Basu, R., Marlier, M., & Kinney, P. (2022). Impacts of fine particulate matter from wildfire smoke on respiratory and cardiovascular health in California. *GeoHealth*, 6(6), e2021GH000578. https://doi.org/10.1029/2021GH000578.

Helm, S. V., Pollitt, A., Barnett, M. A., Curran, M. A., & Craig, Z. R. (2018). Differentiating environmental concern in the context of psychological adaption to climate change. *Global Environmental Change*, 48, 158–167. https://doi.org/10.1016/j.gloenvcha.2017.11.012.

Hobden, B., Freund, M., Rumbel, J. et al. (2023). Mental health and substance use co-occurrence among Indigenous peoples: A scoping review. *International Journal of Mental Health and Addiction.* https://doi.org/10.1007/s11469-023-01114-x.

Horyniak, D., Melo, J. S., Farrell, R. M., & Ojeda, V. D. (2016). Strathdee SA. Epidemiology of substance use among forced migrants: A global systematic review. *PLoS One,* 13;11(7), e0159134. https://doi.org/10.1371/journal.pone.0159134.

Hou, Y., Dan, X., Babbar, M. et al. (2019). Ageing as a risk factor for neurodegenerative disease. *Nature Reviews Neurology*, 15, 565–581.

Huang, K., Jiang, Y., Yuan, Y., & Zhao, L. (2015). Modeling multiple humanitarian objectives in emergency response to large-scale disasters. *Transportation Research Part E: Logistics and Transportation Review*, 75, 1–17. https://doi.org/10.1016/j.tre.2014.11.007.

Internal Displacement Monitoring Centre. (2015). *Global overview 2015: People internally displaced by conflict and violence.* Norwegian Refugee Council. https://www.internal-displacement.org/global-overview-2015.

Jain, N., Prasad, S. Reinis, A. et al. (2022). War psychiatry: Identifying and managing the neuropsychiatric consequences of armed conflicts. *Journal of Primary Care Community Health*, 13, 215013192211066. https://doi.org/10.1177/21501319221106625.

Jaradat, J. H., Altah, B. H., & Jankhout, S. (2024). The burden of neurological diseases in conflict settings: Narrative review Gaza situation. *Qatar Journal of Public Health,* (1), 10.

Jatchavala, C., & Vittayanont, A. (2017). Post-traumatic stress disorder symptoms among patients with substance-related disorders in the restive area of *S. Songklanagarind Medical Journal,* 35(2), 121.

Karaliuniene, R., Campana, A. M., Ori, D., de Filippis, R., Shoib, S., Saeed, F., Mohammed, M., Handuleh, J., Ransing, R., Codati, A., Pinto da Costa, M., Ojeahere, M., Orsolini, L., & Pereira-Sanchez, V. (2022). Effects of the global ecological crisis on the mental health of children and adolescents: An international perspective. *Psychiatria Danubina*, 34(2), 296–298. https://doi.org/10.24869/psyd.2022.296. PMID: 35772140.

Kemkem, S., Ivandić-Zimić, J., Mr Andreas Pichides, et al. (2022). Intercultural responses to drug-related challenges for refugees, migrants and IDPS. *Handbook for professionals working in the field of addictions with refugees, migrants, and internally displaced persons.* Council of Europe.

Keyes, K. M., Hatzenbuehler, M. L., & Hasin, D. S. (2011). Stressful life experiences, alcohol consumption, and alcohol use disorders: The epidemiologic evidence for four main types of stressors. *Psychopharmacology (Berl).* 218(1), 1–17. https://doi.org/10.1007/s00213-011-2236-1.

Khantzian E. J. (1997). The self-medication hypothesis of substance use disorders: A reconsideration and recent applications. *Harvard Review of Psychiatry*, 4(5), 231–244. https://doi.org/10.3109/10673229709030550.

Kim, S. Y., Kim, S. H., Wee, J. H., Min, C., Han, S. M., Kim, S., & Choi, H. G. (2021). Short- and long-term exposure to air pollution increases the risk of ischemic heart disease. *Scientific Reports*, 11(1), 5108. https://doi.org/10.1038/s41598-021-84587-x.

Kirmayer, L. J., Narasiah, L., Munoz, M., Rashid, M., Ryder, A. G., Guzder, J., ... & Pottie, K. (2011). Common mental health problems in immigrants and refugees: General approach in primary care. *CMAJ*, 183(12), E959–E967.

Kovaleva, M., Jones, A., & Maxwell, C. A. (2021). Immigrants and dementia: Literature update. *Geriatric Nursing*, 42(5), 1218–1221. https://doi.org/10.1016/j.gerinurse.2021.04.019.

Kyaw, K. W. Y., Platt, L., Bijl, M. et al. (2023). The effect of different types of migration on symptoms of anxiety or depression and experience of violence among people who use or inject drugs in Kachin State, Myanmar. *Harm Reduction Journal*, 20, 45. https://doi.org/10.1186/s12954-023-00766-1.

Lamaisaard, Y., Sangsuwan, T., Chanyeam, K., & Jatchavala, C. (2021). Depression among first-year university students in restive areas of southern Thailand during COVID-19 pandemic. *Suranaree Journal of Science and Technology*, 29(5), 070054(1–9).

Lemmens, P. H., Dupont, H. B., & Roosen, I. (2017). Migrants, asylum seekers and refugees: An overview of the literature relating to drug use and access to services. https://www.euda.europa.eu/document-library/migrants-asylum-seekers-and-refugees-overview-literature-relating-drug-use-and-access-services_en (Accessed on November 19, 2024).

Li, C., Lawrance, E. L., Morgan, G., Brown, R., Greaves, N., Krzanowski, J., Samuel, S., Guinto, R. R., & Belkin, G., (2022). The role of mental health professionals in the climate crisis: An urgent call to action. *International Review of Psychiatry,* 34(5), 563–570.

Liang, X., Fu, Y., Cao, Wt. et al. (2022). Gut microbiome, cognitive function and brain structure: A multi-omics integration analysis. *Transl Neurodegener,* 11, 49. https://doi.org/10.1186/s40035-022-00323-z.

Lo, J., Patel, P., & Roberts, B. (2015). A systematic review on tobacco use among civilian populations affected by armed conflict. *Tobacco Control*, 25(2), 129–140. https://doi.org/10.1136/tobaccocontrol-2014-052054.

López Jaramillo, A. M., Rangel Gómez, M. G., Morales Chainé, S., López Montoya, A., Lira Chávez, I. A., & Cruz-Piñeiro, R. (2023). Mental, neurological and substance use disorders among the Latino migrant population in the United States who visited the Health Windows and Mobile Health Units in 2021. *Front Public Health*, 11, 959535. https://doi.org/10.3389/fpubh.2023.959535.

Malhi, Y., Franklin, J., Seddon, N., Solan, M., Turner, M. G., Field, C. B., & Knowlton, N. (2020). Climate change and ecosystems: Threats, opportunities and solutions. *Philosophical Transactions of the Royal Society of London. Series B, Biological Sciences*, 375(1794), 20190104. https://doi.org/10.1098/rstb.2019.0104

Mattoo, S., Prasad, S., & Ghosh, A. (2018). Brief intervention in substance use disorders. *Indian Journal of Psychiatry*, 60(8), 466. https://doi.org/10.4103/0019-5545.224352.

McLellan, A. T. (2017). Substance misuse and substance use disorders: Why do they matter in healthcare? *Transactions of the American Clinical and Climatological Association*, 128, 112–130.

National Academies of Sciences, Engineering, and Medicine, Health and Medicine Division, Board on Health Care Services, Board on Health Sciences Policy, Committee on Accelerating Progress in Traumatic Brain Injury Research and Care, Matney, C., Bowman, K., & Berwick, D. (Eds.). (2022). *Traumatic brain injury: A roadmap for accelerating progress: Gaps, challenges, opportunities.* National Academies Press

Okie, S. (2005). Traumatic brain injury in the war zone. *New England Journal of Medicine*, 352(20), 2043–2047

O'Neil, M. E., Carlson, K., Storzbach, D., et al. (2013). *Complications of mild traumatic brain injury in veterans and military personnel: A systematic review.* Department of Veterans Affairs (US).

Omalu, B., Hammers, J. L., Bailes, J., Hamilton, R. L., Kamboh, M. I., Webster, G., & Fitzsimmons, R. P. (2011). Chronic traumatic encephalopathy in an Iraqi War veteran with post-traumatic stress disorder who committed suicide. *Neurosurgical Focus*, 31(5), E3. https://doi.org/10.3171/2011.9.FOCUS11178.

Pagulayan, K. F., Temkin, N. R., Machamer, J. E., & Dikmen, S. S. (2016). Patterns of alcohol use after traumatic brain injury. *Journal of Neurotrauma*, 33(14), 1390–1396. https://doi.org/10.1089/neu.2015.4071.

Palinkas, L. A, & Wong, M. (2020). Global climate change and mental health. *Current Opinion in Psychology*, 32, 12–16. https://doi.org/10.1016/j.copsyc.2019.06.023.

Pörtner, H., Roberts, D., Tignor, M., Poloczanska, E., Mintenbeck, K., Alegría, A., et al. (2022). Climate change 2022: Impacts, adaptation and vulnerability. Working Group II Contribution to the IPCC Sixth Assessment Report Citations of the Intergovernmental Panel on Climate Change. https://doi.org/10.1017/9781009325844.

Priebe, S., Giacco, D., & El-Nagib, R. (2016). Public health aspects of mental health among migrants and refugees: A review of the evidence on mental health care for refugees, asylum seekers and irregular migrants in the WHO European region. WHO Regional Office for Europe, Copenhagen.

Roberts, N. P., et al. (2015). Psychological interventions for post-traumatic stress disorder and comorbid substance use disorder: A systematic review and meta-analysis. *Clinical Psychology, Review,* 38, 25–38.

Ruszkiewicz, J. A., Tinkov, A. A., Skalny, A. V., Siokas, V., Dardiotis, E., Tsatsakis, A., Bowman, A. B., da Rocha, J. B. T., & Aschner, M. (2019). Brain diseases in changing climate. *Environmental Research*, 177, 108637. https://doi.org/10.1016/j.envres.2019.108637.

Sadeghmousavi, S., & Rezaei, N. (2020). COVID-19 and multiple sclerosis: Predisposition and precautions in treatment. *SN Comprehensive Clinical Medicine*, 2(10), 1802–1807. https://doi.org/10.1007/s42399-020-00504-9.

Salama, E. S., Castaneda, A. E., Lilja, E., Suvisaari, J., et al. (2019). Pre-migration traumatic experiences, post-migration perceived discrimination and substance use among Russian and Kurdish migrants—A population-based study. *Society for the Study of Addiction*, 115, 1160–1171. https://doi.org/10.1111/add.14904.

Sarkar, S., Pakhre, A., Murthy, P., & Bhuyan, D. (2020). Brief interventions for substance use disorders. *Indian Journal of Psychiatry*, 62(8), 290–298. https://doi.org/10.4103/psychiatry.indianjpsychiatry_778_19.

Scaramutti, C., Salas-Wright, C. P., Vos, S. R., & Schwrtz, S. J. (2019). The mental health impact of Hurricane Maria on Puerto Ricans in Puerto Rico and Florida. *Disaster Med Public Health Prep*, 30, 1–4. https://doi.org/10.1017/dmp.2018.151.

Schraufnagel, D. E., Balmes, J. R., Cowl, C. T., De Matteis, S., Jung, S. H., Mortimer, K., Perez-Padilla, R., Rice, M. B., Riojas-Rodriguez, H., Sood, A., Thurston, G. D., To, T., Vanker, A., & Wuebbles, D. J. (2019). Air pollution and non-communicable diseases: A review by the Forum of International Respiratory Societies' Environmental Committee, Part 2: Air pollution and organ systems. *Chest*, 155(2), 417–426. https://doi.org/10.1016/j.chest.2018.10.041.

Selten, J. P., Termorshuizen, F., van Sonsbeek, M., Bogers, J., & Schmand, B. (2021). Migration and dementia: A meta-analysis of epidemiological studies in Europe. *Psychological Medicine,* 51(11), 1838–1845. https://doi.org/10.1017/S0033291720000586.

Shetty, S. S., D, D., S, H., Sonkusare, S., Naik, P. B., Kumari, N. S., & Madhyastha, H. (2023). Environmental pollutants and their effects on human health. *Heliyon*, 9(9), e19496. https://doi.org/10.1016/j.heliyon.2023.e19496. Kim, S. Y., Kim, S. H., Wee, J. H., Min, C., Han, S. M., Kim, S., & Choi, H. G. (2021). Short- and long-term exposure to air pollution increases the risk of ischemic heart disease. *Scientific Reports*, 11(1), 5108. https://doi.org/10.1038/s41598-021-84587-x

Sibley, A. L., Colston, D. C., & Go, V. F. (2024). Interventions to reduce self-stigma in people who use drugs: A systematic review. *Journal of Substance Use and Addiction Treatment*, 159, 209284. https://doi.org/10.1016/j.josat.2023.209284.

Teeters, J., Lancaster, C., Brown, D., & Back, S. (2017). Substance use disorders in military veterans: Prevalence and treatment challenges. *Substance Abuse and Rehabilitation*, 8, 69–77. https://doi.org/10.2147/sar.s116720.

Thoele, K., Moffat, L., Konicek, S., et al. (2021). Strategies to promote the implementation of Screening, Brief Intervention, and Referral to Treatment (SBIRT) in healthcare settings: A scoping review. *Substance Abuse Treatment, Prevention, and Policy*, 16(42). https://doi.org/10.1186/s13011-021-00380-z.

Tong T., Ledig C., Guerrero R., Schuh A., Koikkalainen J., Tolonen A., Rhodius H., Barkhof F., Tijms B., Lemstra A.W., et al. (2017). Five-class differential diagnostics of neurodegenerative diseases using random undersampling boosting. *NeuroImage Clinical*, 15, 613–624. https://doi.org/10.1016/j.nicl.2017.06.012.

van Selm, L., White, T. M., Picchio, C. A., Requena-Méndez, A., Busz, M., Bakker, I., Romero, D., Perez Gayo, R., Pouille, A., Vanderplasschen, W., & Lazarus, J. V. (2023). Drug use and access to drug dependency services for vulnerable migrants who use drugs in the European Union: Consensus statements and recommendations from civil society experts in Europe. *International Journal on Drug Policy*, 118, 104087. https://doi.org/10.1016/j.drugpo.2023.104087.

Vasto, S., & Caruso, C. (2004). Immunity & ageing: A new journal looking at ageing from an immunological point of view. *Immun Ageing*, 1, 1. https://doi.org/10.1186/1742-4933-1-1.

Vergunst, F., Berry, H. L., Minor, K., & Chadi, N. (2023). Climate change and substance-use behaviors: A risk-pathways framework. *Perspectives on Psychological Science: A Journal of the Association for Psychological Science*, 18(4), 936–954. https://doi.org/10.1177/17456916221132739.

Virupaksha, H. G., Kumar, A., & Nirmala, B. P. (2014). Migration and mental health: An interface. *Journal of Natural Science, Biology and Medicine*, 5(2), 233–239. https://doi.org/10.4103/0976-9668.136141.

Walinski, A., Sander, J., Gerlinger, G., Clemens, V., Meyer-Lindenberg, A., & Heinz, A. (2023). The effects of climate change on mental health. *Deutsches Arzteblatt International*, 120(8), 117–124. https://doi.org/10.3238/arztebl.m2022.0403.

Walker, K. R., & Tesco, G. (2013). Molecular mechanisms of cognitive dysfunction following traumatic brain injury. *Frontiers in Aging Neuroscience*, 5, 29. https://doi.org/10.3389/fnagi.2013.00029.

Watanabe, Y., & Watanabe, T. (2017). Meta-analytic evaluation of the association between head injury and risk of amyotrophic lateral sclerosis. *European Journal of Epidemiology*, 32(10), 867–879. https://doi.org/10.1007/s10654-017-0327-y.

WHO. (2023). Operational framework for climate resilient and low carbon health systems. World Health Organization. https://www.who.int/publications/i/item/9789240081888. Accessed on 19th November, 2024.

World Migration Report. (2020). International Organization for Migration, Geneva. https://publications.iom.int/system/files/pdf/wmr_2020.pdf. Accessed on November 19th, 2024.

16 Challenges of Neurodegenerative and Substance Use Disorders in Low- and Middle-Income Countries

Joy Louise Gumikiriza-Onoria, Rita Nassanga, Dennis Buwembo, and Kamada Lwere

INTRODUCTION

In regions with limited resources and underdeveloped healthcare systems, addressing substance use disorders (SUDs) and neurodegenerative disorders (NDDs) presents unique public health challenges. This chapter delves into these intersecting conditions in low- and middle-income countries (LMICs), with an emphasis on Sub-Saharan Africa (SSA), where socioeconomic pressures, evolving drug use patterns, and globalization contribute to rising substance use rates. The discussion includes an examination of changing drug use patterns, including the misuse of prescription drugs, alongside the social and economic factors driving these trends. Subsequently, the chapter highlights specific obstacles in diagnosing and managing NDDs in LMICs, where dementia is often stigmatized and misunderstood. A section on SSA further explores the region's unique epidemiological and societal context, noting the scarcity of data on individuals affected by both dementia and drug addiction. The chapter then presents a Uganda case study to illustrate the challenges and potential for culturally relevant interventions. Finally, the chapter proposes recommendations to improve healthcare infrastructure, reduce stigma, and enhance research initiatives aimed at these intersecting health issues. By addressing these needs, LMICs can improve health outcomes for affected populations and better manage the growing burden of co-occurring NDDs and SUDs.

SUBSTANCE USE DISORDERS IN LOW- AND MIDDLE-INCOME COUNTRIES: PATTERNS AND TRENDS

SUDs represent a pressing concern in LMICs, driven by evolving drug use patterns, socioeconomic influences, and the impact of globalization. The prevalence of SUDs has been escalating in LMICs, compounded by several challenges, such as inadequate training among healthcare professionals and limited availability of treatment services due to insufficient public funding. This scarcity is particularly pronounced in rural areas, where treatment gaps can reach up to 95% (Sharma et al., 2024). The associated burden of disease is significant. For instance, compared to high- and middle-income countries, the alcohol-attributable disease burden is disproportionately higher in LMICs (Gosh et al., 2022).

This section delves into the multifaceted issues surrounding substance misuse in LMICs, with particular emphasis on the roles of socioeconomic factors, inadequate healthcare infrastructure, and the broad societal repercussions of these conditions. Additionally, it highlights how limited

DOI: 10.1201/9781003478485-16

resources and the lack of access to effective treatment further complicate the prevention and management of SUDs in these regions, underscoring the urgent need for tailored public health responses.

Evolving Drug Use Patterns and Prescription Drug Misuse

Drug use in LMICs has shifted from traditional substances like cannabis, opium, and coca leaves to synthetic drugs and the non-medical use of prescription medications. The rise of amphetamine-type stimulants, such as methamphetamine, is particularly pronounced in Southeast Asia and Africa, driven by these substances' affordability, accessibility, ease of production, and potency (UNODC, 2021). The misuse of prescription drugs, such as analgesics and sedatives, is also increasing, especially in countries with lax pharmaceutical regulations. The perceived safety and easy availability of these drugs contribute to dependency, addiction, and health complications, particularly in settings with inadequate regulatory frameworks and limited public awareness (Preuss et al., 2023). Polysubstance use further exacerbates these risks, complicating treatment and increasing the likelihood of risky behaviors, medical comorbidity, psychiatric sequelae, and overdose (Were & Nyongesa, 2022). A systematic review of substance use in Kenya noted a significant burden across all age groups, as well as an increased likelihood of adverse mental and physical health outcomes directly attributable to misuse (Jaguga et al., 2022).

Socioeconomic Influences on Drug Use

Socioeconomic conditions, such as poverty, unemployment, and conflict, heavily influence drug use in LMICs. Economic instability and financial stress contribute significantly to addiction, often driving individuals to substance misuse as a coping or survival mechanism in regions where traditional livelihoods and social support networks have collapsed (Stephenson et al., 2023; Sunday & Okereke, 2021; Were et al., 2022). In conflict-affected areas like South Sudan and the Democratic Republic of Congo, SUDs, particularly those involving alcohol and cannabis, are prevalent as coping mechanisms for trauma and instability. Refugee populations in countries such as Kenya, Uganda, and Lebanon also report alcohol and marijuana use as a means of coping with stressors related to war zones, displacement, adjustment, poor living conditions, and material and personal losses (Abbas et al., 2021; Makoha & Denov, 2024).

Income inequality and unemployment further exacerbate the likelihood of SUDs, as individuals may turn to drugs to cope with economic hardship (Gavurova et al., 2021; Manhica et al., 2023). The affordability and accessibility of certain drugs make them more appealing to economically disadvantaged populations, particularly in urban centers. Additionally, studies show that nations with significant income disparities experience higher rates of substance misuse (Were et al., 2022).

Drug addiction in LMICs has far-reaching societal consequences, affecting entire communities. Weak social networks, limited community resources, and inadequate policies exacerbate this problem. Strengthening community-based interventions, such as recreational activities and support groups, is essential in curbing SUDs (Henning et al., 2019). Addressing the socioeconomic barriers to addiction treatment, such as lack of health insurance and absence of culturally sensitive care, requires robust policy frameworks to expand access to care for marginalized populations and prioritize funding for treatment programs (Matsuzaki et al., 2018). Table 16.1 summarizes the main challenges in addressing drug addiction in LMICs and outlines potential responses.

Globalization and Urbanization Effects

Globalization has significantly contributed to the spread of drug trafficking networks into LMICs, making a wide variety of substances, including cannabinoids, cocaine, and narcotics, more accessible. The increased interconnectedness between countries has facilitated the movement of both legal and illicit substances across borders, often outpacing regulatory efforts (Petcu, 2017). Alongside

TABLE 16.1
Challenges in Addressing Drug Addiction in Low- and Middle-Income Countries

Challenge	Recommendation
Healthcare infrastructure	Improve healthcare systems and expand access to addiction treatment, particularly in rural and underserved areas (Simone et al., 2020).
Stigma and public awareness	Implement educational campaigns to reduce stigma and encourage treatment-seeking behaviors. Collaborate with community leaders to change public perceptions (Smith and Peyson, 2020).
Policy and regulations	Develop evidence-based drug policies that balance enforcement with harm reduction, tailored to local contexts (Matthew et al., 2020).
International collaborations	Strengthen cross-border cooperation to tackle transnational drug trafficking and enhance rehabilitation efforts across regions (Luca et al., 2021).

this, urbanization has exacerbated the issue by creating densely populated urban environments that are more conducive to the distribution and procurement of drugs. In addition, globalization and urbanization have made drug use more culturally normalized, mainly through social media, the internet, and print media. These platforms often glamorize drug experimentation and substance use, influencing younger populations, especially in urban centers, to view drug use as a socially acceptable or even desirable behavior (Irwin-Rogers, 2019). As a result, youth in cities are increasingly exposed to and encouraged to engage in substance use, with peer pressure and social media trends amplifying the risk of early initiation and habitual use (Luca et al., 2021). These factors, combined with limited access to prevention programs, pose significant challenges in curbing substance misuse in urban areas of LMICs.

Treatment Strategies

Deb and Gupta (2017) discuss potential prevention strategies for substance dependence in challenging and low-resource settings. As previously noted, due to numerous challenges, LMICs face significant limitations in developing and implementing successful prevention programs. Thus far, preventative measures to reduce alcohol misuse remain the most studied and cost-effective interventions. These include banning alcohol advertisements, restricting access to alcoholic beverage outlets, and implementing random breath tests for drivers (Deb & Gupta, 2017).

Commonly identified LMIC-specific barriers to treatment include low motivation and a lack of perceived problem or need for treatment, while family support and resource availability have been described as facilitators of treatment (Sarkar et al., 2021). Addressing these barriers and developing effective prevention and treatment strategies for addiction in LMICs remain in their infancy. Multi-stakeholder efforts are essential for their successful implementation.

CHALLENGES IN APPROACHING DEMENTIA IN LOW- AND MIDDLE-INCOME COUNTRIES

NDDs present unique challenges in LMICs, particularly in regions like SSA. These challenges include diagnostic difficulties, economic and cultural barriers, limited treatment options, and the pervasive influence of stigma. This section outlines the main obstacles to dementia care in LMICs, providing a broader understanding of the multifaceted barriers that hinder effective diagnosis and treatment (Table 16.2).

In a recent review, Parra et al. discuss various affordable, culturally sensitive, and interdisciplinary approaches that can yield practical and reliable solutions to managing NDDs in LMICs (Parra et al., 2019). Efforts to address the challenges of treatment require coordinated action across multiple fronts:

TABLE 16.2
Key Challenges in Dementia Care in Low- and Middle-Income Countries

Challenge	Description
Diagnosis	Limited public awareness, a shortage of trained healthcare professionals, the prohibitive costs of advanced diagnostic procedures like neuroimaging, the misinterpretation of dementia as a normal part of aging, and rural-urban disparities in access to healthcare services all contribute to delays in diagnosis (Alzheimer's Disease International, 2022; Gharbr-Meliani et al., 2021; Prince et al., 2013; Gureje et al., 2011).
Economic barriers	High costs of diagnostic tests, medications, and caregiving, compounded by a lack of health insurance and social security, force families to bear out-of-pocket expenses (WHO, 2023; Wimo et al., 2014).
Cultural factors	Traditional beliefs that dementia is a natural aspect of aging, coupled with stigma, lead to delayed diagnosis and treatment, as well as reliance on non-medical alternatives (Gureje et al., 2006; Kalula & Petros, 2011).
Stigma	Social isolation of individuals with dementia, coupled with many families hiding the condition to avoid discrimination, prevents timely access to medical care and necessary support (Alzheimer's Disease International, 2016).
Limited treatment options	A scarcity of comprehensive care models and essential medications like cholinesterase inhibitors and memantine results in untreated progressive cognitive decline (Guerchet et al., 2017).
Pursuit of non-medical alternatives	Reliance on traditional healers and spiritual explanations for dementia delays accurate diagnosis and intervention. However, there is potential to integrate traditional practices with modern medical care to improve outcomes (Prince et al., 2013; Kalula & Petros, 2011).

- *Public education and awareness*: Raising awareness about dementia as a medical condition rather than a natural consequence of aging is critical to reducing stigma and encouraging early diagnosis. Educational campaigns tailored to the specific cultural contexts of LMICs can help bridge the gap between traditional beliefs and modern medical practices.
- *Expanding access to care*: Improving healthcare infrastructure, particularly in rural areas, and providing affordable diagnostic and treatment options are crucial steps toward equitable dementia care. Investments in healthcare systems and international cooperation can help fund these initiatives and reduce the care gap between LMICs and high-income countries. In addition, there is a dire need for culturally appropriate tools that allow timely diagnosis of dementia, particularly for populations with high rates of illiteracy (Walker & Paddick, 2019).
- *Training healthcare professionals*: Expanding training programs to facilitate early identification and management of NDDs is essential. Capacity-building initiatives can help improve diagnostic accuracy and preventative measures, enhancing patient care. One recommended avenue includes integrating care for dementia into chronic disease clinics delivered by non-specialist healthcare professionals (Walker & Paddick, 2019).
- *Integrating traditional medicine*: While it can sometimes hinder timely medical interventions, traditional medicine provides psychological comfort and social support. Integrating traditional practices with evidence-based medical treatments like cognitive behavioral therapy and UGA-iSupport—a platform developed to provide psychosocial support for caregivers of individuals with dementia in LMICs—offers a promising strategy to improve dementia care in LMICs, particularly in underserved areas (Logsdon et al., 2007; Gumikiriza-Onoria et al., 2024).

Through these efforts, LMICs can make significant progress in overcoming the numerous barriers to dementia care, ensuring better outcomes for individuals living with dementia and their families.

DEMENTIA AND DRUG ADDICTION IN SUB-SAHARAN AFRICA: CHALLENGES, INSIGHTS, AND FUTURE DIRECTIONS

Limited Epidemiological Data

One of the most significant challenges in addressing dementia and drug addiction in SSA is the lack of comprehensive epidemiological data. This data gap complicates efforts to understand the scope and prevalence of these issues, particularly when they occur as comorbidities. The scarcity of research in this area further impedes the development of targeted interventions.

- *Dementia*: While dementia is recognized as a growing global concern, epidemiological data specific to SSA is limited. This impedes the development of accurate prevalence estimates and regional intervention strategies. Furthermore, genetic research, such as studies on the APOE ε4 allele—a known risk factor for dementia—is sparse in African populations, preventing a clear understanding of genetic and environmental contributors to dementia (Mavrodaris et al., 2013; Prince et al., 2016).
- *Drug addiction*: Data on drug use in SSA is similarly limited, particularly among older adults. While the rising incidence of SUDs in younger populations is more frequently documented, the long-term effects of drug use on mental health and its potential to exacerbate NDDs remain largely unexplored (Nyenga et al., 2024).

Specific Risk Factors in Sub-Saharan Africa

SSA faces unique risk factors that exacerbate both NDDs and SUDs. These include genetic predispositions and rising comorbidities, such as HIV.

- *APOE ε4 allele and dementia*: Genetic studies in SSA have identified the APOE ε4 allele as a significant risk factor for dementia in African populations. However, limited research infrastructure and funding hinder investigations into how this allele interacts with region-specific factors, such as diet, lifestyle, and healthcare access, to influence dementia risk (Adebiyi et al., 2016).
- *HIV and substance use*: The intersection of HIV and SUDs presents a significant health challenge in SSA. Substance misuse, especially the use of injectable drugs, increases the risk of HIV transmission. HIV-associated neurocognitive disorders can then complicate the diagnosis and treatment of dementia and other NDDs. This comorbidity further complicates efforts to address both dementia and addiction within healthcare systems that are already stretched thin (Nyenga et al., 2024).

BROADER CHALLENGES COMMON TO LOW- AND MIDDLE-INCOME COUNTRIES

As in most LMICs, SSA faces several systemic challenges that hinder progress in tackling dementia and drug addiction, especially when co-existing. These challenges include underdeveloped healthcare infrastructure, economic constraints, stigma, and a lack of public awareness and education.

- *Healthcare infrastructure*: Healthcare systems in SSA are primarily focused on acute and infectious diseases, leaving chronic conditions such as NDDs and SUDs underfunded and under-resourced. The shortage of healthcare professionals trained in geriatrics and addiction medicine exacerbates the problem, especially in rural areas where access to specialized care is limited (WHO, 2017).

- *Stigma and public awareness*: Cultural beliefs and stigma surrounding both dementia and drug addiction create significant barriers to care. Dementia is often perceived as a natural consequence of aging or attributed to witchcraft or spiritual possession, while SUDs are commonly regarded as a moral failing. These misperceptions lead to delays in seeking treatment, social exclusion, and an overreliance on traditional healers, all of which hinder effective intervention and exacerbate the challenges in addressing these conditions (Parry et al., 2015).

Case Study: Uganda

Introduction

Uganda, an East African lower-middle-income country with a population of approximately 45 million, faces a growing SUD problem. The nation grapples with addiction, exacerbated by socioeconomic disparities, political instability, and underfunded healthcare services. This section explores the patterns of drug addiction in Uganda, the underlying causes, and the responses from the government and communities, while considering potential links between substance misuse and NDDs.

Drug Use and Dementia in Uganda

In Uganda, commonly misused substances include cannabis, alcohol, khat (a stimulant plant), and, more recently, prescription medications such as tramadol and codeine. Amphetamines and methamphetamines have also gained popularity, particularly among youth in urban areas (Uganda Police Force, 2022). However, data on drug use among older adults, especially in relation to NDDs, remains limited. Research on the cognitive impact of drug use in older adults in Uganda, SSA, and other LMICs is scarce, highlighting the urgent need for further investigation into the long-term effects of addiction on mental health and cognitive decline in these regions.

Challenges and Responses in Uganda

- *Socioeconomic challenges*: Unemployment, poverty, and a lack of opportunities—especially for Uganda's large youth population—are critical drivers of substance misuse. Many young Ugandans turn to drugs as a means of coping with economic difficulties and social pressures. Urbanization, particularly in Kampala, has also contributed to a rise in addiction rates as traditional social structures weaken, leaving individuals more vulnerable to SUDs (Uganda Bureau of Statistics, 2020).
- *Healthcare and legal barriers*: Uganda's healthcare system is under-resourced, with a shortage of specialized drug rehabilitation centers and insufficient mental health services. The stigma surrounding drug addiction further hinders individuals from seeking treatment. The legal framework, led by Uganda's Anti-Narcotics Unit, heavily focuses on punitive measures, often resulting in incarceration rather than rehabilitation. This approach has been criticized for failing to address the root causes of addiction or provide effective long-term recovery solutions (Global Initiative Against Transnational Organized Crime, 2020; Miller et al., 2021; WHO, 2022).
- *Cultural and community responses*: Cultural beliefs and social stigma play significant roles in both the perpetuation of and response to SUDs. In many communities, drug use is frowned upon, and the lack of public awareness campaigns means addiction is often misunderstood. Community-based organizations and non-governmental organizations are stepping in to fill gaps in the healthcare system by providing education, prevention, and treatment services tailored to local needs. Religious and traditional leaders have also been instrumental in changing societal attitudes toward drug use, helping reduce stigma and encouraging individuals to seek help (Scherz et al., 2021; Kitafuna, 2022; Dawhaniuk, 2021).

TABLE 16.3
Challenges and Responses in Uganda

Challenge	Recommendation
Unemployment and poverty	Expand job creation programs and provide social support to vulnerable groups to reduce substance misuse rates.
Underfunded healthcare system	Invest in mental health and addiction treatment facilities, particularly in rural areas, and train healthcare providers.
Punitive legal approach	Shift from punitive legal measures to rehabilitative strategies, prioritizing treatment over incarceration.
Cultural stigma and misconceptions	Launch public awareness campaigns with support from community and religious leaders to reduce stigma.
Limited data on drug use and dementia	Conduct research to explore the cognitive effects of long-term substance misuse and its link to dementia in older populations.

Insights and Future Directions

Uganda's drug addiction crisis intersects with broader health challenges, including mental health issues. A list of challenges and responses in the country is detailed in Table 16.3. Data on drug use among older adults and its potential link to NDDs is scarce. Nevertheless, the increasing prevalence of substance misuse among younger populations suggests that this issue could become more prominent as these individuals age.

Uganda's drug addiction problem is deeply intertwined with socioeconomic challenges, inadequate healthcare infrastructure, and cultural stigmas. While drug use is primarily concentrated among young populations, there is an urgent need for data on the long-term impact of SUDs on older adults, particularly those with NDDs. Addressing Uganda's drug addiction crisis will require a shift toward rehabilitation, improved healthcare services, and community-driven prevention efforts. Understanding Uganda's socioeconomic and cultural context is crucial for designing effective strategies to combat this growing issue.

CONCLUSION

The intersection of NDDs and SUDs in LMICs, particularly in SSA, presents significant hurdles. These include a lack of epidemiological data, underdeveloped healthcare systems, and cultural stigma. Both NDDs and SUDs remain poorly understood and under-addressed, leaving many without proper care. Critical barriers to treatment include misconceptions surrounding dementia and substance misuse, which delay diagnosis and intervention. The shortage of trained professionals, especially in rural areas, exacerbates these issues. Socioeconomic factors, like poverty and unemployment, contribute to higher addiction rates, particularly among younger populations. Meanwhile, the long-term impact of substance misuse on cognitive decline in older adults remains largely unstudied.

A multifaceted and comprehensive approach is essential to address these challenges:

- *Improving public awareness and reducing stigma*: Public awareness campaigns should aim to destigmatize both dementia and SUDs. Utilizing culturally relevant messaging and engaging community influencers, such as religious figures and leaders, can foster openness and encourage treatment-seeking behaviors.
- *Strengthening healthcare systems*: Integrating dementia and addiction services into primary healthcare systems can ensure early detection and intervention. This requires the development of standardized, culturally sensitive protocols for screening and treatment,

alongside increased funding to expand access to care. Additionally, employing task-shifting models, where non-specialist healthcare workers deliver basic care under supervision, can address workforce shortages, particularly in rural areas. Governmental and non-governmental organizations must invest in healthcare infrastructure to establish specialized centers in underserved areas.

- *Enhancing research and data collection*: Fostering international collaboration to enhance research and data collection is imperative. Building regional research networks and partnerships with academic institutions can help bridge the epidemiological data gap. Research priorities should include understanding the prevalence and comorbidity of NDDs and SUDs in LMICs, investigating the long-term effects of substance misuse on cognitive health, and developing region-specific prevention and treatment models.
- *Socioeconomic interventions*: Addressing underlying socioeconomic factors, such as poverty and unemployment, is crucial. Providing vocational training, community support programs, and education on the risks of substance misuse can help mitigate addiction rates among vulnerable populations.
- *Community-driven approaches*: Empowering local communities to take an active role in addressing these challenges is essential. As previously highlighted, engaging religious, cultural, and traditional leaders in health promotion and stigma reduction efforts can build trust and improve community acceptance of seeking treatment for NDDs and SUDs.
- *Rehabilitation over punishment*: Policy reforms should prioritize rehabilitation and harm reduction over punitive approaches to substance use. Implementing evidence-based interventions, such as needle exchange programs and medication-assisted treatment, can improve outcomes and reduce societal costs.

In conclusion, addressing the combined challenges of dementia and drug addiction in LMICs requires region-specific interventions that account for the cultural, economic, and healthcare realities of these countries. By prioritizing public awareness, healthcare integration, infrastructure development, and research, significant progress can be made in improving the quality of life for individuals affected by these conditions. Collaborative efforts among governments, local communities, academic institutions, hospitals, and international agencies will be essential to achieving sustainable change.

REFERENCES

Abbas, Z., Eiden, C., Salameh, P., & Peyriere, H. (2021). Substance use among refugees in three Lebanese camps: A cross-sectional study. *International Journal of Drug Policy*, 94, 103204.

Adebiyi, A. O., Uchendu, O. C., & Fasoro, A. A. (2016). Epidemiology of dementia in Nigeria: A review. *Journal of Clinical Neuroscience*, 25(2), 1–4.

Alzheimer's Disease International. (2016). World Alzheimer Report 2016: Improving healthcare for people living with dementia. https://www.alzint.org/resource/world-alzheimer-report-2016/

Alzheimer's Disease International. (2022). World Alzheimer Report 2022: Journey through the diagnosis of dementia. https://www.alzint.org/resource/world-alzheimer-report-2022/

Deb, K. S., & Gupta, S. K. (2017). Prevention strategies for substance use disorders in low-resource settings. *Indian Journal of Social Psychiatry*, 33(2), 112–117.

Dowhaniuk, N. (2021). Exploring country-wide equitable government health care facility access in Uganda. *International Journal for Equity in Health*, 20, Article 38. https://doi.org/10.1186/s12939-020-01371-5

Gavurova, B., Ivankova, V., & Rigelsky, M. (2021). Income inequality, education, and substance use disorders: Cross-country analysis in low- and middle-income countries. *International Journal of Environmental Research and Public Health*, 18(6), 3082. https://doi.org/10.3390/ijerph18063082

Gharbr-Meliani, C., Guerchet, M., & Prince, M. J. (2021). Risk factors for dementia in Sub-Saharan Africa: A review. *Lancet Neurology*, 20(2), 134–140.

Ghosh, A., Singh, P., Das, N., Pandit, P. M., Das, S., & Sarkar, S. (2022). Efficacy of brief intervention for harmful and hazardous alcohol use: A systematic review and meta-analysis of studies from low middle-income countries. *Addiction (Abingdon, England)*, 117(3), 545–558. https://doi.org/10.1111/add.15613

Global Initiative Against Transnational Organized Crime (2020). Uganda's Narcotic Drugs and Psychotropic Substances Act 2016 has left individuals vulnerable to abuse while failing to deter major drug traffickers. Global Initiative Against Transnational Organized Crime. Risk Bulletin #5. https://riskbulletins.globalinitiative.net/esa-obs-005/04-ugandas-narcotic-drugs-and-psychotropic-substances-act.html

Guerchet, M., Mayston, R., Lloyd-Sherlock, P., Prince, M., Akinyemi, R., Paddick, S. M., Wimo, A., Amoakoh-Coleman, M., Uwakwe, R., Ezeah, P., & Aboderin, I. (2017). *Dementia in Sub-Saharan Africa: Challenges and Opportunities* (Doctoral Dissertation, Alzheimer's Disease International).

Gumikiriza-Onoria, J. L., Mayega, R. W., Nakigudde, J., Giordani, B., Sajatovic, M., Mukasa, M. K., ... & Nakasujja, N. (2024). Feasibility and acceptability of the World Health Organization's iSUPPORT program for dementia caregivers in Uganda, 1(2). https://doi.org/10.20935/MHealthWellB7292

Gureje, O., Ogunniyi, A., Kola, L., & Afolabi, E. (2006). Dementia in Sub-Saharan Africa: The Nigerian context. *Alzheimer's & Dementia*, 2(5), 279–285. https://doi.org/10.1016/j.jalz.2006.08.005

Gureje, O., Ogunniyi, A., Kola, L., & Afolabi, E. (2011). The profile and impact of probable dementia in a Sub-Saharan African community: Results from the Ibadan study of aging. *Journal of Psychosomatic Research*, 71(5), 232–237. https://doi.org/10.1016/j.jpsychores.2011.03.016

Henning, L., & Gunnarsson, E. (2019). Social cohesion and its effects on substance use: A comparative study between rural and urban settings in low-income countries. *Substance Use & Misuse*, 54(8), 1294–1305. https://doi.org/10.1080/10826084.2019.1577305

Irwin-Rogers, K. (2019). Illicit drug markets, consumer capitalism and the rise of social media: A toxic trap for young people. *Critical Criminology*, *27*(4), 591–610.

Jaguga, F., Kiburi, S. K., Temet, E., Barasa, J., Karanja, S., Kinyua, L., & Kwobah, E. K. (2022). A systematic review of substance use and substance use disorder research in Kenya. *PloS One*, 17(6), e0269340.

Kalula, S. Z., & Petros, S. G. (2011). Use of healthcare services by older adults with dementia in Cape Town, South Africa. *Journal of Dementia Care*, 19(6), 32–36. https://doi.org/10.1016/j.dementia.2011.09.001

Kitafuna, K. B. (2022). A critical overview of mental health-related beliefs, services, and systems in Uganda and recent activist and legal challenges. *Community Mental Health Journal*, 58(6), 829–834. https://doi.org/10.1007/s10597-022-00947-5

Logsdon, R. G., McCurry, S. M., & Teri, L. (2007). Evidence-based interventions to improve quality of life for individuals with dementia. *Alzheimers Care Today,* 8(4), 309–318.

Luca, M., Stevenson, A., & Roberts, S. (2021). Globalization, urbanization, and the proliferation of drug trafficking in low- and middle-income countries: Implications for public health. *Global Public Health*, 16(5), 678–694. https://doi.org/10.1080/17441692.2021.1872580

Makoha, G., & Denov, M. (2024). War, forced displacement, and alcohol abuse: Experiences and perceptions of war-affected south Sudanese refugee youth living in Bidibidi refugee settlement in northern Uganda. *Frontiers in Public Health*, 12, 1232504.

Manhica, H., Lundberg, P., & Neupane, S. (2023). Poverty, psychiatric disorders, and substance use in adolescents from low-income countries: A comprehensive review. *Social Psychiatry and Psychiatric Epidemiology*, 58(2), 323–337. https://doi.org/10.1007/s00127-022-02330-5

Matsuzaki, M., Vu, Q.M., Gwadz, M., et al. (2018). Perceived access and barriers to care among illicit drug users and hazardous drinkers: Findings from the Seek, Test, Treat, and Retain Data Harmonization Initiative (STTR). *BMC Public Health*, 18, 366. https://doi.org/10.1186/s12889-018-5291-2

Mavrodaris, A., Powell, J., & Thorogood, M. (2013). Prevalence of dementia and cognitive impairment among older people in Sub-Saharan Africa: A systematic review. *Journal of Geriatric Psychiatry and Neurology*, 26(4), 183–192. https://doi.org/10.1177/0891988713496588

Matthew, S. A., & Stewart, B. (2020). The challenge of implementing international drug control standards in local contexts: The case of LMICs. *International Journal of Drug Policy*, 85, 102934. https://doi.org/10.1016/j.drugpo.2020.102934

Miller, A. P., Ziegel, L., Mugamba, S., Kyasanku, E., Wagman, J. A., Nkwanzi-Lubega, V., Nakigozi, G., Kigozi, G., Nalugoda, F., Kigozi, G., Nkale, J., Watya, S., & Ddaaki, W. (2021). Not enough money and too many thoughts: Exploring perceptions of mental health in two Ugandan districts through the Mental Health Literacy Framework. *Qualitative Health Research*, 31(5), 967–982. https://doi.org/10.1177/1049732320986164

Nyenga, A., Parry, C. D. H., & Myers, B. J. (2024). Drug addiction in Sub-Saharan Africa: Socioeconomic and public health implications. *African Journal of Drug and Alcohol Studies*, 23(1), 1–9.

Parra, M. A., Butler, S., McGeown, W. J., Brown Nicholls, L. A., & Robertson, D. J. (2019). Globalising strategies to meet global challenges: The case of ageing and dementia. *Journal of Global Health*, 9(2), 020310. https://doi.org/10.7189/jogh.09.020310

Parry, C. D. H., Myers, B., & Morojele, N. K. (2015). Substance use and its impact on health in Sub-Saharan Africa: A review. *Addiction*, 110(10), 1571–1581.

Petcu, C. (2017). *Globalization and drug trafficking*. The New School University.

Preuss, C. V., Kalava, A., & King, K. C. (2023). *Prescription of controlled substances: Benefits and risks*. (Updated 29 April 2023). In: StatPearls [Internet]. Treasure Island (FL): StatPearls Publishing; January. Available from: https://www-ncbi-nlm-nih-gov.ezproxy.aub.edu.lb/books/NBK537318/

Prince, M., Bryce, R., Albanese, E., Wimo, A., Ribeiro, W., & Ferri, C. P. (2013). The global prevalence of dementia: A systematic review and meta-analysis. *Alzheimer's & Dementia: The Journal of the Alzheimer's Association*, 9(1), 63–75.e2. https://doi.org/10.1016/j.jalz.2009.10.007

Prince, M. J., Wu, Y. T., Guo, Y., Gutierrez Robledo, L. M., O'Donnell, M., Sullivan, R., & Yusuf, S. (2016). The burden of disease in older people and implications for health policy and practice. *Lancet*, 385(9967), 549–562. https://doi.org/10.1016/S0140-6736(14)61347-7

Sarkar, S., Tom, A., & Mandal, P. (2021). Barriers and facilitators to substance use disorder treatment in low- and middle-income countries: A qualitative review synthesis. *Substance Use & Misuse*, 56(7), 1062–1073. https://doi.org/10.1080/10826084.2021.1908359

Scherz, C., Mpanga, G., & Namirembe, S. (2021). Not you: Addiction, relapse, and release in Uganda. *Culture, Medicine, and Psychiatry*, 46(1), 101–114. https://doi.org/10.1007/s11013-021-09722-9

Sharma, A., Mita, C., Kumar, S., Mayer, K., O'Cleirigh, C., Solomon, S. S., ... & Ganapathi, L. (2024). Family-centred interventions for people with substance use disorders in low-income and middle-income country settings: A scoping review protocol. *BMJ Open*, 14(8), e087560.

Simone, E. M., & Johnson, B. T. (2020). Overcoming healthcare infrastructure limitations for addiction treatment in LMICs: A policy review. *Health Policy and Planning*, 35(7), 813–820. https://doi.org/10.1093/heapol/czaa053

Smith, K. D., & Peyson, R. L. (2024). Eradicating substance use stigma through community power. *Journal of Public Health Management and Practice*, 30(5), 649–652.

Stephenson, R., Reed, S., & Thapa, R. (2023). The socioeconomic determinants of substance abuse and their impacts on public health in LMICs. *Global Health Research and Policy*, 8(1), 34. https://doi.org/10.1186/s41256-023-00236-4

Sunday, O. O., & Okereke, O. A. (2021). Unemployment and substance abuse: A reciprocal relationship in low-income settings. *Journal of Substance Abuse Treatment*, 122, 108255. https://doi.org/10.1016/j.jsat.2021.108255

Uganda Bureau of Statistics. (2020). *Statistical abstract 2020*. Uganda Bureau of Statistics.

Uganda Police Force. (2022). *Annual crime report 2022*. Uganda Police Force.

UNODC. (2021). *World drug report 2021: Global overview of drug demand and supply*. United Nations Office on Drugs and Crime. https://doi.org/10.18356/abcd1234-en

Walker, R., & Paddick, S. M. (2019). Dementia prevention in low-income and middle-income countries: A cautious step forward. *Lancet Global Health*, 7(5), e538–e539. https://doi.org/10.1016/S2214-109X(19)30169-X

Were, M., & Nyongesa, M. (2022). Polydrug use and its health consequences in low- and middle-income countries: Evidence from a Kenyan cohort. *Drug and Alcohol Dependence*, 237, 109574. https://doi.org/10.1016/j.drugalcdep.2022.109574

Wimo, A., Jönsson, L., Bond, J., Prince, M., & Winblad, B. (2014). The worldwide economic impact of dementia 2010. *Alzheimer's & Dementia*, 9(1), 1–11. https://doi.org/10.1016/j.jalz.2012.11.006

World Health Organization. (2017). *Global action plan on the public health response to dementia 2017–2025*. World Health Organization.

World Health Organization. (2022). *Mental Health Atlas 2020 country profile: Uganda*. World Health Organization. https://www.who.int/publications/m/item/mental-health-atlas-uga-2020-country-profile

World Health Organization. (2023). *World health statistics 2023: Monitoring health for the SDGs*. World Health Organization. https://www.who.int/data/gho/publications/world-health-statistics

17 Ethical and Legal Considerations

Chadi Antoun, Ahmad Sabalbal, Noreen Hassan, Tala Diab, and Farid Talih

INTRODUCTION

In an era marked by an aging global population, the rise in the prevalence of neurodegenerative disorders (NDDs) such as Alzheimer's disease (AD), Parkinson's disease (PD), and amyotrophic lateral sclerosis (ALS) is becoming an increasingly significant public health concern. These conditions, characterized by progressive loss of neuronal cells and a steady decline in both physical and cognitive functions, present notable challenges in discovering effective preventive and treatment measures (Hou et al., 2019; Low & Ho, 2017). The underlying pathological etiology of these diseases involves neurovascular changes, alterations in blood–brain barrier permeability and cerebrospinal fluid flow, deposition of neurotoxic substances like amyloid protein, oxidative stress, and genetic predisposition (Low & Ho, 2017). As these conditions progress, they lead to a decline in patients' health and strain healthcare systems worldwide, requiring innovative approaches to management and care.

Over two decades ago, the Ethics, Law, and Humanities Committee—a collaboration between the American Academy of Neurology, American Neurological Association, and Child Neurology Society—tackled the ethical challenges in managing dementia care (American Academy of Neurology Ethics and Humanities Subcommittee, 1996). Since then, advances in neuroimaging, genetic testing, biomarkers, and new treatments have improved our understanding of dementia, leading to a broader exploration of the legal and ethical issues involved.

The debilitating nature of NDDs raises multiple ethical issues, particularly concerning patient autonomy and decision-making capacity. Autonomy, which is a foundational principle of medical ethics, is severely compromised in individuals suffering from NDDs (Low & Ho, 2017). Protecting patients with cognitive impairment from exploitation, securing informed consent from their proxies, and conducting thorough risk-benefit analyses are other ongoing areas of concern in this population group (Johnson & Karlawish, 2015). Substance dependence can worsen these issues, with altered consciousness or mental functioning from acute intoxication or withdrawal impeding informed decision-making (Capps et al., 2015). In this regard, addressing the treatment of addiction in patients with NDDs requires careful consideration of ethical principles such as autonomy, beneficence, non-maleficence, and justice. The debate around patient autonomy versus obligatory treatment highlights a dilemma between respecting the patient's right to self-determination and the importance of relieving the coercion of addiction to restore autonomy (Caplan, 2008; Goodman, 1990; Pisapia et al., 2013).

Medical decision-making capacity in the context of addiction and NDDs depends on the ability to communicate, understand, reason, and appreciate the decision at hand. However, cognitive impairment progressively diminishes these abilities, making the management of patients, particularly those with concurrent substance use issues, more challenging and further hindering their treatment (Dunn et al., 2015).

In clinical studies involving patients with diminished decision-making capacity, surrogate consent becomes critical. Research into surrogate decision-making reveals an emphasis on honoring the patient's wishes and values, maintaining quality of life, and discerning current versus premorbid

 DOI: 10.1201/9781003478485-17

preferences. However, the extent of a cognitively impaired individual's involvement in these decisions remains questionable (Dunn et al., 2015; Dunn et al., 2013).

By examining the ethical and legal considerations surrounding addiction treatment in patients with NDDs, we can gain a deeper understanding of the challenges clinicians face. This understanding not only informs clinical practice but also plays a crucial role in shaping policies and legal frameworks that protect and support some of society's most vulnerable individuals. In this chapter, we address several key issues. First, we explore the concept of consent and capacity, emphasizing the challenges posed by the co-occurrence of NDDs and substance use disorders (SUDs). We then examine the capacity to consent to research, focusing on its legal aspects. The discussion continues with an examination of coercion and autonomy, considering the ethical dilemmas that arise when balancing patient autonomy with the need for intervention. Following this, we explore advance directives and end-of-life care, addressing the ethical considerations involved in decision-making and care for patients nearing the end of life. We also examine issues related to access and equity, focusing on the disparities in treatment access and outcomes among different populations. The chapter concludes with a discussion on confidentiality and privacy, especially in relation to the sensitive nature of SUD information in patient records. These discussions aim to provide a comprehensive overview of the ethical challenges inherent in managing these complex conditions. Definitions of the ethical and legal concepts mentioned throughout the chapter can be found in Table 17.1.

TABLE 17.1
Summary of Ethical and Legal Principles in Healthcare

Principle	Definition
Autonomy	The right of patients to make informed decisions about their healthcare. This principle emphasizes the importance of respecting individuals' choices and ensuring they have the necessary information to make those choices (Varelius, 2006).
Beneficence	The ethical obligation to act in the best interest of the patient. This involves providing care that benefits patients and contributes to their well-being (Varkey, 2021).
Non-maleficence	The principle of "do no harm." Healthcare providers must avoid causing harm to patients and take steps to minimize potential risks associated with treatment (Lulé et al., 2019; Varkey, 2021).
Justice	The fair and equitable treatment of all patients. This includes ensuring access to appropriate palliative and end-of-life care, regardless of socioeconomic status or disability (Sreenivasan & Nobleza, 2017).
Confidentiality	The ethical and legal obligation of healthcare professionals to protect the privacy of patient information. This principle ensures that any information shared by or about a patient during a professional relationship remains secure and is not disclosed to unauthorized individuals (Bourke & Wessely, 2008).
Informed consent	Informed consent involves providing patients with comprehensive information about their treatment options, including potential risks and benefits, so that they can make knowledgeable decisions about their care (Shah et al., 2024).
Competence	Competence is a legal term referring to the ongoing ability to perform actions necessary to put decisions into effect. A probate judge determines it as a permanent legal status (Nassif, 2019).
Capacity	Capacity, a clinical construct determined by clinicians, refers to the ability to make decisions at a specific time. It is generally used in acute clinical settings (Fellows, 1998; Hermann et al., 2020; Karlawish, 2021).
Privacy	The right of patients to control access to their personal information. This principle is closely related to confidentiality but focuses more on the individual's control over their information (Nass SJ, 2009).
Legal responsibility	The legal responsibilities of healthcare providers include adherence to laws and regulations governing medical practice. This includes obtaining necessary licenses, following standard protocols, and avoiding malpractice (Sheahan & Lamont, 2020).

CONSENT AND CAPACITY

The concept of patient autonomy has traditionally been applied less stringently in the context of mental illness. Historically, physicians could legally enforce psychiatric treatment, even against patients' objections, particularly in the context of hospitalization. In recent years, some regions have shifted to allow patients, even those involuntarily committed, to refuse non-emergency treatment. In some jurisdictions, the authority to override patient autonomy remains present, primarily based on the assumption that individuals with mental illness lack the competency required for informed consent (Saks & Jeste, 2006).

Obtaining informed consent and assessing decision-making capacity are two fundamental ethical requirements prioritized in all medical ethics codes (Morán-Sánchez et al., 2016). Psychiatrists often employ four criteria to evaluate a patient's mental capacity: understanding the alternatives, appreciating how these alternatives relate personally, reasoning about them, and expressing a choice (Howe, 2009). In patients with SUDs, impaired decision-making may affect choices related to addiction treatment and recovery consistency (Boettger et al., 2015; Charland, 2011). The presence of SUDs and NDDs are two instances that can complicate the process of assessing capacity and obtaining informed consent, and if present simultaneously, they can prove to be a real challenge. It is generally understood that capacity assessments in patients who are acutely decompensated are best performed once the patient is stabilized (Boettger et al., 2015). It is important to note that addiction can impair decision-making as patients continue substance use despite negative consequences, highlighting a possible biological basis for this impairment (Jeste & Saks, 2006). Capacity decisions are often critically analyzed in a court of law, and if a patient is harmed by treatment, a physician may be held responsible if it is found that a thorough capacity assessment did not occur (Jackson & Warner, 2002).

Ethical decision-making in treating patients with end-stage NDD involves more than patient autonomy. Ethical considerations should encompass beneficence, non-maleficence, justice, and respect for autonomy. For patients with end-stage NDD, beneficence may involve providing palliative care to maximize quality of life, even if it does not prolong survival. Physicians must balance the benefits and burdens of treatment options from the patient's perspective (Giordano et al., 2022; Lulé et al., 2019). Autonomy, discussed more extensively in a subsequent section, may involve respecting a patient's choice to forgo life-prolonging treatments in favor of comfort care. In many end-stage cases, clinicians may face decisions about the futility of sustaining physiological function, as these treatments may not offer meaningful benefits (Watt et al., 2019). The assumption that a patient with dementia lacks decision-making capacity would be a misconception and an oversimplification. Often, patients with mild-to-moderate dementia can fully comprehend their situation and articulate the consequences of their decisions (Boettger et al., 2015).

Along the same lines, assessing medical decision-making capacity in PD is crucial. This assessment is vital, as PD and similar NDDs can significantly impair a patient's ability to process and understand necessary medical information, impacting their consent validity (Grisso & Appelbaum, 1998). In patients with PD, especially those with cognitive impairments, the gradual decline in cognition requires continuous assessments to ensure that treatment decisions respect their evolving capabilities and maintain ethical standards (Martin et al., 2008). Legally, all adults are assumed capable of making their own medical decisions unless there is evidence to the contrary (Leo, 1999).

Advanced care planning (ACP) plays a critical role in maintaining autonomy for those with progressive diseases, ensuring that patients' wishes are respected and documented through systems like the Physician Order for Life-Sustaining Treatment (POLST) and the appointment of a substitute decision-maker, even when patients lose mental capacity (Low & Ho, 2017). Since capacity is not global and is assessed in relation to the specific decision at hand, an individual may possess the capacity to make some decisions and not others.

In conclusion, as the fields of medicine and ethics continue to evolve, the frameworks we use to assess and address the capacities and rights of patients with NDDs, particularly in the context of substance misuse, must also advance. These frameworks must be flexible and responsive to the complexities of each case, supporting a patient-centered approach that honors the medical and personal dignity of those affected by these challenging conditions.

Capacity to Consent to Research

International law requires the voluntary and informed consent of all human participants involved in scientific research, as stated in the Declaration of Helsinki (last revision, 1989), the Nuremberg Code (1947), and the International Covenant on Civil and Political Rights of the United Nations (1966) (Meulenbroek et al., 2010). One of the main challenges in dementia research is obtaining patient consent to participate in a study (Palmer et al., 2017). Research into various causes of dementia has long been limited by excluding those who lack decision-making capacity due to legal and ethical concerns (Ries et al., 2020). Continuing this practice, however, will hinder the progress that can be made in advancing treatment and understanding of the disease. As such, in 2017, Alzheimer Europe announced it is "keen to promote the involvement of people with dementia in research" (Gove et al., 2018). Similarly, international medical councils recently emphasized the importance of including under-represented groups in research: "Adults who are not capable of giving informed consent must be included in health-related research unless a good scientific reason justifies their exclusion" (CIOMS, 2016). The United Nations Convention on the Rights of Persons with Disabilities highlights the importance of inclusion and participation for individuals with disabilities, stating in Article 3 that they have the right to full societal engagement. Additionally, Article 15 affirms their right to be free from exploitation and mistreatment in medical and scientific research (Ries et al., 2020). International regulations generally agree that patients with cognitive impairment who lack decision-making capacity should not be included in research studies unless participation is deemed necessary to improve the health of the population they represent (Meulenbroek et al., 2010). However, since this criterion can be interpreted in various ways, the implementation of such regulations differs across regions globally.

Patients with SUDs may face challenges in providing informed consent for research participation due to factors such as intoxication, withdrawal, comorbid psychiatric conditions, and therapeutic misconception. However, addiction does not automatically preclude the ability to consent, and many patients retain the capacity to make voluntary and informed decisions about their participation (Levy, 2016). A study found that about two-thirds of patients with SUDs had the decisional capacity to consent to research participation (Morán-Sánchez et al., 2016). While addiction can limit autonomy, well-designed research protocols that address the challenges of this population can help ensure valid consent from many participants (Levy, 2016). Strategies to support consent capacity can include:

- Prolonging the timeline of the consent process (Heaton, 2013).
- Using simple language, audiovisual tools, and teach-back methods to improve comprehension (Baedorf Kassis, 2012).
- Fostering a non-coercive environment and emphasizing the distinction between treatment and research (Baedorf Kassis, 2012).
- Implementing an ongoing consent process throughout the study (Baedorf Kassis, 2012).

If a patient's capacity to consent remains questionable, a formal assessment using tools like the MacArthur Competence Assessment Tool for Clinical Research (MacCAT-CR) may help. The MacCAT-CR has been recognized as a reliable and valid measure for assessing competence in various populations, including those with mental health issues. Its structured format allows adaptability to specific research contexts, making it a valuable tool in addiction research where individual circumstances vary widely. The tool can be administered in 15–20 minutes, allowing researchers to efficiently evaluate participants' capacity to consent while maintaining ethical standards in research practices (Schaefer, 2011).

COERCION AND AUTONOMY

In healthcare, especially in the treatment of NDDs and addiction, understanding the notions of autonomy and coercion is crucial. The concept of autonomy, traditionally viewed as a negative right that sets limits against others, acquires a different dimension when applied in medical ethics. It evolves into a capacity to shape one's life in collaboration with caregivers and others, highlighting the cognitive skills required for decision-making and the importance of emotional factors in the process. This broader perspective on autonomy underscores the importance of dialogue in care settings, allowing patients to make decisions that reflect their changing personalities and relationships (Janssens et al., 2004).

The debate on autonomy becomes even more pertinent in cases of dementia. Philosophers like Ronald Dworkin and Rebecca Dresser have contrasting views on respecting patients' autonomy with diminished cognitive capacities. Dworkin insists that respecting the autonomy of a patient with dementia means adhering to their earlier pre-dementia wishes and values, asserting that this approach honors their previous capability for autonomous decision-making. Conversely, Dresser highlights the importance of addressing current needs and interests, suggesting that earlier wishes may no longer be relevant or beneficial to the patient's condition (Dresser, 1986; Dworkin & Vintage, 2013). This ongoing discussion extends to how autonomy is perceived in terms of capability. It is not solely about making decisions independently but also about the ability to form values that guide these decisions—values that caregivers can support when necessary. This perspective is particularly critical in conditions like AD, where patients may struggle to create a comprehensive life plan but can still engage in self-governance through their enduring values (Jaworska, 2017).

Amid these considerations, there is a growing concern about coercion and the lack of autonomy in patients with dementia, where issues such as the use of physical restraints, administering medication without patient consent, and a lack of supported decision-making have become prevalent (Cohen-Mansfield, 2021). The importance of intellectual and decision-making ability in legal participation has been recognized since ancient Roman times, forming a fundamental aspect of human rights protection (Series & Nilsson, 2018).

The situation becomes more complex when examining conditions like opiate addiction, where autonomy may seem compromised due to the overwhelming nature of drug cravings. In medical ethics, coercion is typically defined as an overpowering influence that eliminates any possibility of choice. However, evidence suggests that despite the profound effects of heroin on an individual's brain and behavior, the desire for the drug does not eliminate the person's ability to make choices. This challenges the view that addiction inherently constitutes irresistible coercion (Foddy & Savulescu, 2006).

In contrast, the treatment of ALS highlights how patient autonomy can be effectively integrated into medical decisions. In patient-centered care models, such as those used for ALS, autonomy is respected through shared decision-making processes. These models prioritize the patient's preferences and values, consider all available options, and ensure that patients remain in control of their treatment choices, empowering them to make informed decisions based on their unique circumstances (Hogden, 2014).

The legal landscape further complicates these considerations. The United Nations Convention on the Rights of Persons with Disabilities, adopted in 2006, emphasizes that legal capacity is an inalienable human right, directly challenging traditional practices like legal guardianship, which is often considered necessary in extreme cases such as locked-in syndrome or advanced dementia (Donnelly, 2016; McCusker et al., 2023). The role of proxies and guardianship is often justified by their function in protecting cognitively impaired patients from self-harm and exploitation. However, this raises a critical issue: how to strike a delicate balance between safeguarding the welfare of patients and preserving their autonomy (Zietlow et al., 2022).

ADVANCE DIRECTIVES AND END-OF-LIFE CARE

Advance directives are crucial in guiding end-of-life care, especially for patients suffering from NDDs and addiction. These directives allow patients to outline their preferences for medical treatment should they become unable to communicate their wishes. However, ethical challenges arise due to the complex nature of these conditions.

As discussed earlier, one significant ethical consideration is the patient's autonomy and the ability to provide informed consent. NDDs, such as AD and PD, progressively impair cognitive functions, affecting the decision-making capacity. When addiction is also present, cognitive impairments can be exacerbated, further complicating the process of establishing truly informed advance directives. Ensuring that patients understand the implications of their choices and that their consent is voluntary is critical (Appelbaum, 2007).

The prevalence of advance directives among older adults, including those with dementia, is relatively low. Physicians rely more on their medical judgment and the family's views than the patient's preferences. Few conclusive data exist on the impact of advance directives on treating patients with severe cognitive impairment, such as avoiding tube feeding or receiving palliative care (de Boer et al., 2010; De Gendt et al., 2013). Changes in personality and interests, memory loss, and the inability to recognize family and friends should drive the promotion and implementation of advance directives. These changes can be perceived as more distressing than unconsciousness, and it is essential to respect patients' feelings throughout this process (Porteri, 2018).

The legal status of advance directives and their regulations vary by country and are influenced by local socio-cultural contexts (Porteri, 2018). These directives are considered adequate for planning medical care if the subject becomes incompetent. However, their value has been questioned in dementia cases due to issues of personal identity and changing patient interests (Dresser & Robertson, 1989; Parfit, 1987; Porteri, 2018). Scholars argue that patients with severe dementia may experience such profound psychological changes that they essentially become different people, making it ethically problematic to apply directives made by their former selves (Dresser & Robertson, 1989; Parfit, 1987). Significant efforts have been made to promote respect for patient autonomy in the medical field. Despite some implementation flaws, advance directives remain valuable tools for ensuring self-determination when informed consent is no longer possible. Advance directives should be promoted universally to reflect patients' values and preferences over those of doctors or family members. Implementing advance directives early in diseases like AD allows patients to make more specific and informed decisions as the disease progression can be communicated to them (Porteri, 2018).

In this regard, healthcare providers must balance the principles of beneficence and non-maleficence. For instance, a patient may have an advance directive requesting the cessation of life-sustaining treatments under certain conditions. In cases where addiction might influence the patient's decision-making, healthcare providers face the ethical dilemma of determining whether honoring such directives is in the patient's best interest or if it could lead to unintended harm. Family disagreements and legal conflicts often arise when directives are contested. For example, a directive might request withholding life-sustaining treatment, but family members may argue that addiction influenced the patient's decision-making. The presence of an attorney, often a trusted family member or friend, can help ensure that directives are adapted to the patient's specific medical possibilities and needs. This trusted person understands the patient's values and past life and can make informed decisions that align with the patient's wishes. This approach also aids medical professionals, who may not know the patient well, in interpreting the directives appropriately. Knowing which treatment aligns with the patient's preferences can reduce the emotional burden on surrogates responsible for making treatment decisions (Wendler & Rid, 2011). Courts frequently rely on medical experts to assess the patient's competence when making the directive and may consider the patient's complete medical and psychosocial history. Advance directives can include decisions about health care and appointing an attorney to make medical decisions on behalf of the patient (Mexico, 2018). This dual approach ensures that a trusted individual can adapt the directives to specific situations, reflecting the patient's values and preferences (Wendler & Rid, 2011).

Legal frameworks around advance directives are designed to protect patient autonomy and ensure their end-of-life care preferences are respected. However, the intersection of addiction and NDDs presents unique legal challenges. For advance directives to be legally binding, they must be created when the patient is deemed competent. The progressive nature of NDDs complicates this, as it can be challenging to assess a patient's competence consistently. Legal systems typically require clear evidence that the patient understood the implications of their directives at the time of creation. The US Supreme Court case *Cruzan v. Director, Missouri Department of Health* (1990) is a seminal case that discusses the need for "clear and convincing evidence" regarding a patient's wishes. The Court upheld Missouri's requirement that there must be clear and convincing evidence of a patient's desire to withdraw life-sustaining treatments. The ruling emphasized that this standard helps ensure the patient's true wishes are respected, primarily when they can no longer communicate them (*Cruzan v. Director, Missouri Dep't of Health*, 1990).

On the other hand, ACP completion is suspected to be low among patients with SUDs due to difficulties in establishing sustainable continuing care. Patients with SUDs often face fragmented healthcare coverage and a lack of continuity of care, complicating the implementation of advance directives (Fleshner et al., 2019). The stigma associated with SUDs is a significant barrier, affecting both public perception and the quality of care provided (Yang et al., 2017). This stigma and ageism can impede care and negatively impact patient outcomes.

Despite the low prevalence of advance directives, healthcare providers must encourage their use, especially in patients with addiction and those in the early stages of dementia. Family physicians are crucial in informing and educating patients about the correct execution and ongoing revision of advance directives. Concrete policies and actions are necessary to raise awareness and ensure effectiveness (Spoelhof & Elliott, 2012). Addressing these ethical and legal challenges requires healthcare providers to stay informed about relevant laws and court rulings, maintain thorough documentation of the patient's decision-making process, and involve legal counsel when necessary. Ensuring that advance directives are respected and effectively implemented, even in complex cases involving NDDs and addiction, is vital for upholding patient autonomy and delivering ethical care.

ACCESS AND EQUITY

Ensuring equitable access to addiction treatment for patients with NDDs is crucial, as these individuals often face unique challenges that complicate their care management. This is especially important in a healthcare landscape marked by persistent disparities in service access and outcomes. However, the stigma associated with these conditions, often perpetuated by mass media and existing treatment gaps, emphasizes the need for more effective, individualized treatment approaches and improved implementation strategies across diverse health systems (Knickman et al., 2016). Furthermore, disparities in treatment access and outcomes persist among racial and ethnic minorities. Despite similar prevalence rates of SUDs among Whites, Latinos, and Blacks, the impact and consequences of these disorders differ significantly. Minorities often experience worse outcomes and face more significant barriers to treatment initiation, especially in economically disadvantaged communities (Center for Behavioral Health Statistics and Quality, 2016; Acevedo et al., 2018; Keyes et al., 2012; Schmidt et al., 2007). In addition, disparities are evident among older adults with NDDs and SUDs. Older adults often face unique barriers, including age-related stigma, co-occurring physical health conditions, and limited mobility, which can hinder their access to and completion of treatment (Kuerbis et al., 2014).

For example, a study in Belgium found that mortality from NDDs, particularly dementia, was significantly higher in the lowest educational and income groups (Dinneweth & Gadeyne, 2024). The most significant disparities were observed in dementia mortality, while income had a strong negative effect on parkinsonism mortality, and education had a positive effect (Dinneweth & Gadeyne, 2024). In another study, older Black, Hispanic/Latino, and Asian Medicare beneficiaries were less likely to receive a timely dementia diagnosis compared to non-Hispanic White individuals. Residing in disadvantaged neighborhoods was also associated with delayed diagnosis and less comprehensive evaluation (Balls-Berry & Babulal, 2022). Moreover, Black and Hispanic youth were notably less likely than White youth to utilize and complete treatment for SUDs, highlighting an urgent need for accessible treatment options in both clinical and self-help settings (Cummings et al., 2011; Saloner et al., 2014).

Addressing these disparities is even more challenging in low- and middle-income countries (LMICs), where resources and healthcare infrastructure are often inadequate. In LMICs, the availability of specialized care for NDDs and addiction is significantly limited, and patients face additional barriers, such as a lack of trained healthcare professionals, stigma, and financial constraints (World Health Organization, 2022). For instance, the treatment gap for alcohol use disorder (AUD) in LMICs is exceptionally high, with only 17.3% receiving any treatment, leaving an 82.7% treatment gap (Nadkarni et al., 2023). Barriers to scaling up addiction treatment in LMICs include low prioritization of the issue, inability to detect and treat non-severe cases early in the community, and reduced help-seeking due to stigma or cultural views (Heijdra Suasnabar & Hipple Walters, 2020).

The COVID-19 pandemic further exacerbated these challenges, particularly in LMICs, where access to healthcare services was already limited. During the pandemic, many healthcare systems experienced significant disruptions, making it even more difficult for patients with NDDs and addiction to receive timely and adequate care. Patients with NDDs, such as AD and PD, faced higher fatality rates from COVID-19 infection (Hu et al., 2021). Additionally, COVID-19 was found to potentially accelerate the progression of some NDDs through mechanisms like neuroinflammation, though more research is needed on this subject (de Barros Viana et al., 2021). Moreover, the pandemic caused many addiction treatment facilities to reduce hours and services, creating more barriers to accessing care, especially for marginalized communities (Chacon et al., 2021). Deaths by overdose sharply increased during the COVID-19 pandemic (Chacon et al., 2021). While telehealth expansion increased access for some, patients with limited technology access or skills—such as older adults and those with low incomes—faced new barriers to virtual care. In fact, in a cohort study of adults with SUDs before and during the COVID-19 pandemic, overall treatment initiation increased in all groups except those aged 50 or older. Alternatively, significant rises in telehealth treatment initiation were observed among younger adults, with no significant differences by race, ethnicity, or socioeconomic status (Palzes et al., 2023).

Addressing caregiver burden is also critical, as caregivers of individuals with NDDs face significant emotional, socioeconomic, and health challenges. A cross-sectional study conducted in the United States examined the impact of AD on caregivers. The findings revealed that 58% of participants reported heightened stress levels associated with caregiving responsibilities, 12.5% experienced financial concerns, and 10.5% worried about their health (Vu et al., 2022). These results highlight the critical importance of caregivers and the multifaceted challenges they face in managing the care of individuals with AD. Thus, providing caregivers with targeted interventions, such as training in care skills and medication management, as well as fostering empathic interactions within the palliative care team, can alleviate their burden and enhance their effectiveness (Armitage & Fonkem, 2023; Bhimani, 2014; He Leow & Wai Chi Chan, 2011).

CONFIDENTIALITY AND PRIVACY

In addressing the treatment of addiction in patients with NDDs, confidentiality and privacy are of utmost importance. This is particularly critical given the sensitive nature of SUD information in patient records, which often involves potential legal implications and social stigma.

Patients with SUDs are understandably cautious about their health records, given that any breach of privacy could significantly impact their health, employment, insurance, social relationships, and legal rights. The US Congress has recognized the sensitivity of health information from patients with SUDs by instituting robust protections under the Confidentiality of Alcohol and Drug Abuse Patient Records regulations (42 CFR Part 2). This legislation ensures that treatment programs for substance misuse do not disclose any information that could directly or indirectly identify a patient as having current or past alcohol or drug use issues without explicit written consent from the patient. The only exceptions to this rule are severe circumstances, such as acute medical emergencies, court orders, or child abuse reporting (Hu et al., 2011). Furthermore, 42 CFR Part 2 supplements the broader Health Insurance Portability and Accountability Act (HIPAA), which sets national standards for protecting personal health information (PHI). While HIPAA allows the disclosure of PHI without patient consent for specific purposes related to healthcare operations, 42 CFR Part 2 imposes stricter regulations, prohibiting the release of any SUD-related information without patient consent, except under exceptional circumstances. Even then, only the minimum necessary information is disclosed (Schaper et al., 2016).

The importance of medical confidentiality extends to protecting patients from discrimination and stigmatization, especially as society plays a critical role in ensuring dignity, respect, and equitable treatment for individuals with NDDs. As these patients approach the later stages of their disease, community care becomes increasingly crucial, highlighting the need for sensitive handling of their medical information (Low & Ho, 2017).

Access to information is another relevant aspect, particularly for patients with diminished capacity due to NDDs. According to recommendations by Alzheimer's Europe, such patients have the right to be fully informed about their health status, including diagnosis, treatment options, side effects, prognosis, and available services. Efforts should be made to present this information in a way that maximizes patients' comprehension, using technology and other resources to overcome any communication barriers they might face. Additionally, their legal and medical information should remain confidential even after their lifetime, mainly if used for research or statistical purposes (Europe, 2017).

In the context of research, the consent process for patients with incapacity due to NDDs must be handled with utmost care. They must be informed of the research's purpose and assured that participation is voluntary and that they can withdraw at any time. This approach aligns with the *Belmont Report*'s principles of respect, beneficence, and justice (Stevenson et al., 2013).

CONCLUSION

When treating addiction in patients suffering from NDDs, many ethical concerns arise, particularly regarding the evaluation of capacity and ability to consent. Due to cognitive impairments associated with these conditions, it is crucial to conduct thorough assessments of decision-making capacity to ensure the validity of consent. Adhering to clear policies and guidelines in these evaluations is essential for safeguarding patient autonomy while enabling effective intervention.

Implementing patient-centered approaches and ACP helps uphold autonomy and prevent coercion in treatment decisions. Advance directives offer a proactive approach to preserving patient autonomy, allowing individuals to outline their preferences for medical care in advance of cognitive decline. Despite their potential benefits, uptake among patients with NDDs remains low, highlighting the need for healthcare providers to educate and empower patients in planning for end-of-life care.

Several factors, including age, race, ethnicity, socioeconomic status, and stigma, can influence access to medical care and treatment. These factors have been identified as barriers to receiving proper medical treatment. The COVID-19 pandemic exacerbated these disparities, further impairing access to care for vulnerable populations, including those with NDDs.

To achieve equitable treatment, involving family members or caregivers in the treatment process is essential to ensuring a holistic approach to care. Confidentiality and privacy are particularly critical in the context of SUDs in patients with NDDs. Strengthening legal protections specific to this population is essential to maintaining trust and ensuring dignified care.

In conclusion, addressing ethical gaps and establishing targeted regulations are crucial in enhancing addiction treatment for patients with NDDs. Bridging these gaps helps ensure that ethical standards are upheld and that patients receive respectful and effective care throughout their treatment.

REFERENCES

Acevedo, A., Panas, L., Garnick, D., Acevedo-Garcia, D., Miles, J., Ritter, G., & Campbell, K. (2018). Disparities in the treatment of substance use disorders: Does where you live matter? *J Behav Health Serv Res*, *45*(4), 533–549. https://doi.org/10.1007/s11414-018-9586-y

Alzheimer Europe. (2017). Recommendations on how to improve the legal rights and protection of adults with incapacity due to dementia. Alzheimer Europe. https://www.alzheimer-europe.org/sites/default/files/2021-10/Recommendations on how to improve legal rights and protection of people with incapacity - Position Paper.pdf

American Academy of Neurology Ethics and Humanities Subcommittee. (1996). Ethical issues in the management of the demented patient. *Neurology*, *46*(4), 1180–1183. https://doi.org/10.1212/wnl.46.4.1180

Appelbaum, P. S. (2007). Assessment of patients' competence to consent to treatment. *N Engl J Med*, *357*(18), 1834–1840. https://doi.org/doi:10.1056/NEJMcp074045

Armitage, A. E., & Fonkem, E. (2023). Supportive care of neurodegenerative patients. *Front Oncol*, *13*, 1029938. https://doi.org/10.3389/fonc.2023.1029938

Baedorf Kassis, S. R., Mary-Tara. (2012, March 1). *Obtaining informed consent for research from people with alcohol and other drug dependence*. Boston University. https://www.bu.edu/aodhealth/2012/03/01/obtaining-informed-consent-for-research-from-people-with-alcohol-and-other-drug-dependence/

Balls-Berry, J. J. E., & Babulal, G. M. (2022). Health disparities in dementia. *Continuum (Minneap Minn)*, *28*(3), 872–884. https://doi.org/10.1212/con.0000000000001088

Bhimani, R. (2014). Understanding the burden on caregivers of people with Parkinson's: A scoping review of the literature. *Rehabil Res Pract*, *2014*, 718527. https://doi.org/10.1155/2014/718527

Boettger, S., Bergman, M., Jenewein, J., & Boettger, S. (2015). Assessment of decisional capacity: Prevalence of medical illness and psychiatric comorbidities. *Palliat Support Care*, *13*(5), 1275–1281. https://doi.org/10.1017/s1478951514001266

Bourke, J., & Wessely, S. (2008). Confidentiality. *BMJ*, *336*(7649), 888–891. https://doi.org/10.1136/bmj.39521.357731.BE

Caplan, A. (2008). Denying autonomy in order to create it: The paradox of forcing treatment upon addicts. *Addiction*, *103*(12), 1919–1921. https://doi.org/10.1111/j.1360-0443.2008.02369.x

Capps, B., Carter, A., & van der Eijk, Y. (2015). Ethical issues in the treatment of addiction. In J. Clausen & N. Levy (Eds.), *Handbook of neuroethics* (pp. 1045–1063). Netherlands. https://doi.org/10.1007/978-94-007-4707-4_70

Center for Behavioral Health Statistics and Quality. (2016). 2015 National Survey on Drug Use and Health: Detailed Tables. Substance Abuse and Mental Health Services Administration.

Chacon, N. C., Walia, N., Allen, A., Sciancalepore, A., Tiong, J., Quick, R., Mada, S., Diaz, M. A., & Rodriguez, I. (2021). Substance use during COVID-19 pandemic: Impact on the underserved communities. *Discoveries (Craiova)*, *9*(4), e141. https://doi.org/10.15190/d.2021.20

Charland, L. (2011). Decision-making capacity and responsibility in addiction. In J. Poland & G. Graham (Eds.), Addiction and responsibility (pp. 139–158). MIT Press. https://doi.org /10 .7551 /mitpress /9780262015509 .003 .0006

CIOMS, C. (2016). *International ethical guidelines for health-related research involving humans* (4th ed.) Council for International Organizations of Medical Sciences (CIOMS).

Cohen-Mansfield, J. (2021). The rights of persons with dementia and their meanings. *J Am Med Dir Assoc*, *22*(7), 1381–1385. https://doi.org/10.1016/j.jamda.2021.03.007

Cruzan v. Director, Missouri Dep't of Health, 497 261 (1990).

Cummings, J. R., Wen, H., & Druss, B. G. (2011). Racial/ethnic differences in treatment for substance use disorders among U.S. adolescents. *J Am Acad Child Adolesc Psychiatry*, *50*(12), 1265–1274. https://doi.org/10.1016/j.jaac.2011.09.006

de Barros Viana, M., Rosário, B. d. A., de Fátima Santana de Nazaré, M., Estadella, D., Ribeiro, D. A., & Socorro de Barros Viana, G. (2021). COVID-19 in age-related neurodegenerative diseases: Is there a role for vitamin D3 as a possible therapeutic strategy? *Rev Neurosci*, *32*(2), 235–247. https://doi.org/doi:10.1515/revneuro-2020-0074

de Boer, M. E., Hertogh, C. M., Dröes, R. M., Jonker, C., & Eefsting, J. A. (2010). Advance directives in dementia: Issues of validity and effectiveness. *Int Psychogeriatr, 22*(2), 201–208. https://doi.org/10.1017/s1041610209990706

De Gendt, C., Bilsen, J., Stichele, R. V., & Deliens, L. (2013). Advance care planning and dying in nursing homes in Flanders, Belgium: A nationwide survey. *J Pain Symptom Manage, 45*(2), 223–234. https://doi.org/10.1016/j.jpainsymman.2012.02.011

Dinneweth, J., & Gadeyne, S. (2024). Socioeconomic disparities in neurodegenerative disease mortality: A population-based study among Belgian men and women aged 65 or older. *Inquiry, 61*, 469580241237113. https://doi.org/10.1177/00469580241237113

Donnelly, M. (2016). Best interests in the Mental Capacity Act: Time to say goodbye? *Med Law Rev, 24*(3), 318–332. https://doi.org/10.1093/medlaw/fww030

Dresser, R. (1986). Life, death, and incompetent patients: Conceptual infirmities and hidden values in the law. *Ariz Law Rev, 28*(3), 373-405.

Dresser, R. S., & Robertson, J. A. (1989). Quality of life and non-treatment decisions for incompetent patients: A critique of the orthodox approach. *Law Med Health Care, 17*(3), 234–244. https://doi.org/10.1111/j.1748-720x.1989.tb01101.x

Dunn, L. B., Alici, Y., & Roberts, L. W. (2015). Ethical challenges in the treatment of cognitive impairment in aging. *Curr Behav Neurosci Rep, 2*(4), 226–233. https://doi.org/10.1007/s40473-015-0055-0

Dunn, L. B., Fisher, S. R., Hantke, M., Appelbaum, P. S., Dohan, D., Young, J. P., & Roberts, L. W. (2013). Thinking about it for somebody else: Alzheimer's disease research and proxy decision makers' translation of ethical principles into practice. *Am J Geriatr Psychiatry, 21*(4), 337–345. https://doi.org/10.1016/j.jagp.2012.11.014

Dworkin, R. M., & Vintage, B. (2013). *Life's dominion: An argument about abortion, euthanasia, and individual freedom*. Vintage Books.

Fellows, L. K. (1998). Competency and consent in dementia. *J Am Geriatr Soc, 46*(7), 922–926. https://doi.org/10.1111/j.1532-5415.1998.tb02734.x

Fleshner, M. J., Kennedy, A. J., Veldkamp, P. J., & Childers, J. W. (2019). Would you be surprised if this patient died this year? Advance care planning in substance use disorders. *J Gen Intern Med, 34*(11), 2630–2633. https://doi.org/10.1007/s11606-019-05223-z

Foddy, B., & Savulescu, J. (2006). Addiction and autonomy: Can addicted people consent to the prescription of their drug of addiction? *Bioethics, 20*(1), 1–15. https://doi.org/https://doi.org/10.1111/j.1467-8519.2006.00470.x

Giordano, A., De Panfilis, L., Perin, M., Servidio, L., Cascioli, M., Grasso, M. G., Lugaresi, A., Pucci, E., Veronese, S., & Solari, A. (2022). Advance care planning in neurodegenerative disorders: A scoping review. *Int J Environ Res Public Health, 19*(2), 803. https://doi.org/10.3390/ijerph19020803.

Goodman, A. (1990). Addiction: Definition and implications. *Br J Addict, 85*(11), 1403–1408. https://doi.org/10.1111/j.1360-0443.1990.tb01620.x

Gove, D., Diaz-Ponce, A., Georges, J., Moniz-Cook, E., Mountain, G., Chattat, R., & Øksnebjerg, L. (2018). Alzheimer Europe's position on involving people with dementia in research through PPI (patient and public involvement). *Aging Ment Health, 22*(6), 723–729. https://doi.org/10.1080/13607863.2017.1317334

Grisso, T., & Appelbaum, P. S. (1998). *Assessing competence to consent to treatment: A guide for physicians and other health professionals*. http://hdl.handle.net/20.500.14038/45786

He Leow, M. Q., & Wai Chi Chan, S. (2011). Factors affecting caregiver burden of terminally ill adults in the home setting—A systematic review. *JBI Libr Syst Rev, 9*(45), 1883–1916. https://doi.org/10.11124/01938924-201109450-00001

Heaton, B. (2013, May 1). *Ensuring subject competence to enroll and participate in addiction research*. Boston University. https://www.bu.edu/aodhealth/2013/05/01/ensuring-subject-competence-to-enroll-and-participate-in-addiction-research/

Heijdra Suasnabar, J. M., & Hipple Walters, B. (2020). Community-based psychosocial substance use disorder interventions in low-and-middle-income countries: A narrative literature review. *Int J Ment Health Syst, 14*, 74. https://doi.org/10.1186/s13033-020-00405-3

Hermann, H., Feuz, M., Trachsel, M., & Biller-Andorno, N. (2020). Decision-making capacity: From testing to evaluation. *Med Health Care Philos, 23*(2), 253–259. https://doi.org/10.1007/s11019-019-09930-6

Hogden, A. (2014). Optimizing patient autonomy in amyotrophic lateral sclerosis: Inclusive decision-making in multidisciplinary care. *Neurodegener Dis Manag, 4*(1), 01–03. https://doi.org/https://doi.org/10.2217/nmt.13.71

Hou, Y., Dan, X., Babbar, M., Wei, Y., Hasselbalch, S. G., Croteau, D. L., & Bohr, V. A. (2019). Ageing as a risk factor for neurodegenerative disease. *Nat Rev Neurol*, *15*(10), 565–581. https://doi.org/10.1038/s41582-019-0244-7

Howe, E. (2009). Ethical aspects of evaluating a patient's mental capacity. *Psychiatry (Edgmont)*, *6*(7), 15–23.

Hu, C., Chen, C., & Dong, X. P. (2021). Impact of COVID-19 pandemic on patients with neurodegenerative diseases. *Front Aging Neurosci*, *13*, 664965. https://doi.org/10.3389/fnagi.2021.664965

Hu, L. L., Sparenborg, S., & Tai, B. (2011). Privacy protection for patients with substance use problems. *Subst Abuse Rehabil*, *2*, 227–233. https://doi.org/10.2147/sar.S27237

Jackson, E., & Warner, J. (2002). How much do doctors know about consent and capacity? *J R Soc Med*, *95*(12), 601–603. https://doi.org/10.1258/jrsm.95.12.601

Janssens, M. J., Van Rooij, M. F., ten Have, H. A., Kortmann, F. A., & Van Wijmen, F. C. (2004). Pressure and coercion in the care for the addicted: Ethical perspectives. *J Med Ethics*, *30*(5), 453–458. https://doi.org/10.1136/jme.2002.002212

Jaworska, A. (2017). 274Ethical dilemmas in neurodegenerative disease: Respecting patients at the twilight of agency. In J. Illes (Ed.), *Neuroethics: Anticipating the future*. Oxford University Press. https://doi.org/10.1093/oso/9780198786832.003.0015

Jeste, D. V., & Saks, E. (2006). Decisional capacity in mental illness and substance use disorders: Empirical database and policy implications. *Behav Sci Law*, *24*(4), 607–628. https://doi.org/https://doi.org/10.1002/bsl.707

Johnson, R. A., & Karlawish, J. (2015). A review of ethical issues in dementia. *Int Psychogeriatr*, *27*(10), 1635–1647. https://doi.org/10.1017/s1041610215000848

Karlawish, J. (2021). Assessment of decision-making capacity in adults. Retrieved August 19, 2019, from https://www.uptodate.com/contents/assessment-of-decision-making-capacity-in-adults

Keyes, K. M., Liu, X. C., & Cerda, M. (2012). The role of race/ethnicity in alcohol-attributable injury in the United States. *Epidemiol Rev*, *34*(1), 89–102. https://doi.org/10.1093/epirev/mxr018

Knickman, J., Krishnan, R., & Pincus, H. (2016). Improving access to effective care for people with mental health and substance use disorders. *JAMA*, *316*(16), 1647–1648. https://doi.org/10.1001/jama.2016.13639

Kuerbis, A., Sacco, P., Blazer, D. G., & Moore, A. A. (2014). Substance abuse among older adults. *Clin Geriatr Med*, *30*(3), 629–654. https://doi.org/10.1016/j.cger.2014.04.008

Leo, R. J. (1999). Competency and the capacity to make treatment decisions: A primer for primary care physicians. *Prim Care Companion J Clin Psychiatry*, *1*(5), 131–141. https://doi.org/10.4088/pcc.v01n0501

Levy, N. (2016). Addiction, autonomy, and informed consent: On and off the garden path. *J Med Philos*, *41*(1), 56–73. https://doi.org/10.1093/jmp/jhv033

Low, J. A., & Ho, E. (2017). Managing ethical dilemmas in end-stage neurodegenerative diseases. *Geriatrics*, *2*(1), 8. https://www.mdpi.com/2308-3417/2/1/8 https://www.ncbi.nlm.nih.gov/pmc/articles/PMC6371100/pdf/geriatrics-02-00008.pdf

Lulé, D., Kübler, A., & Ludolph, A. C. (2019). Ethical principles in patient-centered medical care to support quality of life in amyotrophic lateral sclerosis [Mini Review]. *Front Neurola*, *10*. https://doi.org/10.3389/fneur.2019.00259

Martin, R. C., Okonkwo, O. C., Hill, J., Griffith, H. R., Triebel, K., Bartolucci, A., Nicholas, A. P., Watts, R. L., Stover, N., Harrell, L. E., Clark, D., & Marson, D. C. (2008). Medical decision-making capacity in cognitively impaired Parkinson's disease patients without dementia. *Mov Disord*, *23*(13), 1867–1874. https://doi.org/10.1002/mds.22170

McCusker, P., Gillespie, L., Davidson, G., Vicary, S., & Stone, K. (2023). The United Nations Convention on the Rights of Persons with Disabilities and Social Work: Evidence for impact? *Int J Environ Res Public Health*, *20*(20). https://doi.org/10.3390/ijerph20206927

Meulenbroek, O., Vernooij-Dassen, M., Kessels, R. P. C., Graff, M. J. L., Sjögren, M. J. C., Schalk, B. W. M., Hoogsteen-Ossewaarde, M. E., Claassen, J. A. H. R., Melis, R. J. F., & Olde Rikkert, M. G. M. (2010). Informed consent in dementia research. Legislation, theoretical concepts and how to assess capacity to consent. *Eur Geriatr Med*, *1*(1), 58–63. https://doi.org/https://doi.org/10.1016/j.eurger.2010.01.009

Mexico, U. o. N. (2018). Center for Health Law and Ethics. Albuquerque. *Values history. A form to assist you in making health care choices in accordance with your values*. http://hscethics.unm.edu/common/pdf/values-history.pdf

Morán-Sánchez, I., Luna, A., Sánchez-Muñoz, M., Aguilera-Alcaraz, B., & Pérez-Cárceles, M. D. (2016). Decision-making capacity for research participation among addicted people: A cross-sectional study. *BMC Med Ethics*, *17*, 3. https://doi.org/10.1186/s12910-015-0086-9

Nadkarni, A., Gandhi, Y., Bhatia, U., & Velleman, R. (2023). Closing the treatment gap for alcohol use disorders in low- and middle-income countries. *Cambridge Prisms Glob Ment Health*, *10*, e3, Article e3. https://doi.org/10.1017/gmh.2022.57

Nass, S. J., L. L., & Gostin, L. O. (2009). *Beyond the HIPAA privacy rule: Enhancing privacy, improving health through research*. National Academies Press. https://www.ncbi.nlm.nih.gov/books/NBK9579/

Nassif, W. M. (2019). Assessing decisional capacity in patients with substance use disorders. *Curr Psychiatry*, *18*(10), 35–40.

Palmer, B. W., Harmell, A. L., Pinto, L. L., Dunn, L. B., Kim, S. Y., Golshan, S., & Jeste, D. V. (2017). Determinants of capacity to consent to research on Alzheimer's disease. *Clin Gerontol*, *40*(1), 24–34. https://doi.org/10.1080/07317115.2016.1197352

Palzes, V. A., Chi, F. W., Metz, V. E., Sterling, S., Asyyed, A., Ridout, K. K., & Campbell, C. I. (2023). Overall and telehealth addiction treatment utilization by age, race, ethnicity, and socioeconomic status in California after COVID-19 policy changes. *JAMA Health Forum*, *4*(5), e231018–e231018. https://doi.org/10.1001/jamahealthforum.2023.1018

Parfit, D. (1987). *Reasons and persons*. Oxford University Press.

Pisapia, J. M., Halpern, C. H., Muller, U. J., Vinai, P., Wolf, J. A., Whiting, D. M., Wadden, T. A., Baltuch, G. H., & Caplan, A. L. (2013). Ethical considerations in deep brain stimulation for the treatment of addiction and overeating associated with obesity. *AJOB Neurosci*, *4*(2), 35–46. https://doi.org/10.1080/21507740.2013.770420

Porteri, C. (2018). Advance directives as a tool to respect patients' values and preferences: Discussion on the case of Alzheimer's disease. *BMC Med Ethics*, *19*(1), 9. https://doi.org/10.1186/s12910-018-0249-6

Ries, N. M., Mansfield, E., & Sanson-Fisher, R. (2020). Ethical and legal aspects of research involving older people with cognitive impairment: A survey of dementia researchers in Australia. *Int J Law Psychiatry*, *68*, 101534. https://doi.org/10.1016/j.ijlp.2019.101534

Saks, E. R., & Jeste, D. V. (2006). Capacity to consent to or refuse treatment and/or research: Theoretical considerations. *Behav Sci Law*, *24*(4), 411–429. https://doi.org/10.1002/bsl.708

Saloner, B., Carson, N., & B, L. C. (2014). Explaining racial/ethnic differences in adolescent substance abuse treatment completion in the United States: A decomposition analysis. *J Adolesc Health*, *54*(6), 646–653. https://doi.org/10.1016/j.jadohealth.2014.01.002

Schaefer, L. A. (2011). MacArthur Competence Assessment Tools. In J. S. Kreutzer, J. DeLuca, & B. Caplan (Eds.), *Encyclopedia of clinical neuropsychology* (pp. 1502–1505). Springer. https://doi.org/10.1007/978-0-387-79948-3_1613

Schaper, E., Padwa, H., Urada, D., & Shoptaw, S. (2016). Substance use disorder patient privacy and comprehensive care in integrated health care settings. *Psychol Serv*, *13*(1), 105–109. https://doi.org/10.1037/a0037968

Schmidt, L. A., Ye, Y., Greenfield, T. K., & Bond, J. (2007). Ethnic disparities in clinical severity and services for alcohol problems: Results from the National Alcohol Survey. *Alcohol Clin Exp Res*, *31*(1), 48–56. https://doi.org/10.1111/j.1530-0277.2006.00263.x

Series, L., & Nilsson, A. (2018). Wellcome Trust–funded monographs and book chapters Article 12 CRPD: Equal recognition before the law. In I. Bantekas, M. A. Stein, & D. Anastasiou (Eds.), *The UN Convention on the Rights of Persons with Disabilities: A commentary*. Oxford University Press © The Several Contributors 2018.

Shah, P., Thornton, I., Turrin, D., & Hipskind, J. E. (2024). Informed consent. In *StatPearls*. StatPearls Publishing Copyright © 2024, StatPearls Publishing LLC.

Sheahan, L., & Lamont, S. (2020). Understanding ethical and legal obligations in a pandemic: A taxonomy of "duty" for health practitioners. *J Bioeth Inq*, *17*(4), 697–701. https://doi.org/10.1007/s11673-020-10003-0

Spoelhof, G. D., & Elliott, B. (2012). Implementing advance directives in office practice. *Am Fam Physician*, *85*(5), 461–466.

Sreenivasan, V., & Nobleza, C. O. H. S. (2017). Challenges and ethical issues in the course of palliative care management for people living with advanced neurologic diseases. *Ann Palliat Med*, *7*(3), 304–319. https://apm.amegroups.org/article/view/17002

Stevenson, S., Beattie, B. L., Vedan, R., Dwosh, E., Bruce, L., & Illes, J. (2013). Neuroethics, confidentiality, and a cultural imperative in early onset Alzheimer disease: A case study with a First Nation population [Text]. https://open.library.ubc.ca/collections/52383/items/1.0135717

Varelius, J. (2006). The value of autonomy in medical ethics. *Med Health Care Philos*, *9*(3), 377–388. https://doi.org/10.1007/s11019-006-9000-z

Varkey, B. (2021). Principles of clinical ethics and their application to practice. *Med Princ Pract*, *30*(1), 17–28. https://doi.org/10.1159/000509119

Vu, M., Mangal, R., Stead, T., Lopez-Ortiz, C., & Ganti, L. (2022). Impact of Alzheimer's disease on caregivers in the United States. *Health Psychol Res, 10*(3), 37454. https://doi.org/10.52965/001c.37454

Watt, A. D., Jenkins, N. L., McColl, G., Collins, S., & Desmond, P. M. (2019). Ethical issues in the treatment of late-stage Alzheimer's disease. *J Alzheimers Dis, 68*(4), 1311–1316. https://doi.org/10.3233/jad-180865

Wendler, D., & Rid, A. (2011). Systematic review: The effect on surrogates of making treatment decisions for others. *Ann Intern Med, 154*(5), 336–346. https://doi.org/10.7326/0003-4819-154-5-201103010-00008

World Health Organization. (2022). *Follow-up to the political declaration of the third high-level meeting of the General Assembly on the Prevention and Control of Non-Communicable Diseases* (A75/10 Add.4). https://apps.who.int/gb/ebwha/pdf_files/WHA75/A75_10Add4-en.pdf

Yang, L. H., Wong, L. Y., Grivel, M. M., & Hasin, D. S. (2017). Stigma and substance use disorders: An international phenomenon. *Curr Opin Psychiatry, 30*(5), 378–388. https://doi.org/10.1097/yco.0000000000000351

Zietlow, K., Dubin, L., Battles, A., & Vitale, C. (2022). Guardianship: A medicolegal review for clinicians. *J Am Geriatr Soc, 70*(11), 3070–3079. https://doi.org/10.1111/jgs.17797

Index

For Product Safety Concerns and Information please contact our EU
representative GPSR@taylorandfrancis.com
Taylor & Francis Verlag GmbH, Kaufingerstraße 24, 80331 München, Germany

www.ingramcontent.com/pod-product-compliance
Ingram Content Group UK Ltd.
Pitfield, Milton Keynes, MK11 3LW, UK
UKHW050931290725
461302UK00004B/23
* 9 7 8 1 0 3 2 7 6 2 4 2 5 *